Innovative Approaches in Infrastructure Design, Resilience, and Maintenance

Innovative Approaches in Infrastructure Design, Resilience, and Maintenance

Guest Editor

Xu-Yang Cao

MDPI

Basel • Beijing • Wuhan • Barcelona • Belgrade • Novi Sad • Cluj • Manchester

Guest Editor
Xu-Yang Cao
College of Civil and
Transportation Engineering
Hohai University
Nanjing
China

Editorial Office
MDPI AG
Grosspeteranlage 5
4052 Basel, Switzerland

This is a reprint of the Special Issue, published open access by the journal *Designs* (ISSN 2411-9660), freely accessible at: https://www.mdpi.com/journal/designs/special_issues/SD202YKSEM.

For citation purposes, cite each article independently as indicated on the article page online and as indicated below:

Lastname, A.A.; Lastname, B.B. Article Title. *Journal Name* **Year**, *Volume Number*, Page Range.

ISBN 978-3-7258-7995-3 (Hbk)
ISBN 978-3-7258-7996-0 (PDF)
https://doi.org/10.3390/books978-3-7258-7996-0

Contents

About the Editor

Xu-Yang Cao

Xu-Yang Cao is currently an associate professor at the Institute of Structural Engineering in Hohai University. Dr. Cao's preliminary research focused on the performance-based risk analysis and resilience strategy of existing structures under extreme hazards. A series of research results have been achieved, and more than 70 high-level academic papers have been published, including 50 SCI papers in international authoritative journals, such as *ASCE Journals*, *Mechanical System and Signal Processing*, and *Structural Safety*. The previous research results were written into three industry regulations, including the standard "T/CECS-4-2021 Risk Design Standard for Connections of Prefabricated Frames" in China. Dr. Cao is serving as a guest editor and invited reviewer for many prestigious journals, confirming his competencies recognized by international colleagues.

designs

MDPI

Editorial

Progress in Resilience Design: Innovative Approaches for Civil Infrastructure from Maintenance Aspect

Li-Jia Peng, Zhuo Yang, Tian-Le Jin and Xu-Yang Cao *

College of Civil and Transportation Engineering, Hohai University, Nanjing 210024, China; 2404010111@hhu.edu.cn (L.-J.P.); 2404010108@hhu.edu.cn (Z.Y.); 2404010101@hhu.edu.cn (T.-L.J.)
* Correspondence: caoxy@hhu.edu.cn

1. Introduction

At present, global climate change is intensifying and extreme natural disasters are frequent. The corresponding impacts show cross-domain and cross-regional systemic risk characteristics, which poses new major challenges to infrastructure security. Therefore, it is urgent to evaluate the realistic basis and development trend of China's infrastructure security resilience construction and build a future-oriented technological development strategic framework [1–3]. Resilient infrastructure can better resist natural disasters (such as earthquakes and floods) while ensuring rapid recovery of key services (transportation, energy and water supply) after disasters, thereby minimizing the loss of life and property caused by disasters [4–6]. At present, the research on infrastructure resilience is fragmented in various professional fields, and related innovation achievements are scattered in different directions such as structural design, material science and intelligent monitoring. To this end, this paper integrates 13 high-impact research results in the key areas of infrastructure design, resilience and operation and maintenance, analyzes innovative methods and unsolved problems in structural optimization, material research and development, multi-disaster response, intelligent monitoring, etc., comprehensively expounds the latest research progress in the field of resilient infrastructure, and proposes future research directions.

2. Development of Resilience Design and Maintenance

2.1. Structural Optimization for Enhanced Performance

This section focuses on the optimization design of infrastructure structure, analyzes the key research progress oriented to improving the resilience of engineering structure, and systematically combs the innovative methods and practical results at the component and node levels, aiming to provide theoretical and practical basis for readers to understand how to achieve structural performance breakthrough through optimization design [7–9].

This section consists of the following four articles:

The first article (contribution 1) 'Research on mechanical properties and bearing capacity calculation methods of concrete filled steel tubular members under axial compression' explores the axial compression bearing capacity characteristics of self-stressing concrete filled steel tubular columns. In this study, the researchers used nine specimens (one traditional specimen, eight self-stressing concrete-filled steel tubular specimens) to carry out axial compression tests. The study found that within a specific expansion agent content range, self-stressing can significantly improve the axial compression bearing capacity of the component. Aiming at the optimization design pain point of this widely used structural member, a modified simplified theoretical calculation method of bearing capacity is proposed. This innovation not only improves the accuracy of bearing capacity prediction, but also provides a practical tool for engineering and technical personnel to optimize design parameters in construction projects.

The second article (contribution 2) 'Optimization design of seismic beam variable cross-section I-shaped beam members' constructs and verifies the optimal design scheme for the variable cross-section I-shaped beam members in the three-story steel frame. The researchers carried out a comparative analysis using a beam with a dog-bone weakened section. It is found that the variable cross-section beam can not only effectively avoid brittle failure, but also maintain the stiffness and bearing capacity of the component, and achieve the performance balance between structural safety and material utilization efficiency. The research results provide a feasible alternative for traditional beam design, and have important application value in earthquake-prone areas with the core goal of improving seismic resilience.

The third article (contribution 3) 'Prediction of shear strength of anisotropic structural plane considering size effect' focuses on the prediction of shear strength of anisotropic structural plane considering size effect. By constructing models with different roughness (joint roughness coefficient), different sizes and different normal stresses, the key influencing factors of shear strength are analyzed, and the parameter correlation range is given. Based on the numerical test data, a back propagation neural network prediction model with multiple input parameters is constructed. The model has high prediction accuracy and can effectively predict the peak shear strength, which provides a scientific reference for the evaluation of engineering rock mass stability.

In the fourth article (contribution 4) 'Innovative Design of Railway Sleepers', it is pointed out that in the heavy haul railway infrastructure, the new sleeper reinforcement system (new-tie) can effectively solve the problems of high operation and maintenance costs and subgrade diseases caused by excessive stress transferred from sleepers to ballast. The modular "'Z'-shaped" structure can be compatible with the existing sleepers. Through the multi-sleeper collaborative force, the load diffusion range is expanded, and there is no need to carry out large-scale transformation of the existing track infrastructure. The finite element analysis shows that the system can reduce the effective stress of ballast by 24%, the maximum displacement of sleeper by 12%, and the deformation of ballast by 2%, which significantly improves the durability of track. At the same time, the engineering case verification in the soft soil area of the Black Rock Desert concludes that compared with the traditional geotechnical improvement scheme and the slab track scheme, the new sleeper reinforcement system has more advantages in economy, construction convenience and environmental friendliness.

2.2. Multi-Hazard Resilience and Risk Management

This section focuses on the key issues in the field of civil engineering structure and disaster prevention and mitigation, and systematically reviews the relevant research progress from four dimensions: multi-disaster resilience, structural seismic design, material properties and traditional building protection [10–13].

This section consists of the following five articles:

The first article (contribution 5) 'Modeling and Comparison of Design Features of Pendulum and Radial Micro Hydropower Stations Considering the Variability of Design Parameters' found that in the low-speed water energy collection scenario, the torque output per unit specific area of the pendulum micro hydropower station is increased by up to 12 times compared with the traditional scheme by adjusting the blade area and the lever length. Although the pendulum design is a cyclic operation mode, it is superior to the radial system in terms of low-speed water energy conversion efficiency, which provides the possibility for efficient power generation in underutilized low-flow waters.

https://doi.org/10.3390/designs10020025

The second article (contribution 6) 'Evaluation of seismic isolation design method: Comparison of simplified linear method and nonlinear time history analysis method' By comparing the seismic isolation design methods in the American 'Building Seismic Code' (ASCE/SEI 7-22) [14] and the European 'Seismic Code' (EN 1998-1) [15], a six-story steel structure building with lead rubber isolation bearings (LRBs) and curvature self-centering isolation bearings (CSSs) was used as a case to verify the reliability of the two methods. In this study, 20 ground motion records were used to carry out nonlinear time history analysis, and the key performance indexes of the isolated structure were verified. This study clarifies the advantages and disadvantages of the two standard systems: EC8 is more reliable for the design guidance of lead rubber isolation bearings, while ASCE 7-22 is too conservative for the design requirements of curvature self-centering isolation bearings. This provides guidance for engineers and technicians to choose a reasonable seismic isolation design method, and helps to achieve a balance between safety and cost-effectiveness in structural design in earthquake-prone areas.

The third article (contribution 7) 'Discussion on the calculation formula of additional pressure for seismic design of circular silos in European Code EN1998-4' points out and corrects a key technical defect in the European Code for Seismic Design of Circular Silos (EN 1998-4) [16]. The correction scheme makes the calculation formula of the specification fit with the classical physical model of Trahail theory, so as to ensure that the calculation results can reflect the actual dynamic mass of granular materials under an earthquake. The correction eliminates the problem of overestimation of the original calculated value in the specification. On the premise of ensuring the safety reserve set by the specification, it avoids the over-design of the silo wall, and can greatly save the cost of silo construction in the medium and high seismic intensity area.

The fourth article (contribution 8) 'Influence of the coupling form of the spherical steel bearing and the thickness of the anti-friction interlayer on the structural performance' takes the L-100 spherical steel bridge bearing as the research object, and systematically explores the coupling form of the anti-friction interlayer and the lower steel plate and the influence of the thickness of the interlayer on the structural performance of the bearing. The results of the finite element analysis show that the simulation effect of viscoelastic model is significantly better than that of elastic–plastic model. In all kinds of coupling forms, the fully bonded form can minimize the stress and strain level of the bearing. Increasing the thickness of the interlayer will expand the plastic deformation range of the bearing. This study provides an accurate and efficient engineering reference for geometric optimization, material selection and numerical modeling of bridge bearings.

The fifth article (contribution 9) 'Evaluation of road reconstruction potential based on historical road accident data' proposes a data-driven road reconstruction potential evaluation method based on the historical road accident data of Greece from 2016 to 2022. The key factors affecting road safety are analyzed, including road conditions, intersection design, traffic signs and lighting, road geometric characteristics, and so on. This evaluation method breaks through the limitations of traditional research relying on driving behavior analysis, provides a data-driven scientific basis for road reconstruction planning under limited funds, and has strong cross-regional applicability.

2.3. Intelligent Technologies for Monitoring and Maintenance

This section integrates the latest research results of intelligent monitoring, defect detection, operation and maintenance strategy in the field of railway engineering and geotechnical engineering, and shows the deep integration of technological innovation and engineering practice [17–19].

This section consists of the following four articles:

https://doi.org/10.3390/designs10020025

The first article (contribution 10) 'Intelligent inversion analysis of surrounding rock parameters and deformation characteristics of water diversion surge shaft' focuses on intelligent monitoring and operation and maintenance of tunnels, and constructs an intelligent early warning model of surrounding rock deformation based on coded particle swarm optimization (PSO) and support vector machine (SVM). The model improves the prediction accuracy of surrounding rock deformation. The verification results show that the model can effectively control the deformation of surrounding rock and reduce the risk of large deformation. In addition, the proposed adaptive algorithm and reinforcement learning framework can improve the long-term reliability of railway infrastructure operation and maintenance under complex geological conditions.

The second article (contribution 11) 'Innovative road operation and maintenance: the application of intelligent technology in local infrastructure' takes road operation and maintenance technology as the research object, and compares the operation efficiency of traditional detection methods (flatness index method and artificial visual detection) and innovative technical applications (roadroid mobile phone application and Movi Pro 2 detection equipment). The research shows that the digital detection technology shortens the detection time by 60%, and provides an extensible implementation framework for the digital transformation of pavement operation and maintenance while maintaining high detection accuracy.

The third research review article (contribution 12) 'Railway Network Security: Systematic Bibliometric Analysis' comprehensively analyzes the research status of railway network security. Through bibliometric analysis and systematic review, this study answers the core issues in the field of railway networks, sorts out the research trends, hotspots and emerging directions in this field, and puts forward specific suggestions for improving the resilience of railway network security and promoting technological iteration.

In the fourth paper (contribution 13), 'Prediction of clay shrinkage cracking and shrinkage strength factors based on artificial intelligence', a data-driven method is proposed to predict the cracking and shrinkage strength characteristics of clay due to shrinkage. Four prediction models were evaluated: multiple linear regression (MLR), conditional random field (CRF), adaptive neuro-fuzzy inference system (ANF), and Gaussian process (GP). Among them, the artificial neural network (ANN) model based on Bayesian regularization performs best. The model provides a reliable tool for the risk assessment of clay subgrade, soft soil foundation and expansive soil foundation, which reduces the demand for physical tests and improves the safety of geotechnical engineering. This section systematically synthesizes the latest research in railway and geotechnical engineering, spanning intelligent monitoring, defect detection, and maintenance strategies, to demonstrate the deep integration of technological evolution and engineering practice.

3. Conclusions

This study systematically integrates 13 innovative research results in the field of infrastructure design, resilience and operation and maintenance, and comprehensively expounds the latest research progress in this field. Based on the 13 research results sorted out in this paper, infrastructure resilience design and intelligent operation and maintenance are the core directions for the future development of civil engineering. The importance of infrastructure resilience design and intelligent operation and maintenance is mainly reflected in the following three dimensions:

First, it can effectively break through the inherent limitations of traditional civil engineering in terms of insufficient multi-hazard resistance, low life cycle efficiency, and limited technical adaptability, and accurately respond to the dual strategic needs of global multi-hazard frequency and sustainable development.

https://doi.org/10.3390/designs10020025

Second, relying on structural optimization, new material research and development, intelligent operation and maintenance innovation technology integration, etc., it can systematically improve the safety, durability and economy of geotechnical engineering, transportation infrastructure, underground structure and other core field projects, and provide scientific and feasible technical guidance for engineering practice.

Third, it can help the civil engineering industry to transform to intelligent and low-carbon, enhance the resilience of the whole life cycle of infrastructure, and lay a solid foundation for the high-quality development of the industry.

At present, researchers have a clear understanding and technical countermeasures on the mechanism of a single disaster or single environmental action, the degradation mechanism and recovery method of structural material performance under a single environmental action, and the disaster response and control of a single structure under a single disaster [20–22]. However, the catastrophic mechanism of civil engineering infrastructure under the action of multiple disasters and complex environments is more complex and the improvement of resilience is more difficult. It mainly faces the following four challenges:

(1) The universality of the resilience evaluation system needs to be further improved to adapt to different types of infrastructure and regional conditions.
(2) The large-scale application of new materials and new structures is restricted by cost barriers and lack of technical specifications.
(3) Multi-disaster response strategies need to be further optimized to solve the complex spatio-temporal correlation between disasters.
(4) It is necessary to establish a long-term coordination mechanism among various stakeholders to promote the deep integration of resilient infrastructure construction with community and urban development.

In the future, it is necessary to further study the disaster mechanism and resilience improvement method of civil engineering infrastructure systems under the action of multiple disasters and complex environments, develop a global infrastructure resilience improvement innovation method, and accelerate the application of infrastructure resilience improvement technology [23–25].

Funding: This research received no external funding.

Conflicts of Interest: The authors declare no conflicts of interest.

List of Contributions:

1. Liu, X.; Hu, J.; Zheng, Y. Study on the Mechanical Properties and Calculation Method of the Bearing Capacity of Concrete-Filled Steel Pipes under Axial Pressure Load. *Designs* **2024**, *8*, 90. https://doi.org/10.3390/designs8050090.
2. Palizzolo, L.; Vazzano, S.; Benfratello, S. Seismic Design of Steel Frames with Protected Connections. *Designs* **2024**, *8*, 91. https://doi.org/10.3390/designs8050091.
3. Ma, W.-B.; Zou, W.-H.; Zhang, J.-L.; Li, G. Prediction of Shear Strength in Anisotropic Structural Planes Considering Size Effects. *Designs* **2025**, *9*, 17. https://doi.org/10.3390/designs9010017.
4. Karakouzian, M.; Crow, M.; Van Vlerin, W.; Whitton, P.; Karami, M. An Innovative Design of Railroad Crossties. *Designs* **2025**, *9*, 127. https://doi.org/10.3390/designs9060127.
5. Zhilkashinova, A.; Ocheredko, I.; Azamatov, B.; Nurbaev, M.; Dogadkin, D.; Abilev, M. Modeling and Comparison of Design Features of Pendulum and Radial Micro-Hydropower Plants Considering the Influence of Variable Design Parameters. *Designs* **2024**, *8*, 101. https://doi.org/10.3390/designs8050101.
6. Yaacoub, E.; Nascimbene, R.; Furinghetti, M.; Pavese, A. Evaluating Seismic Isolation Design: Simplified Linear Methods vs. Nonlinear Time-History Analysis. *Designs* **2025**, *9*, 34. https://doi.org/10.3390/designs9020034.

7. Mansour, S.; Silvestri, S. Cautionary Note on the Current EN1998-4 Formula of the Additional Pressure in the Seismic Design of Circular Silos. *Designs* **2025**, *9*, 89. https://doi.org/10.3390/designs9040089.
8. Kamenskikh, A.A.; Bogdanova, A.P.; Nosov, Y.O.; Kuznetsova, Y.S. Influence of the Pattern of Coupling of Elements and Antifriction Interlayer Thickness of a Spherical Bearing on Structural Behavior. *Designs* **2025**, *9*, 117. https://doi.org/10.3390/designs9050117.
9. Gkyrtis, K.; Pomoni, M. Use of Historical Road Incident Data for the Assessment of Road Redesign Potential. *Designs* **2024**, *8*, 88. https://doi.org/10.3390/designs8050088.
10. Zou, X.-W.; Zhou, T.; Li, G.; Hu, Y.; Deng, B.; Yang, T. Intelligent Inversion Analysis of Surrounding Rock Parameters and Deformation Characteristics of a Water Diversion Surge Shaft. *Designs* **2024**, *8*, 116. https://doi.org/10.3390/designs8060116.
11. Jáuregui Gallegos, L.F.; Gamarra Tuco, R.; Espinoza Vigil, A.J. Innovative Road Maintenance: Leveraging Smart Technologies for Local Infrastructure. *Designs* **2024**, *8*, 134. https://doi.org/10.3390/designs8060134.
12. Abudu, R.; Bridgelall, R.; Quayson, B.P.; Tolliver, D.; Dadson, K. Railroad Cybersecurity: A Systematic Bibliometric Review. *Designs* **2025**, *9*, 23. https://doi.org/10.3390/designs9010023.
13. Baghbani, A.; Choudhury, T.; Costa, S. Artificial-Intelligence Based Prediction of Crack and Shrinkage Intensity Factor in Clay Soils During Desiccation. *Designs* **2025**, *9*, 54. https://doi.org/10.3390/designs9030054.

References

1. Bruneau, M.; Chang, S.E.; Eguchi, R.T.; Lee, G.C.; O'Rourke, T.D.; Reinhorn, A.M.; Shinozuka, M.; Tierney, K.; Wallace, W.A.; Von Winterfeldt, D. A framework to quantitatively assess and enhance the seismic resilience of communities. *Earthq. Spectra* **2003**, *19*, 733–752. [CrossRef]
2. Li, D.; Jiang, B.; Li, G.; Zhu, C. Design and Stability Evaluation of Slopes in the Sejiang Deformable Body Region Based on Experimental Data. *Designs* **2025**, *9*, 143. [CrossRef]
3. Roohi, M.; Ghasemi, S.; Sediek, O.; Jeon, H.; van de Lindt, J.W.; Shields, M.; Hamideh, S.; Cutler, H. Multi-disciplinary seismic resilience modeling for developing mitigation policies and recovery planning. *Resilient Cities Struct.* **2024**, *3*, 66–84. [CrossRef]
4. Welsh-Huggins, S.J.; Liel, A.B. Evaluating multiobjective outcomes for hazard resilience and sustainability from enhanced building seismic design decisions. *J. Struct. Eng.* **2018**, *144*, 04018108. [CrossRef]
5. Guerrero-Bustamante, O.; Camargo, R.; Duque, J.; Martinez-Arguelles, G.; Polo-Mendoza, R.; Acosta, C.; Murillo, M. Designing Sustainable Asphalt Pavement Structures with a Cement-Treated Base (CTB) and Recycled Concrete Aggregate (RCA): A Case Study from a Developing Country. *Designs* **2025**, *9*, 65. [CrossRef]
6. Chen, X.; Shen, J.; Bao, X.; Wu, X.; Tang, W.; Cui, H. A review of seismic resilience of shield tunnels. *Tunn. Undergr. Space Technol.* **2023**, *136*, 105075. [CrossRef]
7. Biondini, F.; Camnasio, E.; Titi, A. Seismic resilience of concrete structures under corrosion. *Earthq. Eng. Struct. Dyn.* **2015**, *44*, 2445–2466. [CrossRef]
8. French, E.L.; Birchall, S.J.; Landman, K.; Brown, R.D. Designing public open space to support seismic resilience: A systematic review. *Int. J. Disaster Risk Reduct.* **2019**, *34*, 1–10. [CrossRef]
9. Shang, Q.; Wang, T.; Li, J. A quantitative framework to evaluate the seismic resilience of hospital systems. *J. Earthq. Eng.* **2022**, *26*, 3364–3388. [CrossRef]
10. Shafieezadeh, A.; Burden, L.I. Scenario-based resilience assessment framework for critical infrastructure systems: Case study for seismic resilience of seaports. *Reliab. Eng. Syst. Saf.* **2014**, *132*, 207–219. [CrossRef]
11. Alipour, A.; Shafei, B. Seismic resilience of transportation networks with deteriorating components. *J. Struct. Eng.* **2016**, *142*, C4015015. [CrossRef]
12. Feng, D.C.; Wang, Z.; Cao, X.Y.; Wu, G. Damage mechanics-based modeling approaches for cyclic analysis of precast concrete structures: A comparative study. *Int. J. Damage Mech.* **2020**, *29*, 965–987. [CrossRef]
13. Joyner, M.D.; Puentes, B.; Gardner, C.; Steinberg, S.; Sasani, M. Multi objective optimization of building seismic design for resilience. *J. Struct. Eng.* **2022**, *148*, 04022006. [CrossRef]
14. *ASCE/SEI 7-22*; Minimum Design Loads and Associated Criteria for Buildings and Other Structures. American Society of Civil Engineers: Reston, VA, USA, 2022.
15. *EN 1998-1*; Design of Structures for Earthquake Resistance—Part 1: General Rules, Seismic Actions and Rules for Buildings. European Committee for Standardization: Brussels, Belgium, 2004.

16. *EN 1998-4*; Design of Structures for Earthquake Resistance—Part 4: Silos, Tanks and Pipelines. European Committee for Standardization: Brussels, Belgium, 2006.
17. Shen, D.; Li, M.; Yang, Q.; Wen, C.; Liu, C.; Kang, J.; Cao, X. Seismic performance of earthquake-damaged corroded reinforced concrete beam-column joints retrofitted with basalt fiber-reinforced polymer sheets. *Struct. Infrastruct. Eng.* **2024**, *20*, 1451–1467. [CrossRef]
18. Cao, X.Y.; Feng, D.C. A direct analytical derivation of the multi-dimensional fragility spaces of structures under nonstationary mainshock-multi-aftershock sequences. *Probabilistic Eng. Mech.* **2024**, *76*, 103630. [CrossRef]
19. Cao, X.Y.; Shen, D.; Ji, K.; Qu, Z.; Wang, C.L. Recovery resilience framework of replaceable AB-BRB for seismic strengthening during the aftershock stage. *Thin-Walled Struct.* **2024**, *205*, 112389. [CrossRef]
20. Burton, H.V.; Deierlein, G.; Lallemant, D.; Lin, T. Framework for incorporating probabilistic building performance in the assessment of community seismic resilience. *J. Struct. Eng.* **2016**, *142*, C4015007. [CrossRef]
21. Joyner, M.D.; Kurth, M.H.; Pumo, I.; Linkov, I. Recovery-based design of buildings for seismic resilience. *Int. J. Disaster Risk Reduct.* **2021**, *65*, 102556. [CrossRef]
22. Fang, C.; Wang, W.; Qiu, C.; Hu, S.; MacRae, G.A.; Eatherton, M.R. Seismic resilient steel structures: A review of research, practice, challenges and opportunities. *J. Constr. Steel Res.* **2022**, *191*, 107172. [CrossRef]
23. Lagos, R.; Lafontaine, M.; Bonelli, P.; Boroschek, R.; Guendelman, T.; Massone, L.M.; Saragoni, R.; Rojas, F.; Yañez, F. The quest for resilience: The Chilean practice of seismic design for reinforced concrete buildings. *Earthq. Spectra* **2021**, *37*, 26–45. [CrossRef]
24. Xu, J.G.; Cao, X.Y.; Shi, J.; Wang, Z. A comparative study of the novel externally-attached precast SRC braced-frames for seismic retrofitting under near-field spectrum-compatible non-stationary stochastic earthquake. *Structures* **2023**, *50*, 200–214. [CrossRef]
25. Mata, R.; Nuñez, E.; Forcellini, D. Seismic resilience of composite moment frames buildings with slender built-up columns. *J. Build. Eng.* **2025**, *111*, 113532. [CrossRef]

https://doi.org/10.3390/designs10020025

designs

MDPI

Review

Railroad Cybersecurity: A Systematic Bibliometric Review

Ruhaimatu Abudu [1], Raj Bridgelall [1,*], Bright Parker Quayson [1], Denver Tolliver [2] and Kwabena Dadson [1]

[1] Department of Transportation and Supply Chain, College of Business, North Dakota State University, P.O. Box 6050, Fargo, ND 58108-6050, USA; ruhaimatu.abudu@ndsu.edu (R.A.); bright.quayson@ndsu.edu (B.P.Q.); kwabena.dadson@ndsu.edu (K.D.)

[2] Upper Great Plains Transportation Institute, North Dakota State University, P.O. Box 6050, Fargo, ND 58108-6050, USA; denver.tolliver@ndsu.edu

* Correspondence: raj@bridgelall.com or raj.bridgelall@ndsu.edu

Abstract: Cybersecurity challenges are increasing in the rail industry because of constant technological evolution that includes the Internet of Things, blockchains, automation, and artificial intelligence. Consequently, many railroads and supply chain stakeholders have implemented strategies and practices to address these challenges. However, the pace of cybersecurity implementation in the railroad industry is slow even as cyberthreats escalate. This study uniquely integrates bibliometric analysis with a systematic literature review to provide a holistic view of cybersecurity trends in rail freight. The study analyzes 70 articles focusing on cybersecurity practices in the rail freight industry, structured around four research questions relating to: (1) challenges, (2) measures, (3) emerging trends, and (4) innovations. Key findings are that implementing cybersecurity practices in the rail freight industry comes with numerous challenges and risks. The study concludes that new threats will constantly emerge with technological advancements. Therefore, there is a need for continuous human training, collaboration, and coordination with stakeholders. This study also highlights research gaps and recommends how stakeholders can most appropriately execute cybersecurity strategies and best coordinate them with the various technological functions in the rail freight industry.

Keywords: bibliometric analysis; cyberthreat; cybersecurity; rail freight; risk; safety; security

1. Introduction

The railroad industry, a crucial backbone of the global supply chain, has been grappling with complex security issues. These complexities are related to the physical infrastructure, human resources, and the intricacies of modern technologies that support modern rail operations. These include Internet of Things (IoT) devices, global positioning systems (GPS), and intelligent sensors. For instance, the recent addition of a positive train control (PTC) system has been a critical achievement toward improving safety in this industry [1]. However, these technologies have also opened new vulnerabilities for potential cyberthreats [2]. Cybersecurity in rail freight transportation has become crucial because of the industry's growing dependence on digitalization, connected systems, and automation. For example, malware attacks in the railroad industry have increased by more than 200% globally in the last five years [3]. Although extant studies have explored several areas of railroad cybersecurity, including signaling [4] and cryptographic protocols [5], most of these studies fail to provide a comprehensive perspective on technological, human, and operational risks in railroad cybersecurity [6]. Several of these studies also fail to provide practical mitigation strategies for holistic adaptation. This study aims to bridge this gap by synthesizing findings and recommendations across various domains by offering a comprehensive framework while providing a practical strategy for the growing discourse. Advancing technology in the rail industry heightens the need for robust cybersecurity measures to ensure continuous operations [7]. In addressing these issues, studies focused on various aspects of cybersecurity practices for the rail freight industry, including sustainability [8,9], risk assessment and management [10], cybersecurity framework [3], and technology innovation [11]. The Transportation Security Administration (TSA) recently implemented a new rule for cybersecurity evaluation of the major modes of transportation, including rail freight [12]. Although this regulation aims to enhance cyber resilience, some industry stakeholders have raised concerns about its implementation and the associated compliance costs [13].

In the past, transportation security risks mostly stemmed from physical attacks. After 9/11, there was a notable change in focus toward securing transportation systems from potential terrorist threats, leading to comprehensive policies implemented at both national and international levels. This age saw notable advancements in transportation security due to incorporating digital technologies, which brought new vulnerabilities [14]. The railway industry has encountered difficulties in modifying current safety protocols to effectively manage these cybersecurity threats [15]. The literature has identified signaling and communication networks in railway systems as being susceptible to a range of cyberthreats. Other studies highlight vulnerabilities in the sector, including malware, wireless attacks, and unauthorized system access [16]. Initiatives like Shift2Rail [15] and standards developed by international bodies like the International Organization for Standardization [17,18] and National Institute of Standards and Technology [18,19] have aimed to enhance cybersecurity resilience in railroad operations. These efforts emphasize the importance of digital security in sustaining operational efficiency and environmental compliance [20]. Demiridis and Pyrgidis (2022) highlighted the significance of cybersecurity in the rail freight sector for sustainable development [21]. The study shows the importance of improving essential services in this industry, such as ensuring high reliability and punctuality to increase competitiveness with other modes of transportation. Aminzadegan et al. (2022) highlighted the environmental friendliness of rail as compared with other modes of transportation [22], making it a priority for government and other policymakers. Rail freight transportation is advancing by integrating digital platforms such as online exchanges [23], making cybersecurity crucial for protecting data and transactions.

This research provides a comprehensive overview of current cybersecurity practices in the rail freight industry through a combined systematic literature review (SLR) and a bibliometric analysis. The primary contributions of this study are as follows: (1) synthesizing key cybersecurity challenges and solutions in the rail freight industry, (2) systematically categorizing cyber threats and mitigation strategies using bibliometric insights, (3) identifying research gaps in cybersecurity preparedness for rail freight, and (4) offering actionable recommendations for industry stakeholders to enhance cybersecurity resilience. By addressing these aspects, this review establishes a foundation for future research and provides practical guidance for policymakers and practitioners. These insights will strengthen the rail freight sector's defenses against current and emerging cyber hazards and open new research opportunities to enhance the industry's cybersecurity. Given the continuous increase in cybersecurity issues in rail freight transportation, this review provides a multifaceted exploration by answering the following four research questions:

1. Challenges: What are the primary cybersecurity threats and vulnerabilities identified in rail freight transportation systems?
2. Measures: How do existing cybersecurity measures address the risks associated with rail freight transportation?

3. Emerging Trends: What are the emerging trends in cybersecurity technologies and practices for enhancing the safety and security of rail freight transportation?
4. Innovation: How effective are current cybersecurity practices in mitigating the risk of cyberattacks in rail transportation networks?

The subsequent sections of this paper are as follows. Section 2 describes the SLR methodology. Section 3 offers answers to the research questions based on the reviewed literature. Section 4 discusses the results while offering implications. Section 5 concludes the SLR, highlighting the gaps identified, limitations, and recommendations for future research.

2. Methodology

Undertaking a literature review is a complex process, with many methods available to compare and contrast the knowledge and gaps in existing literature for a particular set of research questions. These methods range from a traditional literature search, forward or snowball sampling of systematic reviews, and meta-analysis to distill trends. However, the most common contemporary guideline used is the "preferred reporting items for systematic reviews and meta-analyses" (PRISMA), which provides a degree of standardization as compared with others [24]. The authors chose the SLR method over others like narrative, critical, and theoretical reviews for its ability to minimize bias and limitations [25]. The SLR methodology along with the PRISMA checklist systematically identifies research gaps and provides a comprehensive and transparent review approach to evidence synthesis [26]. The PRISMA method comprises 27 items across seven sections, which this study divides into four stages: the pre-literature collection, literature (publication) selection, systematic review, and data analysis. The authors selected PRISMA to ensure rigor and minimize biases in synthesizing a comprehensive analysis of existing studies. By following the structured stages, the study identifies critical gaps in railroad cybersecurity literature by integrating a comprehensive overview of relevant studies and narrowing it down to the ones that address the above research questions.

2.1. SLR Workflow

Figure 1 illustrates the framework developed in this study for the SLR. The main stages were as follows:

1. Pre-literature collection: Define the research questions, and their aim and scope. State the specific methodology or tools incorporated. This stage followed the PRISMA framework, which uses combinations of keywords and search phrases to produce potentially relevant results.
2. Literature selection: Conduct an electronic search of different databases using the defined search phrases that focus on the scope of the study. Screen the search results to eliminate irrelevant literature based on titles, abstracts, publication dates, language, keywords, and other criteria.47
3. Systematic review: Analyze the full text of relevant studies and classify the corpus based on the research questions and themes for further analysis.
4. Data analysis: Interpret the analysis by comparing and contrasting the literature analyzed.

2.2. Pre-Literature Collection Stage

This stage focused on identifying literature that addresses the four core research questions: challenges, protective measures, emerging trends, and innovations in cybersecurity within rail transportation. The research questions were critical in shaping the scope, emphasizing cybersecurity in the rail freight industry. The literature search included only journal and conference papers to ensure an elevated level of rigor.

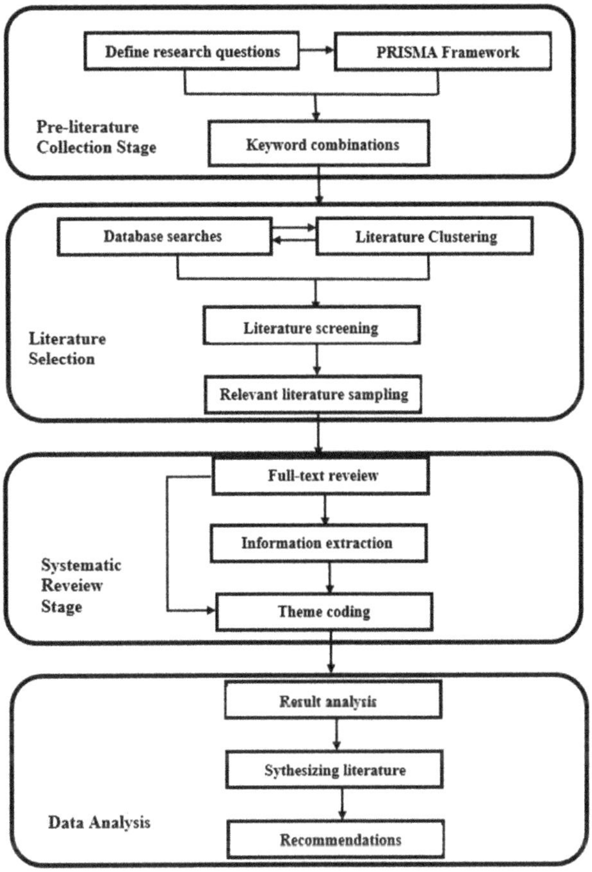

Figure 1. Framework of the study.

Following the PRISMA framework, this stage carefully selected keywords and search phrases to target literature directly aligned with the study's scope. The search strategy aimed to encompass a broad range of relevant studies while systematically narrowing the results to those that best addressed the research questions. The authors then evaluated each identified source for its contribution to understanding the unique cybersecurity concerns in rail freight, as well as the industry's adaptation to emerging threats and technologies. This approach ensured a focused and comprehensive set of literature for the analysis, emphasizing both breadth and depth in addressing the complexities of rail freight cybersecurity.

2.3. Literature Selection

Guided by the research questions, the authors developed Boolean search commands to narrow the literature search and focus on relevant publications. The central premise for selecting relevant literature was to include scope for both rail freight transportation and the implementation of cybersecurity practices. Hence, the selected keywords and Boolean operators were ("rail freight" OR "railway security" OR "train transportation") AND ("cybersecurity" OR "cyberattacks" OR "cyber resilience" OR "cybersecurity practices"). The authors also considered regional variations of key words. For example, "train transportation" and "rail transportation" refer to the same scope reviewed in this study.

The searches utilized the Google Scholar search engine and two databases: Web of Science and Science Direct. The authors selected these databases because they encompass a broad spectrum of interdisciplinary studies.

Table 1 summarizes the Boolean search commands and results from each tool. Search criteria included language, publication year, accessibility to text, uniqueness, and publication type. Table 2 summarizes the categories and criteria considered for the selected publications. The authors excluded reports that did not address the research questions. Excluding non-English language articles ensured accuracy in cybersecurity terminology, consistency in bibliometric analysis, and replicability of the study.

Table 1. Search results from the selected databases.

Database	Search Phrase and Format	Results
Google Scholar	("rail freight" OR "railway security" OR "train transportation") AND ("cybersecurity" OR "cyberattacks" OR "cyber resilience" OR "cybersecurity practices")	532
Science Direct	("rail freight" OR "railway security" OR "train transportation") AND ("cybersecurity" OR "cyberattacks" OR "cyber resilience" OR "cybersecurity practices")	33
Web of Science	(TS = (rail freight transportation) OR TS = (railway security) OR TS = (train freight security)) AND (TS = (cybersecurity) OR TS = (cyberattacks) OR TS = (cyber resilience) OR TS = (cybersecurity practices))	4658

Table 2. Criteria for relevant literature.

Category	Inclusion Criteria	Exclusion Criteria
Source type	Peer-reviewed journal articles and conference papers	Reports, dissertations, news, clinical trials, grants, etc.
Text accessibility	Full-text access to relevant publications	Availability of only title and/or abstract.
Language	English language	Non-English articles.
Search phrases	Based on the selected keywords	Keywords outside the chosen keywords.
Database	Relevant literature published in Google Scholar, Web of Science, and Science Direct	Published in other databases.
Uniqueness	Non-duplicate relevant studies from the selected databases	Duplicate publications from other databases or the selected databases.
Focus	Relevant studies focus on cybersecurity practices in the rail freight industry.	Studies outside the defined scope.
Publication date	January 2017 to November 2024	Studies published before January 2017 and after November 2024.

2.4. Bibliometric Assessment

The bibliometric analysis assessed the temporal and geographic distribution of the selected body of literature to gain insights into the trends of discourse in railroad cybersecurity. The authors also utilized the NVivo software (version 14) [27] to categorize articles by analyzing their full text for relevance to the research questions, and to provide visualizations by a word cloud. The software provided a visualization of key terms, emphasizing gaps in extant studies. The authors also utilized the VOSviewer software (version 1.6.20) to visualize co-occurrence networks of important terms and concepts extracted from the selected body of literature [28]. The visualizations included term co-occurrence from the titles and abstracts to highlight the bibliographic coupling of key themes and concepts. Hence, the bibliographic networks provided a qualitative assessment and validation of the subject matter focus on the selected literature, highlighting the effectiveness of the SLR workflow and screening.

3. Results

The subsections that follow present the results of the literature screening, the bibliometric analysis to visualize and validate the screening results, and an assessment of the findings with respect to addressing the four research questions, and an evaluation of current industry practices.

3.1. Literature Screening

Figure 2 shows the results of the PRISMA workflow. The initial search yielded 5223 studies across Google Scholar, Science Direct, and Web of Science. This result included peer-reviewed journal articles and conference papers from January 2017 to November 2024. To streamline the search for all the databases, the authors utilized the "refine function" for each database to limit the literature search to only journal and conference articles. Applying the inclusion and exclusion criteria in Table 2 reduced the relevant studies to 253 for further scrutiny. The initial screening phase identified and removed 2251 duplicates and excluded an additional 2701 studies that focused solely on either cybersecurity or rail freight transportation, without addressing both areas.

The authors removed an additional 18 studies because the abstracts and keywords did not align with the research area. Of the 253 articles screened, 99 studies lacked comprehensive data on cybersecurity framework in rail freight transportation, leaving 154 articles for further analysis. There were additional exclusions based on full text unavailability—some of the studies had only their abstracts in English with the full text in other languages, and some studies were not journal or conference papers. The final review yielded 70 articles.

3.2. Bibliometric Analysis

Figure 3 shows the temporal distribution of the publications. The number of publications steadily increased throughout the years, peaking in 2023. This reflects a growing trend in the integration of digital systems and autonomous operations in transportation, which has in turn increased vulnerability to cyber threats. Figure 4 shows the publication distributions by country. Sweden and France had the most publications on the topic, followed by Italy, the United States, and China. This reflects Sweden and France taking a lead role in railway cybersecurity research, stemming from Europe's continual investments in projects like Shift2Rail [15]. The Shift2Rail initiative is essential as it employs blockchain technology to ensure the security of data communication within rail systems. Nonetheless, the initiative does not adequately represent other regions like Africa and various areas in Asia, even as they increasingly embrace railroad infrastructure. This geographical disparity results in a significant oversight for these areas, largely attributed to limited resources, ineffective regulatory structures, or a deficit in understanding how to address intricate cybersecurity challenges within the railroad industry. In order to bridge this gap, there is a need for information that highlights a comprehensive approach to integrating technological innovation with operational strategies for effective policy implementations. Relevant studies highlighted in this study, therefore, aim to address advanced cybersecurity practices across diverse economies and infrastructure.

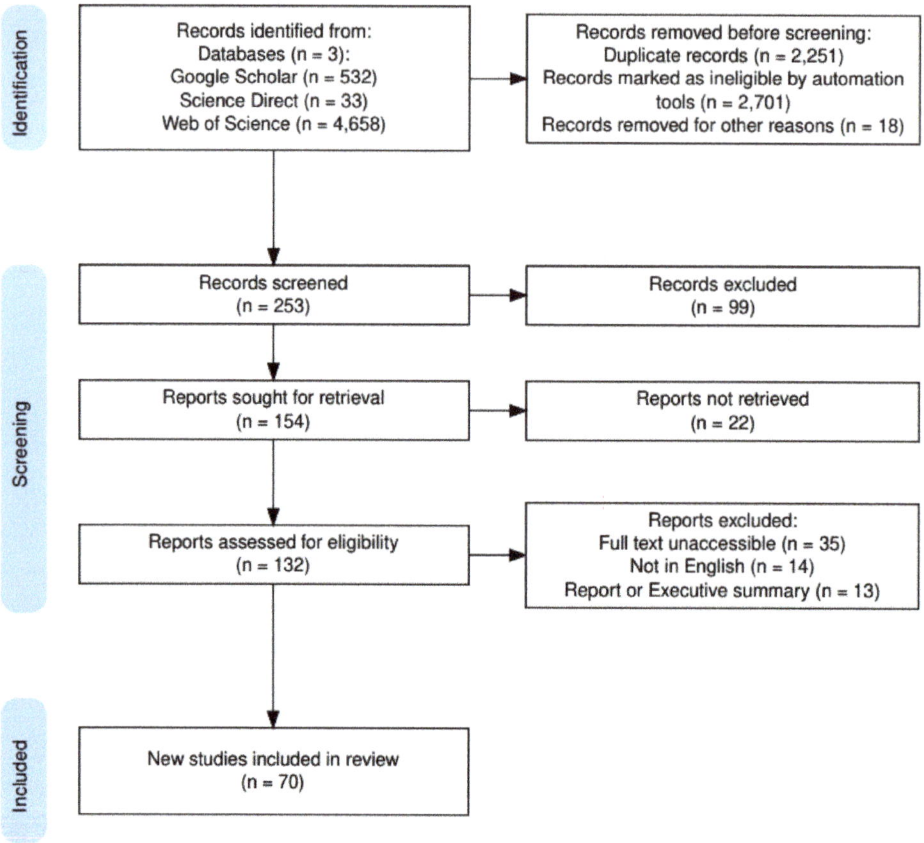

Figure 2. PRISMA workflow results.

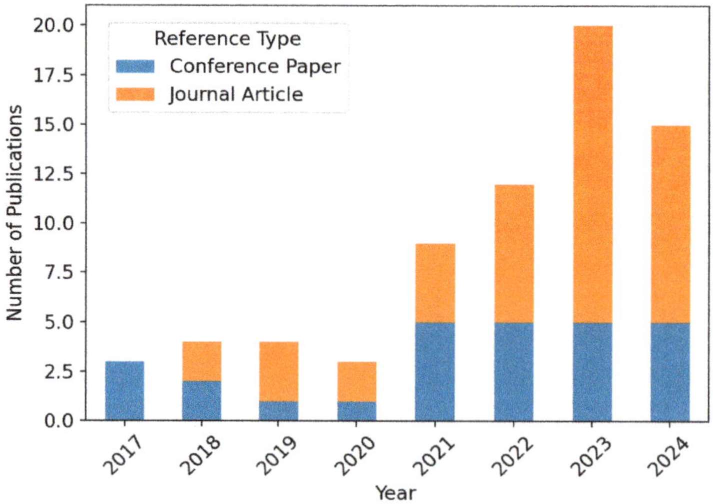

Figure 3. Distribution of publication by year and reference type.

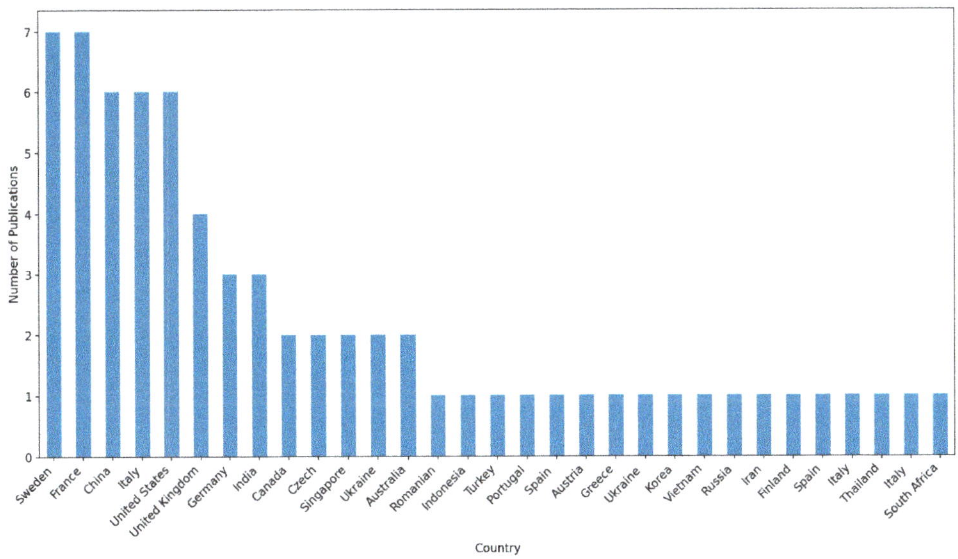

Figure 4. Number of publications by country.

Figure 5 shows a word cloud of the most frequently used terms in the selected corpus where the font size reflects their relative count. This word cloud validates the quality of the literature selection with the most frequent words being "railway", "train", "system", "safety", "attacks", "cybersecurity", and "security".

Figure 5. Word cloud for the relevant studies using NVivo.

Figure 6 shows the term co-occurrence network. The size of the circles represents the relative frequency of the term in the corpus. The thickness of the lines connecting terms represents the relative frequency of their co-occurrence in the corpus. The distinct colors represent cluster terms based on their co-occurrence and relationship within document subsets.

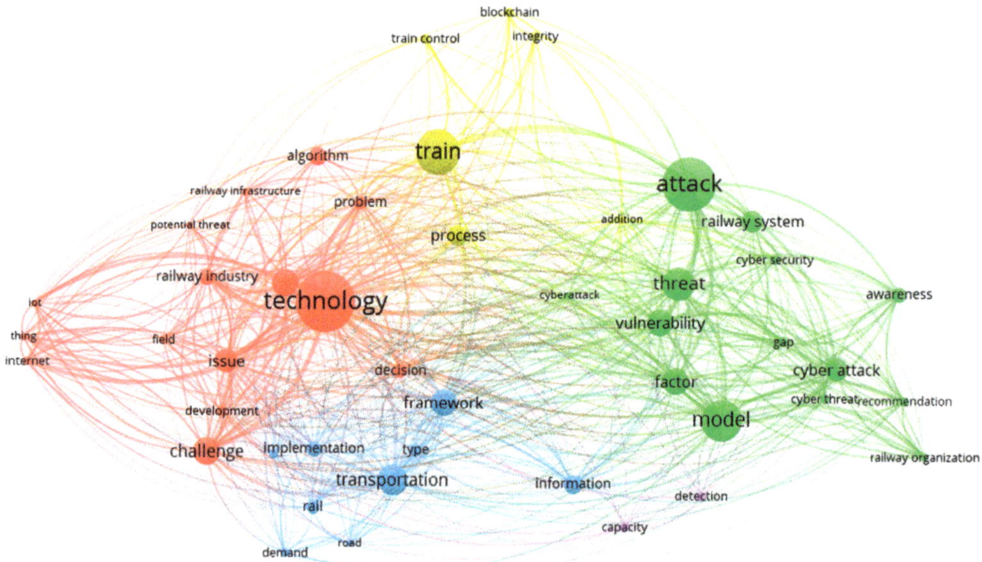

Figure 6. Term co-occurrence network.

These clusters represent thematic areas in the field. The largest cluster in red color, centered on "technology", emphasizes the pervasive role of technological advancements in addressing cybersecurity issues in rail freight systems. The other prominent cluster in green color focus on "threat", "attack", and "vulnerability", indicating concentrated research on vulnerabilities, cyberattacks, and operational impacts. Terms in the blue cluster like "transportation", "framework", and "implementation" reflect ongoing efforts to develop strategic and analytical approaches to mitigate risks in transportation systems. The connections among clusters demonstrate the interdisciplinary nature of cybersecurity, linking technological innovation with operational and strategic measures. This visualization highlights critical research areas and interdependencies, offering insights into emerging trends and gaps in the field.

Figure 7 show the co-authorship network which highlights the collaborative structure among authors contributing to cybersecurity research in the rail industry. The network shows 25 documents where each author has a minimum of two co-authored publications. Distinct clusters represent research groups working closely together, suggesting strong collaboration patterns within specific subfields. The dispersed nature of some authors indicates their independent contributions or limited collaborations within the dataset.

Figure 8 shows the citation network that contains a subset of 39 documents, each having a minimum of five citations. The network center shows 15 documents forming interconnected citation relationships. The most influential works, as indicated by larger nodes, suggest their foundational role in shaping cybersecurity and bibliometric research methodologies. Their prominence indicates frequent citations because of their systematic literature reviews and evidence synthesis. Smaller clusters suggest specialized research areas in the field of rail cybersecurity. The presence of disconnected nodes indicates that some relevant studies remain isolated in citation patterns.

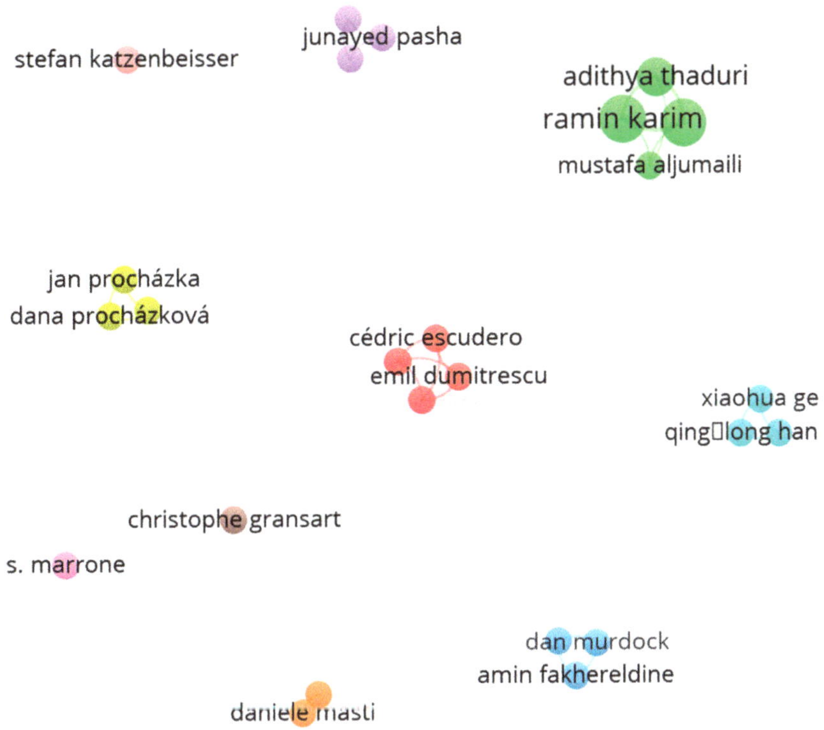

Figure 7. Co-authorship network.

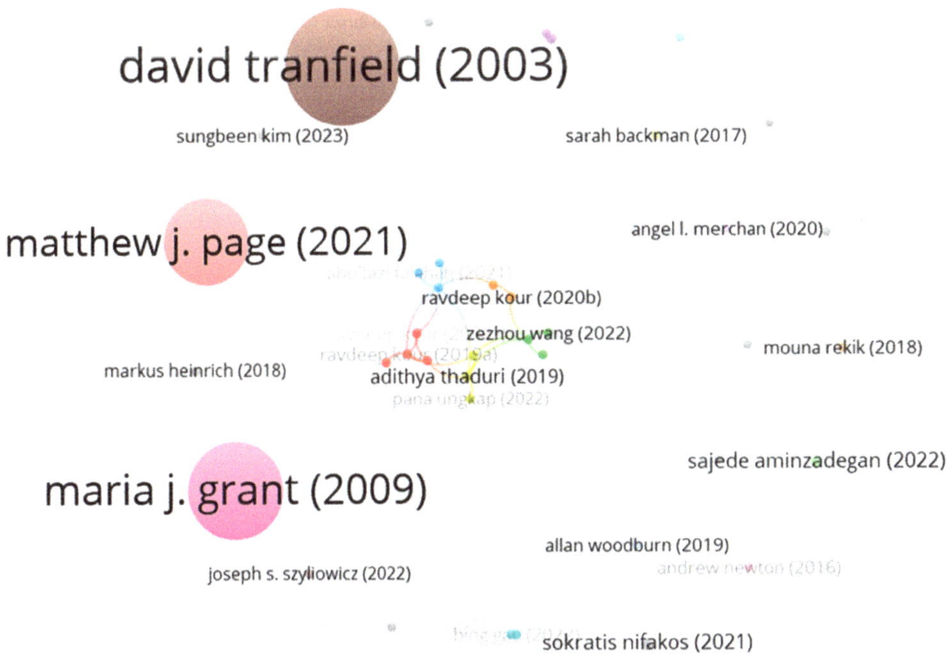

Figure 8. Citation network.

Table 3 classifies representative studies into the five categories and 12 themes shown.

Table 3. Categorization of the relevant studies.

Category	Category Explanation	Themes	References
Resilience and control systems	Studies that focus on resilient control systems for rail freight networks in terms of communication protocols and cyber monitoring systems	Control strategies	[18,29–33]
		Autonomous monitoring	[2,8,17,19,29,30,34–36]
		Resilience framework	[37–44]
Cryptographic mitigation and management	The application of cryptographic mitigation strategies for protecting data within the freight industry with the adoption of advanced encryption algorithms and frameworks	Data protection techniques	[45–54]
		Encryption strategy	[18,31,38,55–59]
Risk assessment	Identifying various rail freight cyber risks and assessing their impact on the sector through various risk assessment tools and evaluation metrics	Risk identification	[5,20,35,60–65]
		Risk evaluation	[44,52,54–56,66]
		Risk impact	[4,17,32,42,50,67]
Security frameworks	Includes studies that implement comprehensive frameworks for managing cyber threats within the rail industry by focusing on cybersecurity standards and policy frameworks	Cybersecurity standards	[37,39,41,48,51,56,68]
		Framework development	[18,19,60,69–71]
Technology adoption	Explores the advancements in rail freight cybersecurity frameworks over the years in terms of technology adoption and deployment	Technology advancement	[39–42,72,73]
		Practical implementations	[20,37,45,55,57,74–76]

3.3. *Types of Cyberattacks in the Rail Industry*

There were 23 articles that focused on the types of cyberattacks common in the rail freight industry. The authors categorized them into five main types: anomaly detection, phishing, software vulnerability, malware, and denial of service, as shown in Table 4. Identifying the type of cyberattack is the first step in fully understanding the mitigation strategy to implement. All the studies asserted that these types of cyberattacks pose vulnerability to railroad operations. Even though various organizations have implemented numerous mitigation strategies, attackers often find other ways to disrupt operations [18]. Even though all these attacks increase vulnerability in the rail freight industry, their severity level differs [38].

Table 4. Types of cyberattacks on rail freight transportation in the reviewed studies.

Attack Types	Description	Literature
Anomaly detection	Anomaly detection in cybersecurity involves recognizing patterns in a dataset that deviate from expected behavior. Identifying unauthorized access or malicious activity that differs from regular operations is crucial.	[5,29,34,36,69,77]
Phishing page identification	Phishing is deceiving people into divulging confidential information, such as credit card numbers, usernames, and passwords, by seeming to be a reliable source in online correspondence. Phishing can also introduce malware into the network, which perpetrators can activate later to interfere with operations.	[18,30,36,38,56,78,79]
Software vulnerability diagnosis	This entails identifying and fixing software vulnerabilities that attackers could exploit to obtain unauthorized access or produce malicious results. Vulnerabilities can result in data breaches, impacting the confidentiality and integrity of critical business and operational data.	[19,29]
Malware identification	Identifying malware entails detecting and eliminating malicious software programs created to damage or exploit any programmable device, service, or network. It is critical to identify and mitigate malware attacks to preserve the confidentiality and integrity of system data.	[18,29,38,58,77]

Table 4. *Cont.*

Attack Types	Description	Literature
Denial of Service (DoS)	Perpetrators design a denial-of-service attack to render a machine or network resource inaccessible to its intended users by temporarily or permanently disrupting the services of a host connected to the Internet. The rail freight industry must establish protocols to identify and address denial-of-service attacks to maintain rail freight transportation networks' uninterrupted and safe functioning.	[5,18,38,69,80,81]

There are direct relationships between the research categories identified in Table 3 and the cyberattack types in Table 4. For instance, resilience frameworks and autonomous monitoring systems align with anomaly detection techniques to identify system irregularities and mitigate DoS attacks. Cryptographic mitigation strategies directly counter phishing attempts and software vulnerabilities by strengthening encryption protocols. Risk assessment methodologies help evaluate malware threats, while security frameworks provide industry-wide protective measures against all categories of cyberattacks. Lastly, the adoption of advanced technologies such as AI-driven cybersecurity tools and Zero Trust Architecture (ZTA) enables proactive defenses against emerging cyber threats in the rail freight sector.

Based on an analysis of 1496 security breaches within a railway infrastructure from January to April 2021, the authors concluded that denial of service, phishing, and malware were the most frequent attack types, accounting for over 40% of incidents. Malware and denial-of-service attacks affected operational systems, leading to delays [81]. For instance, in 2016, the San Francisco rail system experienced a ransomware attack where passengers could ride for free [82]. In October 2017, a distributed denial of service (DDoS) attack targeted Sweden's transportation network, causing a system failure to track rail locations, which led to disruptions in road traffic maps [83]. These incidents underscore the importance of implementing layered defense strategies that include anomaly detection and training programs for employees to reduce these risks effectively. Phishing was by far the most common type of cyberattack in the rail freight industry [79], with five of the 23 papers discussing it. Such attacks involve sending deceptive emails to employees to gain unauthorized access to sensitive information within the business or sector. Some articles also highlighted other attack types, such as SQL injection, watering holes, cross-site scripting, malware, and brute force [18]. Despite phishing being the predominate type of attack in the railroad industry, affecting more of the recorded incidents in the sector [81], most of the suggested mitigation strategies focus on attack detection and network vulnerabilities rather than a layered strategy that combines real-time anomaly detection and awareness tailored to bridge the gap of human contribution to those risks. This disparity highlights a critical gap the study seeks to address.

3.4. Challenges of Cybersecurity in the Rail Industry

Research question 1 focused on different themes for the types of cyberattacks commonly reported in the rail freight industry, the vulnerability of the railway systems, and the risk assessment methodologies currently implemented to mitigate these challenges. The increase in automation across the transportation industry hinges on the reliance on digitalization and connectivity. While this initiative offers to increase efficiency and reduce cost, these systems become more vulnerable to cyberattacks. With the constant evolution of technology, the rail industry has evolved significantly from using coal-powered steam locomotives to integrating sophisticated technologies like PTC, reflecting significant technological advancements [1].

Seventeen of the relevant papers focused on network vulnerabilities in the rail freight industry. The studies asserted that the use of wireless communications makes these systems vulnerable to cyberattacks [5,29,80]. For example, Kiviharju et al. (2022) highlighted the need for an extensive evaluation of the communication systems in the European Rail Traffic Management System [5]. In addressing the issue of vulnerabilities in this industry, the authors identified three broad categories of vulnerabilities: (1) operational, (2) technical, and (3) interconnected. Table 5 characterizes the distinct types of vulnerabilities with their corresponding relevant studies. For example, while acknowledging the railway systems as a critical infrastructure in the transportation industry, Chan (2022) discussed the operational vulnerabilities within the sector [84]. On the other hand, Gaggero et al. (2024) highlighted the technical vulnerability within the industry, focusing on infrastructure attacks, especially the communication network [30]. The authors concluded by highlighting the gaps within the wireless communication industry. Soderi et al. (2023) highlighted that some of the vulnerabilities exist due to the industry's lack of a safety and security mindset during the manufacturing process [4]. Kour et al. (2020) highlighted the importance of becoming aware of attacks and vulnerabilities when evaluating cybersecurity programs in the rail freight industry [18].

Table 5. Types of vulnerabilities in the rail freight industry in the reviewed studies.

Vulnerabilities	Description	Relevant Literature
Operational	This vulnerability stems from human errors, uncontrolled environmental conditions, or supply chain discrepancies that affect the rail freight industry's operations. Examples are incorrect data entry, miscommunication, rail freight supply chain failures, extreme weather conditions, and natural disasters.	[4,5,29,84–86]
Technical	This vulnerability affects the rail freight industry's infrastructure and equipment. It includes derailments, mechanical failures, and aging technology, making the system more prone to failure and inefficiency.	[18,19,30,35,36]
Interconnected	Due to the complex nature of the rail freight network and its interconnectedness, a failure in one node might affect the entire system. These vulnerabilities may include data breaches, communication interruptions, and network intrusions. They may lead to data inaccuracy and system unreliability.	[29,34,69,80]

Many relevant studies have presented steps for implementing risk assessment in the rail freight industry. Rekik et al. (2018) presented critical steps in assessing risks for train control and monitoring systems: (1) identification of critical assets, (2) assessing their vulnerabilities to cyberattacks, and (3) evaluating their likelihood and consequences of an attack [2]. The industry has also utilized tools like simulation and predictive models to assess potential cyberattack scenarios. For example, Kour et al. (2020) adopted the cyber kill chain model in a simulation scenario to assess potential cyberattacks [81]. The authors iterated the seven stages of this model developed by Lockheed Martin (2009), focusing more on the initial stages. Most studies asserted that the assessment should consider the cyber-physical nature of the constant evolution of technological advancements in rail freight systems to address the need for a cross-domain threat.

3.5. Risk Management in the Rail Industry

Research question 2 focused on existing cybersecurity measures in the rail freight industry through the lens of security frameworks and their components. The studies analyzed presented a broad range of mitigation methodologies for addressing risks in the rail freight industry. Each method presented a different approach to distinct types of cyberattacks. Each measure attempted to address an existing cyber security issue or to simulate a cyberattack. The authors delineated these measures into soft and hard strategies to reflect prevalent strategies implemented in the field. Soft strategies encompass measures that do not necessitate infrastructural changes but involve alterations in human training and operational methodologies within the rail freight industry. These strategies include training programs, adopting operational practices, and implementing regulations and standards. An example of a soft strategy is the new TSA regulation [12] implemented to help safeguard the transportation sector, including the railroad industry, against cyber-related disruptions. Conversely, hard strategies involve measures that require modifications to the physical infrastructure, such as upgrading security systems, enhancing network infrastructure, and implementing real-time monitoring systems and robust cyber-physical systems.

More than half of the studies suggested a soft strategy approach to address various rail freight industry risks. A significant body of literature highlighted the importance of continuous employee training as a cornerstone of cybersecurity. Thaduri et al. (2019) highlighted the shift toward digitalization in the rail freight industry [8]. They hinted at various soft strategies such as human factor training, policy development and compliance, information sharing and collaboration, incident response and management, and continuous auditing to protect sensitive information within the railway freight industry. Thron et al. (2024) recommended a dynamic strategy for educating employees on cybersecurity [37]. Rekik et al. (2018) implied undertaking a comprehensive threat landscape assessment like the one proposed in the ROLL2RAIL project to determine unmitigated cyber risks [2]. The other half of the studies adopted the hard strategies approach. The most common strategy was an improved firewall and intrusion detection system, which is essential for early detection within the systems [29,80,84]. Overall, the rail freight industry can enhance its defenses against cyberattacks by conducting cyber-physical security risk assessments and implementing customized security measures.

3.6. Emerging Trends in the Rail Freight Industry

This section focuses on research question 3 to explore trends in network security management and future directions in rail cybersecurity. This industry is evolving rapidly by integrating modern technologies and methods to improve safety and cybersecurity. For example, electric-powered trains are emerging to establish greener, more energy-efficient rail freight systems [87]. Enhancing cybersecurity awareness and providing training is crucial for increasing the reliability and robustness of railway systems, emphasizing the beneficial influence of a cybersecurity culture in businesses [18]. The emergence of Industry 4.0 has also assisted in the convergence of information technology in the railway industry, resulting in reliability and operational efficiency [58]. The evolution of technology in the rail freight industry has warranted constant reliance on communication systems for signaling, thereby creating growing concerns due to system vulnerabilities [88]. Therefore, in ensuring rail freight transportation safety, there should be comprehensive traffic management within the networks in addition to infrastructure management [89]. Another emerging technology is the use of blockchains to provide a decentralized security framework in managing transactions, improving reliability and data distribution across various networks and components [46]. Adopting a zero-trust architecture (ZTA) is another emerging model in the rail freight industry to ensure strict access and continuous authentication to minimize inside threats and cyberattacks [90]. There has been a significant shift toward predictive strategies where systems are proactive to threats and provide mitigation strategies based on patterns and modern technology [18]. Overall, the future of rail freight transportation should incorporate sustainable practices and advanced technologies while simultaneously adopting relevant cybersecurity measures.

3.7. Current Practices in the Rail Freight Industry

This section explores research question 4 in terms of the extent that current cybersecurity practices mitigate cyberattack risks in rail transportation and research gaps. As a recurring theme, the reviewed studies described the constant evolution of technology as a significant contributing factor to cyber vulnerability in the rail freight industry. For example, Kour et al. (2019) asserted that the increased vulnerability of railroads to cyberattacks stems from the integration of advanced technologies during maintenance processes [58]. Falahati and Shafiee (2022) discussed the use of machine learning algorithms and fuzzy logic systems in detecting and responding to cyberattacks in rail freight transportation [61]. Likewise, Singh et al. (2021) highlighted the implementation of advanced cybersecurity measures in autonomous rail operations to safeguard communication systems and operational data [31]. Thaduri et al. (2019) shed light on the importance of comprehensive training programs as a risk mitigation tool [8].

4. Discussion

The subsections that follow link the original research questions to the themes exposed in the results for a comprehensive understanding of cybersecurity in the rail freight industry.

4.1. Assessment of Rail Freight Mitigation Strategies

Numerous studies argued that due to the constant evolution of technology within the railway industry, a continuous and adaptive cyber mitigation strategy is essential. This suggests the need to cross-integrate some cyber assessment strategies from other sectors to tackle the issue within the railway industry. Stakeholders within the railway industry must explore benchmark practices within adjacent industries such as aviation and connected vehicles using vehicle-to-anything communications and adapt their strategies. This adoption becomes especially crucial with a greater reliance on automation, control, and communications technology to improve safety and security. Using a robust quantitative approach such as predictive models for multi-stage cyberattack simulation can be effective [81]. Another key strategy is cybersecurity awareness and training. Railroads can cultivate a cyber hygiene culture [33], especially where employees maintain the security and integrity of digital systems and data [91]. This practice has been in the healthcare sector and other areas, and it has been beneficial in safeguarding sensitive electronic information [92,93]. Additionally, stakeholders within the railway sector should leverage technologies such as blockchains to enhance the security and integrity of their transactions [94].

4.2. Technological Adaptation in the Rail Industry

It is evident that technological evolution is crucial in streamlining processes in the rail industry. These technological solutions will require continuous human improvement and education, organizational measures and methodologies, and improved digitalization with advanced networks. Human improvement should emphasize dynamic, scenario-based programs that replicate real-world issues, promoting a proactive cybersecurity culture within the workforce. The rail freight industry has begun adopting a ZTA to build a reliable system that combats ambiguous security boundaries within a complex network. For instance, Feng et al. (2023) proposed a theoretical use of blockchains and the Merkle tree [90]. ZTA has demonstrated its effectiveness in implementing stringent authentication measures and restricting unauthorized access to a network by continuously verifying access requests. However, the practical implementation of this approach may result in significant challenges, including regulatory compliance, human factors, inherent security concerns, and interoperability. To effectively implement ZTA, the industry must develop targeted strategies to minimize potential risks. There will be a need for constant skills development, hands-on training, and real-life experiences through optimal training delivery methods. This research highlights the need to expand the use of the ZTA to more effectively align with predictive models, including AI-based real-time threat assessment and blockchain-supported decentralized data exchange. These improvements would bolster the resilience of rail systems, especially in tackling emerging threats within interconnected networks. Future research should focus on enhancing cybersecurity resilience through adaptive AI-driven anomaly detection, predictive threat modeling, and blockchain-based secure communication frameworks. Additionally, the application ZTA in railway cybersecurity remains underexplored, requiring further empirical studies to assess its practical implementation and effectiveness in mitigating both insider and external threats.

4.3. Mitigating Cybersecurity Gaps in Rail Freight

This section highlights plausible recommendations to fill strategic gaps within the rail freight industry regarding cybersecurity. A permanent solution in cybersecurity is impossible because threats continuously evolve. Hence, implementing fewer targeted strategies to enhance cybersecurity measures may improve efficiencies. In addressing the issue of human errors, the deployment of autonomous trains could be a significant step in improving safety in the rail industry [31]. However, this also comes with a significant risk as autonomous systems possess different vulnerabilities within the rail network due to their interconnectedness for real-time information access. This may require advanced encrypted techniques to mitigate potential threats. Combining advanced technological strategies and taking a proactive organizational approach can significantly enhance the industry's resilience against cyber threats. Further studies should explore the integration of cybersecurity by design principles in railway infrastructure development. As digitalization increases, future work must also examine the regulatory and compliance challenges of cybersecurity frameworks across different rail networks globally. Additionally, interdisciplinary collaboration between cybersecurity experts and rail industry professionals will be essential to developing sector-specific cybersecurity risk assessment models.

4.4. Enhancing Cybersecurity Resilience in Rail Freight

Existing works on cybersecurity in the rail freight industry have focused on balancing operating efficiency and incorporating advanced technology. This review highlights the significance of maintaining a robust security framework to accommodate technological changes that avoid operation disruptions [30,36,61,84]. The theoretical work of Thron et al. (2024) highlighted the balance between human factors and cybersecurity risks [37]. The study highlighted that many industries, including the aviation sector, also face similar challenges in balancing automation and human error [95]. In particular, ongoing human resource training and development initiatives are necessary to reduce human error, which is a severe cybersecurity vulnerability. Furthermore, utilizing advanced technology for threat detection and response, such as agent-based modeling [96] and artificial intelligence [61], can aid in addressing changing cybersecurity issues. Current research indicates that collaboration and information sharing among stakeholders significantly improves cybersecurity in the rail industry [94]. International cooperation and public-private partnerships can enhance the exchange of threat intelligence and best practices. A critical direction for future research involves developing standardized cybersecurity benchmarks for rail freight operations that align with industry regulations. The increasing reliance on real-time data for train operations also highlights the need for quantum-safe cryptographic techniques to ensure long-term cybersecurity resilience. Future investigations should assess the feasibility of implementing decentralized security models in rail freight networks to reduce single points of failure.

While soft strategies like human training may provide a more economical approach, their potential for scalability and long-term effectiveness has yet to be evaluated. On the other hand, hard strategies, such as firewalls and anomaly detection systems, tackle technical risks but do not provide solutions for mitigating human-centric threats like phishing. The railroad industry needs to learn from related sectors that emphasize the use of predictive analytics. For optimal effectiveness, blockchain-enabled decentralized data sharing should support these frameworks to improve resilience across interconnected networks while responding to ongoing technological advancements.

4.5. Recommendations for Policymakers

Improving cybersecurity measures in the railroad sector demands specific approaches that involve all parties, including policymakers and industry practitioners. They are essential in developing the regulatory framework and securing the funding to tackle various issues, such as cyber threats. As a result, governance must implement compulsory cybersecurity regulations while ensuring that current policies are adhered to. Frameworks such as ISO 27001 or NIST should be adopted in countries with a limited understanding of railroad cybersecurity. This will help maintain consistency among operators and ensure adherence to international best practices [19,50]. As mentioned earlier, incorporating ZTA [61,90] is essential for improving overall authentication and access management while reducing the likelihood of security breaches.

Public-private partnership between rail operators, cybersecurity firms, and academicians can foster innovation. The Shift2Rail initiative [15,41] serves as a successful example of how collaboration can effectively tackle cybersecurity issues in the railway sector, demonstrating its ability to address vulnerabilities within rail systems. In order for all the proposed strategies to be successful, policymakers must also budget for innovation in this area. Innovative technologies like AI-driven predictive analytics [81] and blockchain data structures [46] have demonstrated effectiveness in tackling cybersecurity issues, so it is essential to allocate specific funds for their implementation.

The authors accounted for both scalability and collaboration to guarantee the effective execution of these recommendations. The strategies proposed in this study are flexible for rail operators of different sizes and they are also combinable with other models to tackle the challenge of interoperability. Given that the railroad industry consists of interconnected systems, many of the proposed strategies and recommendations are relevant for achieving unified global standards and minimizing regional disparities.

5. Conclusions

This work summarized the research on cybersecurity practices in the rail freight industry by focusing on the growing literature published in peer-reviewed journals and conference papers. To focus the review, the authors posed four research questions within the following categories: (1) challenges, (2) measures, (3) emerging trends, and (4) cybersecurity practices. The study collated findings from 70 relevant peer-reviewed articles and conference papers. In addressing the first research question, the study highlighted that technological evolution and human error significantly impact cybersecurity practices. Technological evolution in the rail freight industry revolves around the increasing utilization of advanced digital systems to improve operational efficiency and safety. On the other hand, the literature attributes human error to insufficient training, a lack of situational awareness, and the complexities of managing recent technologies.

This review identified common threats such as anomaly detection, phishing, software vulnerabilities, malware, and denial-of-service categorized into operational, technical, and interconnectedness vulnerabilities. The second research question focused on measures implemented to enhance cybersecurity, differentiating between hard strategies, which involve physical infrastructure changes, and soft strategies focusing on operational and training adjustments. The third research question addressed emerging trends, including automation, zero-trust architecture, blockchain, and machine learning, highlighting concerns over system vulnerabilities due to technological advancements. The research emphasized current cybersecurity practices and their effectiveness in risk reduction.

The authors identified obstacles and recommendations for future research, including the need for standardized cybersecurity protocols, investment in human resources, and stakeholder collaboration for sharing best practices and threat intelligence. Despite these strategies, uncertainty remains about whether specific strategies will effectively address ongoing cybersecurity challenges in the rail freight industry and how to best coordinate these strategies with technological functions. This study makes a significant contribution by providing a structured synthesis of cybersecurity research in rail freight transportation, a field that has received limited attention despite growing digitalization. The integration of bibliometric and systematic review methodologies ensures a comprehensive assessment of trends, gaps, and future directions. The findings highlight the urgent need for standardized cybersecurity protocols, enhanced human resource training, and greater collaboration among industry stakeholders to mitigate evolving threats.

This study posits that additional research must fill substantial gaps. For example, there is a need for additional investigation into integrating cybersecurity measures with existing operational technology and information technology systems while maintaining seamless functionality. Additionally, establishing a unified defense mechanism necessitates the development of standardized cybersecurity protocols specifically designed to meet the unique requirements of the rail industry.

This study also emphasized the importance of taking a proactive and unified approach to cybersecurity in the railroad sector, highlighting the need to address technical and human-related vulnerabilities. The swift development of technology and changing industry trends present substantial opportunities and challenges for cybersecurity within the rail freight industry. This research highlights various potential areas for future investigation that aim to fill existing gaps in the literature and prepare for anticipated emerging threats. Future research should aim to create adaptable frameworks that blend AI-powered predictive tools, which are gradually becoming the foundation of cybersecurity mitigation strategies across industries, with established cybersecurity standards, especially in regions often underrepresented. Policymakers and industry stakeholders should make these initiatives a priority to protect global rail systems from emerging cyber threats, ensuring the safety of critical supply chains, and promoting sustainable development around the world.

Funding: This research received funding from the United States Department of Transportation, Center for Transformative Infrastructure Preservation and Sustainability (CTIPS), Funding Number 69A3552348308.

Data Availability Statement: This article includes the data presented in the study.

Conflicts of Interest: The authors declare no conflicts of interest.

References

1. U.S. DOT. *PTC Communications: Cybersecurity Technology Review and Concept of Operations*; U.S. DOT: Washington, DC, USA, 2023.
2. Rekik, M.; Gransart, C.; Berbineau, M. Cyber-physical Threats and Vulnerabilities Analysis for Train Control and Monitoring Systems. In Proceedings of the 2018 International Symposium on Networks, Computers and Communications (ISNCC), Rome, Italy, 19–21 June 2018; pp. 1–6.
3. Soderi, S.; Masti, D.; Lun, Y.Z. Railway cyber-security in the era of interconnected systems: A survey. *IEEE Trans. Intell. Transp. Syst.* **2023**, *24*, 6764–6779. [CrossRef]

4. Soderi, S.; Masti, D.; Hämäläinen, M.; Iinatti, J. Cybersecurity Considerations for Communication Based Train Control. *IEEE Access* **2023**, *11*, 92312–92321. [CrossRef]
5. Kiviharju, M.; Lassfolk, C.; Rikkonen, S.; Kari, H. A Cryptographic and Key Management Glance at Cybersecurity Challenges of the Future European Railway System. In Proceedings of the 2022 14th International Conference on Cyber Conflict: Keep Moving! (CyCon), Tallinn, Estonia, 31 May–3 June 2022; pp. 265–284.
6. Umar, A.M.; Lazi, M.K.A.M.; Hassan, S.A.; Hashim, H.I.C.; Zhang, Y. A Bibliometric Analysis of Railway Safety Research: Thematic Evolution, Current Status, and Future Research Directions. *J. Traffic Transp. Eng. (Engl. Ed.)*, 2025; *in press.*.
7. Woodburn, A. Rail network resilience and operational responsiveness during unplanned disruption: A rail freight case study. *J. Transp. Geogr.* **2019**, *77*, 59–69. [CrossRef]
8. Thaduri, A.; Aljumaili, M.; Kour, R.; Karim, R. Cybersecurity for eMaintenance in railway infrastructure: Risks and consequences. *Int. J. Syst. Assur. Eng. Manag.* **2019**, *10*, 149–159. [CrossRef]
9. Singh, P.; Elmi, Z.; Meriga, V.K.; Pasha, J.; Dulebenets, M.A. Internet of Things for sustainable railway transportation: Past, present, and future. *Clean. Logist. Supply Chain.* **2022**, *4*, 100065. [CrossRef]
10. Wang, Z.; Liu, X. Cyber security of railway cyber-physical system (CPS)—A risk management methodology. *Commun. Transp. Res.* **2022**, *2*, 100078. [CrossRef]
11. López-Aguilar, P.; Batista, E.; Martínez-Ballesté, A.; Solanas, A. Information Security and Privacy in Railway Transportation: A Systematic Review. *Sensors* **2022**, *22*, 7698. [CrossRef] [PubMed]
12. Freight Waves. TSA Proposes New Cybersecurity Rule for Some Railroads, Other Transit Systems. 2024. Available online: https://www.freightwaves.com/news/tsa-proposes-new-cybersecurity-rule-for-some-railroads-other-transit-systems (accessed on 12 November 2024).
13. The Record. 2024. Available online: https://therecord.media/railroad-cyberthreats-tsa-regulations (accessed on 13 November 2024).
14. Newton, A. Multimodal transport security: Frameworks and policy applications in freight and passenger transport. *Secur. J.* **2017**, *30*, 328–330. [CrossRef]
15. Masson, É.; Gransart, C. *Cyber Security for Railways—A Huge Challenge-Shift2Rail Perspective*; Springer International Publishing: Cham, Switzerland, 2017; pp. 97–104.
16. THALES. Cybersecurity for Rail. 2019. Available online: https://www.thalesgroup.com/en/worldwide/transport/news/cybersecurity-rail (accessed on 8 February 2024).
17. Prochazka, J.; Novobilsky, P.; Prochazkova, D.; Kertis, T. Certification Cycles of Train Cyber Gateway. In Proceedings of the 29th European Safety and Reliability Conference (ESREL), Hannover, Germany, 22–26 September 2019; pp. 1973–1978.
18. Kour, R.; Karim, R.; Thaduri, A. Cybersecurity for railways—A maturity model. *Proc. Inst. Mech. Eng. Part F J. Rail Rapid Transit* **2020**, *234*, 1129–1148. [CrossRef]
19. Bellini, E.; Marrone, S.; Marulli, F. Cyber Resilience Meta-Modelling: The Railway Communication Case Study. *Electronics* **2021**, *10*, 583. [CrossRef]
20. Szyliowicz, J.; Zamparini, L. Freight transport security and the robustness of global supply chains. *Transp. Rev.* **2022**, *42*, 717–724. [CrossRef]
21. Demiridis, N.; Pyrgidis, C. Getting freight trains back on track–How railway undertakings, infrastructure owners and regulators can navigate the main dilemmas in freight business to drive sustainable growth. *Front. Sustain.* **2022**, *3*, 903945. [CrossRef]
22. Aminzadegan, S.; Shahriari, M.; Mehranfar, F.; Abramović, B. Factors affecting the emission of pollutants in different types of transportation: A literature review. *Energy Rep.* **2022**, *8*, 2508–2529. [CrossRef]
23. Jain, A.; van der Heijden, R.; Marchau, V.; Bruckmann, D. Towards Rail-Road Online Exchange Platforms in EU-Freight Transportation Markets: An Analysis of Matching Supply and Demand in Multimodal Services. *Sustainability* **2020**, *12*, 10321. [CrossRef]
24. Page, M.J.; McKenzie, J.E.; Bossuyt, P.M.; Boutron, I.; Hoffmann, T.C.; Mulrow, C.D.; Shamseer, L.; Tetzlaff, J.M.; Akl, E.A.; Brennan, S.E.; et al. The PRISMA 2020 statement: An updated guideline for reporting systematic reviews. *Int. J. Surg.* **2021**, *88*, 105906. [CrossRef] [PubMed]
25. Tranfield, D.; Denver, D.; Smart, P. Towards a methodology for developing evidence-informed management knowledge by means of systematic review. *Br. J. Manag.* **2003**, *14*, 207–222. [CrossRef]
26. Grant, M.J.; Booth, A. A typology of reviews: An analysis of 14 review types and associated methodologies. *Health Inf. Libr. J.* **2009**, *26*, 91–108. [CrossRef] [PubMed]
27. Nvivo. 2024. Available online: https://lumivero.com/product/nvivo/ (accessed on 15 November 2024).
28. Eck, N.J.V.; Waltman, L. VOSviewer. 2024. Available online: https://www.vosviewer.com (accessed on 30 June 2024).
29. Miller, N.; Satsrisakul, Y.; Faist, K.; Fehling-Kaschek, M.; Crabbe, S.; Poliotti, M.; Naderpajouh, N.; Setunge, S.; Ergün, S.; Kanak, A.; et al. A Risk and Resilience Assessment Approach for Railway Networks. In Proceedings of the 31st European Safety and Reliability Conference (ESREL 2021), Angers, France, 19–23 September 2021; pp. 2071–2078.

30. Gaggero, G.B.; Marchese, M.; Girdinio, P. A survey on wireless-communication vulnerabilities of ERTMS in the railway sector. *J. Surveill. Secur. Saf.* **2024**, *5*, 52–61. [CrossRef]
31. Singh, P.; Dulebenets, M.A.; Pasha, J.; Gonzalez, E.D.R.S.; Lau, Y.-Y.; Kampmann, R. Deployment of Autonomous Trains in Rail Transportation: Current Trends and Existing Challenges. *IEEE Access* **2021**, *9*, 91427–91461. [CrossRef]
32. Fakhereldine, A.; Zulkernine, M.; Murdock, D. Detecting Intrusions in Communication-Based Train Control Systems. In Proceedings of the ICC 2022—IEEE International Conference on Communications, Seoul, Republic of Korea, 16–20 May 2022; pp. 4193–4198.
33. Bal, O. Formation and management of safety culture in the railway industry: Best practices and strategies. *Transp. Syst. Technol.* **2023**, *42*, 69–80. [CrossRef]
34. Sharma, G.; Sherif, E.; He, M.; Boiten, E. Analysis of Cyber-Attacks for Modern Digital Railway System Using Cyber Range. In Proceedings of the 2022 IEEE Conference on Interdisciplinary Approaches in Technology and Management for Social Innovation (IATMSI), Gwalior, India, 21–23 December 2022; pp. 1–6.
35. Xiong, S.-H.; Qiu, M.-R.; Li, G.; Zhang, H.; Chen, Z.-S. Balancing the signals: Bayesian equilibrium selection for high-speed railway sensor defense. *Inf. Sci.* **2024**, *661*, 120196. [CrossRef]
36. Abdellaoui, S.; Dumitrescu, E.; Escudero, C.; Zamaï, E. Cyber Threat Assessment in Monitoring Turnout Railway Systems. In Proceedings of the 2023 IEEE 28th International Conference on Emerging Technologies and Factory Automation (ETFA), Sinaia, Romania, 12–15 September 2023; pp. 1–8.
37. Thron, E.; Faily, S.; Dogan, H.; Fr, M. Human factors and cyber-security risks on the railway—The critical role played by signalling operations. *Inf. Comput.* **2024**, *32*, 236–263. [CrossRef]
38. Nebaba, A.N.; Savvas, I.K.; Butakova, M.A.; Chernov, A.V.; Shevchuk, P.S. Improving Multiclass Classification of Cybersecurity Breaches in Railway Infrastructure using Imbalanced Learning. In Proceedings of the ESSE 2021: 2021 2nd European Symposium on Software Engineering, Larissa, Greece, 19–21 November 2021; Volume 4, pp. 100–105.
39. Hoang, A.T.; Nguyen, X.K. Managing Technological Security of Smart Environment Monitoring Systems: Study Of a coastal province in Vietnam. *Int. J. Crit. Infrastruct.* **2023**, *19*, 383–403. [CrossRef]
40. Prochazka, J.; Novobilsky, P.; Procházková, D. Mobile Cyber Gateway Security Control. In Proceedings of the 31st European Safety and Reliability Conference, Angers, France, 19–23 September 2021; pp. 3199–3206.
41. Varvarigou, D.; Espes, D.; Bersano, G. Orchestrator for ensuring interdependency between safety and cybersecurity in railway control systems. *Int. J. Rail Transp.* **2023**, *12*, 749–764. [CrossRef]
42. Wang, Z.; Hei, X.; Ma, W.; Wang, Y.; Wang, K.; Jia, Q. Parallel anomaly detection algorithm for cybersecurity on the highspeed train control system. *Math. Biosci. Eng.* **2022**, *19*, 287–308. [CrossRef]
43. Bearfield, G.; Gulijk, C.V.; Thomas, R.J. Redefining rail systems verification and validation: The safety/security STAIRCASE model. *Proc. Inst. Mech. Eng. Part F J. Rail Rapid Transit* **2023**, *237*, 266–274. [CrossRef]
44. Ge, X.; Wu, Q.; Han, Q.-L.; Zhang, X.-M. Resilient Virtual Coupling Control of Automatic Train Convoys with Intermittent Communications. *IEEE Trans. Veh. Technol.* **2023**, *73*, 6183–6195. [CrossRef]
45. Weaver, G.A. Scientific Data Management for Interconnected Critical Infrastructure Systems. In Proceedings of the 2021 ACM/IEEE Joint Conference on Digital Libraries (JCDL), Champaign, IL, USA, 27–30 September 2021; pp. 192–201.
46. Kim, S.; Kim, D. Securing the Cyber Resilience of a Blockchain-Based Railroad Non-Stop Customs Clearance System. *Sensor* **2023**, *23*, 2914. [CrossRef]
47. Purwanto, Y.; Ruriawan, M.F.; Alamsyah, A.; Wijaya, F.P.; Husna, D.N.; Kridan, A.; Fakhrudin, F.N.A.; Itqon, M.; Febrianta, M.Y.; Widiyanesti, S.; et al. Security Architecture for Secure Train Control and Monitoring System. *Sensors* **2023**, *23*, 1341. [CrossRef] [PubMed]
48. Hatzivasilis, G.; Fysarakis, K.; Ioannidis, S.; Hatzakis, I.; Vardakis, G.; Papadakis, N.; Spanoudakis, G. SPD-Safe: Secure Administration of Railway Intelligent Transportation Systems. *Electronics* **2021**, *10*, 92. [CrossRef]
49. Backman, S.; Rhinard, M. The European Union's capacities for managing crises. *J. Contingencies Crisis Manag.* **2018**, *26*, 261–271. [CrossRef]
50. Schmittner, C.; Tummeltshammer, P.; Hofbauer, D.; Shaaban, A.M.; Meidlinger, M.; Tauber, M.; Bonitz, A.; Hametner, R.; Brandstetter, M. Threat Modeling in the Railway Domain. In *Reliability, Safety, and Security of Railway Systems. Modelling, Analysis, Verification, and Certification: Third International Conference, RSSRail 2019, Lille, France, 4–6 June 2019; Proceedings 3*; Springer International Publishing: Cham, Switzerland, 2019; pp. 261–271.
51. Romero, V.; Fernández, E.B. Towards a Security Reference Architecture for Cyber- Physical Systems. In Proceedings of the 15th LACCEI International Multi-Conference for Engineering, Education, and Technology: "Global Partnership for Development and Engineering Education", Boca Raton, FL, USA, 19–21 July 2017; pp. 261–271.
52. Marrone, S. Towards a Unified Definition of Cyber and Physical Vulnerability in Critical Infrastructures. In Proceedings of the 2017 IEEE European Symposium on Security and Privacy: Workshops (EuroS&PW), Paris, France, 26–28 April 2017; pp. 167–173.

53. Babeshko, I.; Illiashenko, O.; Giandomenico, F.D. Towards Effective Safety and Cybersecurity Co-engineering in Critical Domains. In Proceedings of the 2023 13th International Conference on Dependable Systems, Services and Technologies (DESSERT), Athens, Greece, 13–15 October 2023; pp. 1–8.
54. Aktouche, S.R.; Sallak, M.; Bouabdallah, A.; Schön, W. Towards Reconciling Safety and Security Risk Analysis Processes in Railway Remote Driving. In Proceedings of the 2021 5th International Conference on System Reliability and Safety (ICSRS), Palermo, Italy, 24–26 November 2021; pp. 148–154.
55. Agirre, I.; Calderon, A.J.; Yarza, I.; Mugarza, I.; Garcia, D.; Borracci, L. UP2DATE software updating framework compliance with safety and security regulations and standards. In Proceedings of the 2023 26th Euromicro Conference on Digital System Design (DSD), Golem, Albania, 6–8 September 2023; pp. 399–406.
56. Yekta, A.R.; Spychalski, D.; Yekta, E.; Yekta, C.; Katzenbeisser, S. VATT&EK: Formalization of Cyber Attacks on Intelligent Transport Systems—A TTP based approach for Automotive and Rail. In Proceedings of the CSCS '23: Computer Science in Cars Symposium, Darmstadt, Germany, 5 December 2023; pp. 1–17.
57. Wang, Q.; Yu, W.; Huang, D.; Guo, Y. Weighted Train-to-Train Communication-Based Data-Driven Consensus Tracking of Multiple HSTs Subject to Deception Attacks. *IEEE Trans. Veh. Technol.* **2023**, *73*, 6346–6356. [CrossRef]
58. Kour, R.; Al-Jumaili, M.; Karim, R.; Tretten4, P. eMaintenance in railways: Issues and challenges in cybersecurity. *Proc. Inst. Mech. Eng. Part F J. Rail Rapid Transit* **2019**, *233*, 1012–1022. [CrossRef]
59. Kour, R.; Karim, R. Cybersecurity workforce in railway: Its maturity and awareness. *J. Qual. Maint. Eng.* **2021**, *27*, 453–464. [CrossRef]
60. Chakrabarty, S.; Sikdar, B. Detection of Cyber Attacks on Railway Autotransformer Traction Power Systems. *IEEE Trans. Ind. Appl.* **2023**, *59*, 7188–7200. [CrossRef]
61. Falahati, A.; Shafiee, E. Improve Safety and Security of Intelligent Railway Transportation System Based on Balise Using Machine Learning Algorithm and Fuzzy System. *Int. J. Intell. Transp. Syst. Res.* **2022**, *20*, 117–131. [CrossRef]
62. Nunes, J.; Cruz, T.; Simões, P. Railway Infrastructure Cybersecurity: An Overview. In Proceedings of the 23rd European Conference on Cyber Warfare and Security, Jyväskylä, Finland, 27–28 June 2024.
63. Abdellaoui, S.; Dumitrescu, E.; Escudero, C.; Zamaï, E. Temporal assessment of malicious behaviors: Application to turnout field data monitoring. In Proceedings of the International Conference on Control, Automation and Diagnosis (ICCAD24), Paris, France, 15–17 May 2024.
64. Avcı, I.; Koca, M. A Novel Security Risk Analysis Using the AHP Method in Smart Railway Systems. *Appl. Sci.* **2024**, *14*, 4243. [CrossRef]
65. Niekerk, B.v. Vulnerability of South African Commodity Value Chains to Cyber Incidents. *Sci. Mil. S. Afr. J. Mil. Stud.* **2023**, *51*, 161–186. [CrossRef]
66. Zoppi, T.; Mungiello, I.; Ceccarelli, A.; Cirillo, A.; Sarti, L.; Esposito, L.; Scaglione, G.; Repetto, S.; Bondavalli, A. Safe Maintenance of Railways using COTS Mobile Devices: The Remote Worker Dashboard. *ACM Trans. Cyber-Phys. Syst.* **2023**, *7*, 1–20. [CrossRef]
67. Ge, X.; Han, Q.-L.; Zhang, X.-M. Secure virtual coupling control of connected train platoons under cyber attacks. *Veh. Syst. Dyn.* **2025**, *63*, 93–120. [CrossRef]
68. Predescu, S.G.; Savu, D.; Badea, V.E. Cybersecurity in the Railway Sector. *Rom. Cyber Secur. J.* **2022**, *4*. [CrossRef]
69. Fakhereldine, A.; Zulkernine, M.; Murdock, D. CBTCset: A Reference Dataset for Detecting Misbehavior Attacks in CBTC Networks. In Proceedings of the 2023 IEEE 34th International Symposium on Software Reliability Engineering Workshops (ISSREW), Florence, Italy, 9–12 October 2023; pp. 57–62.
70. ter Beek, M.H.; Fantechi, A.; Gnesi, S.; Lenzini, G.; Petrocchi, M. Can AI Help with the Formalization of Railway Cybersecurity Requirements? In Proceedings of the International Symposium on Leveraging Applications of Formal Methods, Crete, Greece, 27–31 October 2024; Springer: Berlin/Heidelberg, Germany, 2024.
71. Voronko, I. The security of IoT systems in railway transport. *Transp. Syst. Technol.* **2024**, *43*, 90–99. [CrossRef]
72. Vayadande, K.; Bhoyar, A.; Gorave, A.; Behare, H.; Bhale, Y.; Bhojane, O. Secure Stations: Revolutionizing Railway Security Using AI. In *International Conference on Innovations and Advances in Cognitive Systems*; Springer: Cham, Switzerland, 2024.
73. Janani, S.; Vijayaram, S.; Vijayganesh, R.C. IoT-SafeRails: Revolutionizing Railway Collision Avoidance Technology. In Proceedings of the 2024 International Conference on Inventive Computation Technologies (ICICT), Lalitpur, Nepal, 24–26 April 2024; pp. 1666–1673.
74. Heinrich, M.; Vieten, J.; Arul, T.; Katzenbeisser, S. Security Analysis of the RaSTA Safety Protocol. In Proceedings of the 2018 IEEE International Conference on Intelligence and Security Informatics (ISI), Miami, FL, USA, 9–11 November 2018; pp. 199–204.
75. Dai, M.; Li, G.; Du, W. Research on data sharing and exchange technology in the railway industry based on blockchain. In Proceedings of the 7th International Conference on Traffic Engineering and Transportation System (ICTETS 2023), Chongqing, China, 22–24 September 2023.
76. Kongsap, P.; Kaewunruen, S. Agent-based modelling of high-speed railway interdependent critical infrastructures facing physical and cyber threats. *Front. Built Environ.* **2024**, *10*, 1249584. [CrossRef]

77. Zou, B.; Choobchian, P.; Rozenberg, J. *Cyber Resilience of Autonomous Mobility Systems: Cyber-Attacks and Resilience-Enhancing Strategies*; World Bank: Washington, DC, USA, 2021; pp. 137–155.
78. Ungkap, P.; Daengsi, T. Cybersecurity Awareness Modeling Associated with Influential Factors Using AHP Technique: A Case of Railway Organizations in Thailand. In Proceedings of the 2022 International Conference on Decision Aid Sciences and Applications (DASA), Chiangrai, Thailand, 23–25 March 2022; pp. 1359–1362.
79. Osliak, O.; Saracino, A.; Martinelli, F.; Mori, P. Cyber threat intelligence for critical infrastructure security. *Concurr. Comput. Pract. Exp.* **2023**, *35*, e7759. [CrossRef]
80. Gao, B.; Bu, B.; Wang, X. A Comprehensive Resilient Control Strategy for CBTC Systems Through Train-to-Train Communications Under Malicious Attacks. *IEEE Trans. Intell. Transp. Syst.* **2022**, *23*, 21015–21033. [CrossRef]
81. Kour, R.; Thaduri, A.; Karim, R. Predictive model for multistage cyber-attack simulation. *Int. J. Syst. Assur. Eng. Manag.* **2020**, *11*, 600–613. [CrossRef]
82. MIT Technology Review. Ransomware Took San Francisco's Public Transit for a Ride. 2016. Available online: https://www.technologyreview.com/2016/11/28/69496/ransomware-took-san-franciscos-public-transit-for-a-ride/ (accessed on 1 December 2024).
83. Barth, B. DDoS Attacks Delay Trains, Stymie Transportation Services in Sweden. 2017. Available online: https://www.scworld.com/news/ddos-attacks-delay-trains-stymie-transportation-services-in-sweden (accessed on 1 December 2024).
84. Chan, R. *A Security Framework for Railway System Deployments*; Springer International Publishing: Cham, Switzerland, 2022; pp. 247–253.
85. Rahiminia, S.; Mehrabi, A.; Pourseyed Aghaee, M.; Jamili, A. Adopting a Bi-level Optimization Method for the Freight Transportation Problem: A Multi-objective Programming Approach. *Transp. Res. Rec. J. Transp. Res. Board* **2023**, *2677*, 490–504. [CrossRef]
86. Valdivia, L.J.; Adin, I.; Arrizabalaga, S.; Anorga, J.; Mendizabal, J. Cybersecurity-The Forgotten Issue in Railways: Security Can Be Woven into Safety Designs. *IEEE Veh. Technol. Mag.* **2018**, *13*, 48–55. [CrossRef]
87. Merchan, A.A.; Belboom, S.; Léonard, A. Life cycle assessment of rail freight transport in Belgium. *Clean Technol. Environ. Policy* **2020**, *22*, 1109–1131. [CrossRef]
88. Steele, H.; Roberts, C.; Hillmansen, S. Railway smart grids: Drivers, benefits and challenges. *Proc. Inst. Mech. Eng. Part F J. Rail Rapid Transit* **2019**, *233*, 526–536. [CrossRef]
89. Kanis, J.; Dolinayova, A.; Kmetik, M. Draft of strategy of the European rail freight corridor Amber. *MATEC Web Conf.* **2018**, *235*, 00023. [CrossRef]
90. Feng, Y.; Zhong, Z.; Sun, X.; Wang, L.; Lu, Y.; Zhu, Y. Blockchain enabled zero trust based authentication scheme for railway communication networks. *J. Cloud Comput.* **2023**, *12*, 62. [CrossRef]
91. Kour, R.; Thaduri, A.; Karim, R. Railway Defender Kill Chain to Predict and Detect Cyber-Attacks. *J. Cyber Secur. Mobil.* **2019**, *9*, 47–90. [CrossRef]
92. Nifakos, S.; Chandramouli, K.; Nikolaou, C.K.; Papachristou, P.; Koch, S.; Panaousis, E.; Bonacina, S. Influence of Human Factors on Cyber Security within Healthcare Organisations: A Systematic Review. *Sensors* **2021**, *21*, 5119. [CrossRef] [PubMed]
93. Alkhaledi, R.; Hawamdeh, S. Electronic Health Records and Cyber Hygiene: A Qualitative Study of the Awareness, Knowledge, and Experience of Physicians in Kuwait. *Proc. Assoc. Inf. Sci. Technol.* **2023**, *60*, 21–30. [CrossRef]
94. Laiton-Bonadiez, C.; Branch-Bedoya, J.W.; Zapata-Cortes, J.; Paipa-Sanabria, E.; Arango-Serna, M. Industry 4.0 Technologies Applied to the Rail Transportation Industry: A Systematic Review. *Sensors* **2022**, *22*, 2491. [CrossRef] [PubMed]
95. Miranda, A.T. Understanding Human Error in Naval Aviation Mishaps. *Hum. Factors* **2018**, *60*, 763–777. [CrossRef] [PubMed]
96. Savadatti, S.G.; Srinivasan, K.; Hu, Y.-C. A Bibliometric Analysis of Agent-Based Systems in Cybersecurity and Broader Security Domains: Trends and Insights. *IEEE Access* **2024**, *13*, 90–119. [CrossRef]

designs

MDPI

Article

Use of Historical Road Incident Data for the Assessment of Road Redesign Potential

Konstantinos Gkyrtis [1,*] and Maria Pomoni [2]

[1] Department of Civil Engineering, Democritus University of Thrace (D.U.Th.), 67100 Xanthi, Greece
[2] Senior Doctoral Pavement Researcher, 15772 Athens, Greece
* Correspondence: kgkyrtis@civil.duth.gr

Abstract: Drivers' safety and overall road functionality are key triggers for deciding on road interventions. Because of the socioeconomical implications of traffic incidents, either fatal or no, continuous research has been dedicated over the previous decades on the assessment of factors contributing to crash potential. Apart from the behavioral aspects of driving, which are commonly studied through simulation and advanced modelling techniques, the road infrastructure status is of equal or even higher significance. In this study, an approach is presented to discuss the road redesign potentials based on the evaluation of network-level historical incident records from road crashes in Greece. Based on total and fatal crash records, the following infrastructure-related aspects were assessed as critical for the discussion of the road redesign potential needs: the status of road's surface (i.e., dry, wet, etc.), the issue of improving driving conditions near at-grade intersections, the presence and suitability of signage and/or lighting, and the consideration of particular geometric design features. Overall, it is deemed that intervention actions for at least one of these pillars should aim at enhancing the safety and functionality of roadways.

Keywords: crash records; road surface status; road redesign; signage; lighting; drivers' safety; road functionality

1. Introduction

1.1. Background on the Social and Enviromental Implications of Roadways

Prior to the construction of safe roads, these are geometrically designed in a way that comprises a three-dimensional layout that balances out three distinct elements, i.e., the horizontal alignment, the vertical alignment, and the cross-sectional profile. Selecting appropriate design values for tangents, curve radii, longitudinal grades, sight distances, lane widths, transverse slope, etc., guarantees adequate traffic flow at typical operating speeds along with functionality, safety, and an environmental-aesthetic adaptability of roads [1,2].

In the meantime, road collisions continue to be a huge global problem with great economic and social implications, despite tremendous efforts and pertinent advances in improving road safety. The World Health Organization's most recent assessment of global road safety states that traffic crashes claim the lives of 1.35 million drivers worldwide each year, ranking them as the eighth most common cause of death for people of all ages [3]. Significant disparities exist in the performance of road safety throughout the various world regions, with Europe recording the lowest numbers of deaths per population [3]. There are notable variations among the regions of a nation as well, which may be related to a number of socioeconomic issues [4].

Enhancing the safety of the European road network is a priority for the European Commission (EC). In order to achieve this goal, the EC approved some years ago the Road Safety Program, which sought to reduce traffic fatalities by half by 2020 compared to 2010. Nevertheless, as the recorded fatality reduction was equivalent to 37%, this aggregate aim was not fulfilled everywhere [5]. Greece was the only member of the European Union (EU) to meet this goal, with a performance of −54%. In addition to the fact that Greece experienced a major economic recession, there could be one major reason for the notable decline in road fatalities in the country over the past ten years. This relates to the fact that the main road network was significantly improved, with 2200 km of motorways in 2018 compared to 750 km in 2007 [5]. Greece, however, was placed 23rd out of 27 EU nations in 2023 for road fatalities per million people, which is far higher than the average for the EU, which is 46 per million people [6]. As a result, it appears that sustained and coordinated efforts are continuously needed to solve this significant issue of traffic accidents.

From a different perspective, roadways encompass environmental and sustainability-related components. When vehicles move on a roadway, they consume fuel and, thus, emit pollutants [7]. The contribution of the transportation sector, and in particular road transportation activities, to environmental pollution is well-known so far. Based on data from the International Energy Agency, the global transport sector produced 8 million tons of CO_2 emissions in 2019, or roughly 1/4 (24.2%) of all CO_2 emissions worldwide. Of these, 6.5 billion tons were produced by the road transportation sector, making up 81% of the transportation sector as a whole [8]. Hence, it is a big challenge for road engineers and vehicle industries to seek for countermeasures to meet sustainability principles in roadways, like eco-friendlier vehicle technologies [9,10] and peculiar characteristics of road design and/or pavement surface condition that could ensure less fuel consumption [7,11]. In parallel to the environmental component of vehicle movement on safely designed roadways, speed choice conforming to speed limits is one of the core pillars to meet sustainability principles in traffic safety, thereby preventing road deaths, serious injuries, and permanent injury by systematically reducing the underlying risks of the entire traffic system.

1.2. Background on the Intervention Potential

While there is enough margin for improvements in the national and international road network, it is a matter of fact that modern road planning, construction, and maintenance repeatedly call for the wise and efficient use of scarce financial resources. In the meantime, the majority of the road network and corridors are completed or are under construction at both the national and international level [12]. In other words, planning for completely new roadway paths might be almost absent, especially in the developed countries. As such, it can be confidently stated that during the preceding decades, the concerns of roadway engineers have been directed toward the upkeep and restoration of the existing road network by slight interventions in the geometric design and by more frequent interventions because of poor pavement condition [13]. The main reason for that is again the budgetary constraints of road entities.

On these grounds, effective road maintenance guidelines are now regarded as being just as significant as sound initial construction and building techniques. A quite frequent and first-decision action when planning interventions in a roadway is to focus on the restoration of road surface condition. Of course, a prerequisite is that the pavement's structural and mechanical behavior is adequate [14–16]. Moreover, this is aligned with the concept of long-life, or perpetual, pavements, where structural soundness is guaranteed in the long term and only surface interventions to improve road serviceability take place [17]. Among the road surface characteristics, tire–road friction is drastically related to safe driving conditions and can be a contributing factor to road incidents [18]. Indeed, nearly 120,000 individuals lost their lives on U.S. roads between 2010 and 2013 [19]. Even in the event of non-fatal incidents, the economic implications are unquestionably dominant in addition to the potential for societal disturbance due to unfriendly and hazardous road infrastructures [20].

Other minor treatments that can be chosen before deciding for redesign interventions include improvements in signage, road markings, and/or lighting conditions especially during night hours on roadways with high incident risk potential [21]. The presence of road safety barriers is another critical safety factor of a hazardous roadway section. If none of the above can contribute to a safer and more sustainable driving mode, then modification or improvement of road design aspects during safety interventions in existing parts of the road network could be opted for. Of course, potential restraints because of the terrain configuration and the available space must be considered, which is usually a matter of concern, especially in urban or suburban areas [7].

Towards better addressing the needs for potential road redesign, the necessity of using various forms of qualitative assessments and analysis is emphasized in the EU Directive on Safety Management of Road Infrastructures [22], including safety audits for the improvement of existing roads and inspections of existing roads to identify safety issues in several road design aspects, and safety audits to suggest intervention improvements [23].

Key common triggers for interventions on the roadways are road safety and the provided functionality for drivers. Because of the socioeconomical implications of traffic incidents, either fatal or no, continuous research has been dedicated over the previous decades on the assessment of records of traffic volumes and crash rates [24]. Driving behavior under various sensitivity analyses of road design parameters, traffic and weather scenarios has been extensively investigated through the use of a driving simulator that can mimic real traffic conditions, in order to retrofit road design procedures [25]. Although the latter seems promising, its use needs caution because of the deviations between the predicted and the observed driving performance [26]. Further to this, the behavioral aspects of driving are commonly studied through simulation and advanced modelling techniques, whereas infrastructure status is of equal or even higher significance.

1.3. Objective

Thus, in this study an approach is presented to discuss road redesign potential based on the evaluation of network-level historical incident records from road crashes in the Greek road network for a seven-year period, i.e., 2016–2022. Publicly available data from the website of the Hellenic Statistical Authority reflecting different aspects of road infrastructure status [27] were analyzed to identify major contributing factors that can be thereafter assessed in order to strategically draw potential intervention plans. Besides the behavioral component of road safety, which is extensively addressed in the literature, the discussion provided in this study has to do with infrastructure-related components that are of equal, if not higher, significance, and are almost overlooked or sparsely investigated in the international literature.

The rest of the paper is organized as follows: Section 2 provides remarks and findings about common existing practices during road intervention planning. Section 3 includes the processing of the available data from the incident records together with supportive discussion points. Section 4 provides the concluding remarks of the current study.

2. Aspects to Consider for Driving Conditions Improvement and/or Road Redesign

2.1. Pavement Surface Characteristics

The condition of a pavement's surface directly affects driving safety. Pavement friction is the most known pavement characteristic that guarantees efficient vehicle maneuvering when brakes are applied, a quite necessary action especially under wet-weather conditions. The water film on the pavement surface reduces the contact area between the tire and the road surface, thereby increasing the potential for skidding events or even hydroplaning [28,29].

The behavioral attributes of skid resistance are well reported elsewhere [30,31]. Herein, for the sake of completeness, it is worth recalling its seasonal variation performance. According to Figure 1, after a prolonged dry period without rainfall, dust and other contaminants are expected to be present on the pavement's surface. Thus, skid resistance tends to be reduced as a result of the combined effect of "pollutants" and extended dry spells. Practically, a fine layer of contamination is built up on road surfaces, which can be particularly intense on heavily trafficked highways, because of the repeated tire–road wear process [31]. This layer can consist of particles of rubber, bitumen, road material, dust, and other debris that are all mixed together to form a kind of pollution.

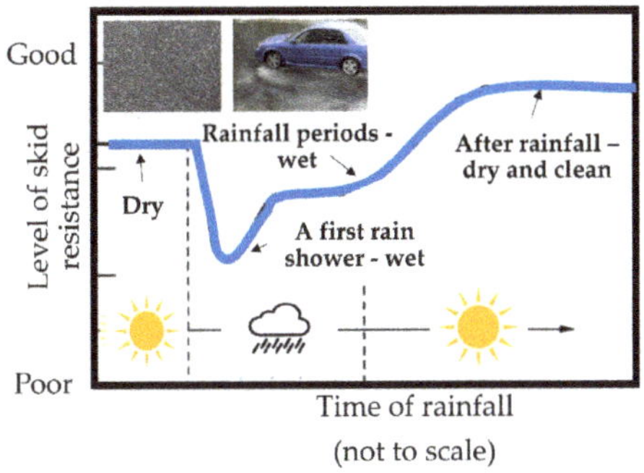

Figure 1. Illustration of friction level variation because of weather changes (adapted from [31,32]).

Rainfall is normally expected to wash the contaminants away. But after a prolonged dry period, the previously mentioned fine layer is additionally mixed with a small amount of water from the first and short rain shower events and the result is an extremely slippery road surface (also called the "summer ice" phenomenon), which is usually perceived by experienced drivers as a risky condition. Once rainfall persists for a long period, the "washing and cleaning" phenomenon takes place and as a result, skid resistance appears increased after a long period of rainfall with usually higher levels compared to the available friction in dry periods (right part of Figure 1) [31,32]. This is why many highway entities may tend to measure skid resistance twice per year; once at drought periods and once after rainfall. It is noted that there are multiple methods for measuring skid resistance, like the static methods (e.g., British Pendulum Number), and continuous methods, like locked-wheel or fixed-slip systems [32]. A common standard for friction measurements is ASTM E2340.

There is much documentation supporting the liaison between road incidents on wet surfaces and the level of provided skid resistance. Indeed, improved pavement friction has the potential to prevent or reduce around 70% of wet pavement crashes, according to research undertaken by the Federal Highway Administration (FHWA) and the National Transportation Safety Board [33]. Using a before-and-after analysis, Lyon et al. [34] assessed the safety impact of several pavement surface modifications in California, North Carolina, Pennsylvania, and Minnesota. The interventions appear to have reduced wet-road collisions in a statistically significant way, based on the combined results for all treatment types. From a different research perspective, Geedipally et al. [35] characterized the link between crash frequency and traffic, geometry, and pavement characteristics for horizontal curves in a southern state of the United States and created safety performance functions. It was discovered that the skid number had a considerable impact on all incidents, with rainy weather run-off-road crashes exhibiting the highest safety benefit.

Lyon et al. [36] recently assessed the safety impact of High Friction Surface Treatments (HFSTs) in various U.S. States. A specialized pavement treatment called HFST is a thin layer applied on the pavement surface to improve or restore friction, consisting of high-quality materials. Local small-scale treatments of curves, ramps, intersections, and steep grades—where the need for friction is greater than what can be met by traditional paving materials—are frequently carried out with it. The results demonstrated the contribution to lane-keeping driving performance when cars navigated curves as well as the remarkable statistically significant crash reductions at curve sites. As such, skid-resistant pavements have been shown to be an effective strategy to target lowering run-off-the-road collisions [37].

Irregularities in the road surface that could lead to a driver losing control of their vehicle are another significant factor that could potentially be dangerous for moving vehicles [38]. Unevenness (i.e., irregularities with wave-length between 50 m and 0.5 m) or roughness (Figure 2) is the profile of the road surface that impacts vehicle dynamics and water drainage. It is quantified using the International Roughness Index (IRI), expressed in m/km and calculated using a road profiler with a quarter-car math model at a standard speed of 80 km/h [39]. Because of the bilateral relationship of road distresses and roughness, already acknowledged in the international literature [40,41], rough surfaces can have the tendency to worsen pavement fatigue status, impose surface deformations, and raise vertical stresses in the pavement [40]. This could result in excessive surface distress, so that a driver may lose directional stability and increase stopping distances [37].

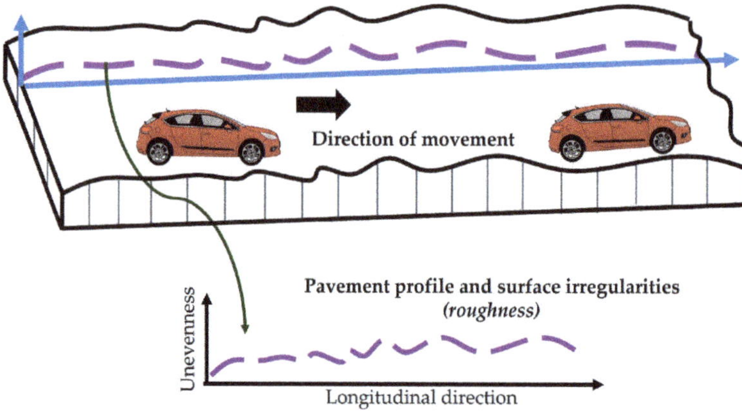

Figure 2. Illustration of a pavement surface with roughness issues (adapted from [39]).

The IRI is frequently used as the foundation for decisions on pavement maintenance because it is regularly gathered by numerous highway authorities across the globe and has been shown to accurately reflect the sense of road quality held by users. Indeed, Chan et al. [42] used long-term performance data from Tennessee to study the impact of pavement conditions (such as the International Roughness Index (IRI) and other indices) on collisions. They discovered a positive impact of the IRI on collisions, demonstrating that more frequent crashes are linked to poor pavement conditions. Nemtsov et al. [43] discovered most recently that there is a statistically significant correlation between an increase in IRI and a rise in crashes along two-lane rural highways.

Nevertheless, the surface characteristic that is most often used to trigger maintenance actions in favor of safe driving is skid resistance. In the case of pavement resurfacing, there can be indirect benefits, even short-term, in terms of restoring the ride quality as well.

2.2. Road Geometry

Once surface interventions to restore the pavement condition status seem to be insufficient, then one should more carefully consider the road design elements and evaluate the safety efficiency of the existing road network. Indeed, numerous studies have demonstrated how geometric design significantly affects traffic safety. AASHTO 2010's Highway Safety Manual (HSM) uses a number of crash modification factors (CMFs) to measure how changes in geometric design parameters from baseline conditions affect crash frequency [37]. The most commonly used geometric design features that are routinely investigated include lane width, shoulder width and type, horizontal curve superelevation, curve radius and length, existence of spiral transitions, and vertical grade [37].

It is a commonly accepted fact that safety effects on driving performance rarely depend solely on a single design factor. Researchers suggest that the impact of geometric design is multi-parametric and safe driving is greatly influenced by the overall highway alignment and the location of any geometric element in relation to the overall alignment and the distance from the other geometric elements [44,45]. In the same context, Cafiso et al. [37] stated that a highway design that guarantees that subsequent elements are coordinated in a way that produces harmonious and homogeneous driver performance along the road is regarded as consistent and safe.

This justifies why it is quite impossible to improve safety conditions by marginal interventions in the road geometry, which are indeed costly and thus undesirable. Further to that, safety interventions at existing road sections are usually restrained because of potential constraints resulting from the layout of the terrain configuration and the available space [7], especially in urban or suburban areas, or in mountainous areas with sharp relief and increased height deviations that could require much more expensive technical works.

However, reality may usually contradict safety design principles. If the horizontal and vertical alignment of a roadway deviate from the safety criteria of road design, then sharp speed changes could be anticipated, something that can increase the likelihood of crashes. The directional changes in a horizontal alignment should be smooth, so that a driver can be alerted and increase his/her degree of attention and expectations about the future alignment of the road [44,46]. As such, a driver can properly adjust the vehicle speed. For instance, on a relatively straight road, a sharp curve is much more dangerous than one on a winding route because the driver might not anticipate it and thus needs more time to adjust the vehicle speed [47]. Moreover, it is not so unusual to observe more crashes on straight roads of high length [48,49], because of the increased speeds, something that can be magnified once the pavement is distressed enough (as explained earlier), and/or warning signage is rather absent.

Finally, the impact of visibility and sight distance is another crucial parameter of road design, as crash sensitivity for a given roadway section with erroneous design elements might be more intense and detrimental at night, thereby requiring additional countermeasures (e.g., lighting, signage). A combination of various design elements that seems to be functional for driving in daytime may not maintain this functionality during nighttime [50]. The role of lighting and signage is further explained in Section 2.4.

2.3. Intersections Versus Roundabouts

A roadway intersection is the area where two or more roads join or cross at-grade. One of the most dangerous places in the road transportation network is an intersection, where there is an increased likelihood of fatalities and severe injuries [51]. In both developing and developed nations, intersection-related crashes make up roughly 20–45% of all recorded crashes [52–54]. There are many different and complex circumstances that lead to crashes involving intersections. In most cases, drivers pay little attention to the complex environment of an intersection and may be prone to illegal maneuvering because of misconception and erroneous understanding of the possible driving directions [55]. Crash types, as well as their frequency and severity, are influenced by a variety of factors, including intersection kinds, vehicle control, and road geometry [56]. For instance, there are up to four times as many crashes at four-legged junctions than at simple T-intersections. Rear-end collision rates are higher at signalized intersections than at stop-controlled intersections [55]. However, compared to signal-controlled crossings, stop-controlled intersections see a higher number of angle crashes.

An innovative solution to cope with the increased incidents at intersections is to transform them into roundabouts. This action has shown that a sufficient degree of road safety and traffic capacity can be achieved [57]. Drivers need to slow down as they approach a roundabout in order to navigate it easily and exit it without being involved in an incident. Without traffic signals, traffic delays are therefore streamlined or even reduced. As a matter of fact, studies have shown that converting a signalized intersection into a roundabout reduces traffic delays by 89% and complete vehicle stops by 56% [58]. Roundabouts also have a special opportunity to raise drivers' safety. The explanation is straightforward: once the emphasis is on urban crossings, conflict points are drastically reduced (Figure 3), for both cars and pedestrians, in comparison to conventional intersections. The same does also apply for roundabouts on rural roadways. Drivers are compelled to slow down, which makes it much easier to manage their risk of becoming involved in an accident with a pedestrian or another car [59].

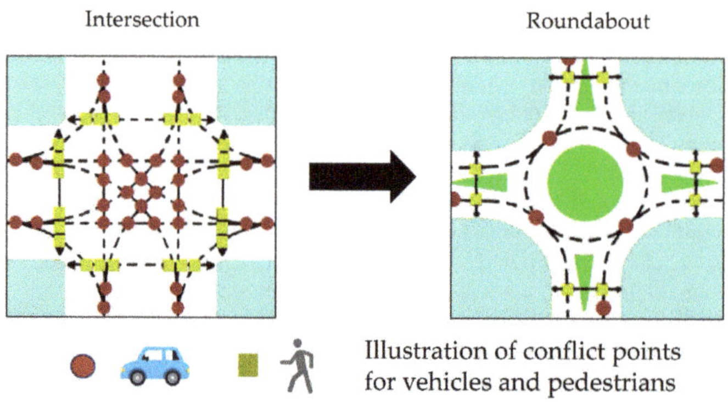

Figure 3. Comparison of the number of conflict points at intersections and roundabouts, including cross conflicts, merge conflicts, diverge conflicts, and pedestrian cross-walks.

Of course, the selection of an optimal type of roundabout (e.g., mini-roundabout, single-lane or double-lane roundabouts, etc.) with proper values for its geometric elements aims at reaching a balance between drivers' safety and traffic capacity. The drivers' perception of danger, which is influenced by their driving performance and experience, is undoubtedly influenced by geometric design components [60]. In order to ensure a smoother transition pattern at this type of cyclic intersection, vehicles should typically yield to the oncoming traffic rather than being forced to a complete stop at roundabout entry positions [61]. Because of the reduced vehicle speeds, Burdett et al. [62] found a 38% reduction in fatal injuries and severe collisions at roundabouts. Within the same framework, De Brabander et al. [63] found that the average rate of reduction for the total number of injury accidents at roundabouts was 38% for serious injury accidents.

However, it is still unclear how roundabouts affect the quantity of crashes that result in no injuries [64]. Polders et al. [65] claim that accidents still happen even with roundabouts' contribution to traffic safety. From a geometrical perspective, the number of the available lanes of a roundabout is critical. Overall, the likelihood of a light, non-injury crash increases with the number of lanes [57]. Indeed, Johnson [66] has recently noted a considerable rise in property-damage-only crashes at multi-lane roundabouts.

Nevertheless, the goal of interventions at intersections is to make the road infrastructure better and easily understandable for drivers with emphasis on vulnerable road users. Moreover, a rational and sustainable vision for safer roads comprises little or no fatal incidents, or the ambitious goal of "zero deaths" along roads [67,68]. Non-injuries crashes seem affordable, as it is impossible or at least unrealistic to fully eliminate them based on the randomness of human driving behavior even under ideal infrastructure conditions.

2.4. Additional Road Equipment

Even when a road's geometric design meets all safety criteria, there are still some actions needed to guarantee safe driving conditions. These include proper equipment to account for: (a) comprehensive signage and road markings, (b) adequate lighting, and (c) necessary road safety barriers in critical areas.

According to Cho et al. [69], there is no doubt that elements connected to the road environment, such as missing road markings, broken signs or traffic lights, and inadequate lighting, can have a direct impact on traffic incidents and/or fatal accidents. In summary, traffic safety signs alert drivers to possible dangers such as sharp curves, junctions, and construction zones and help them become gradually adaptable to the road environment.

They also offer guidance on how to handle these circumstances through safe navigation. In other words, drivers can become capable of higher cognitive activity once proper signs and markings are present [21]. Rightfully, road signs continue to be an essential public tool to ensure driving safety [70,71]. The likelihood and frequency of traffic accidents are decreased when traffic signs are appropriately obeyed. The kind and placement of the sign as well as the behavioral actions of drivers influence how much road signs reduce the frequency of traffic accidents. Important elements include traffic law enforcement, driver education, and safe road and highway design [72].

In the same context, road safety is greatly enhanced by clear and well-maintained road markings. Visibility is aided by the difference between the road surface and the road markings, particularly at night or in adverse weather. Numerous studies demonstrate a relationship between road markings and the lateral position of the vehicle within the lanes [73,74]. In particular, when the width of the road markings increases, drivers shift their vehicle's lateral position and bring it closer to the side of the road, lowering the possibility of a head-on collision [73,74]. The application of vibration markings, road curve signs, transverse warning markings, and markers result in a notable decrease in the incidence of centerline crossings.

Road illumination at night does, in fact, contribute significantly to traffic safety too. According to research by the International Commission on Illumination, road lighting decreased nighttime accidents in 15 different nations by 13–75% [75]. Road lighting in Belgium, Britain, and Sweden decreased the number of crashes that occurred at night by 23% [76]. As a result, assessing the quality of road lighting is important to increase urban traffic safety and the quality of the road lighting at night [77].

Of course, maintaining adequate lighting everywhere at night is impossible as this is translated into environmental costs. This is why intelligent street lighting has appeared as an alternative sustainable option, so that efficient and safe driving at night is no longer compromised. With a privileged domain of implementation in cities, smart lighting technologies, like the one illustrated in Figure 4, include motion detectors to activate illumination only when necessary. This can contribute towards the minimization of road crashes because of nighttime limited road visibility by simultaneously ensuring road functionality and energy efficiency.

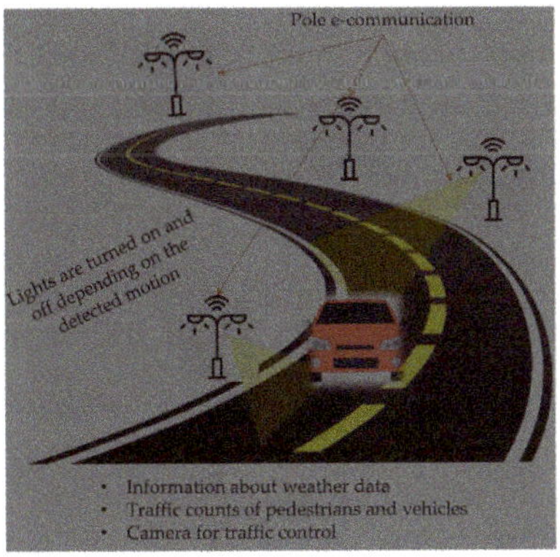

Figure 4. Conceptualization of a smart illumination system of roadways at nighttime.

Finally, standard safety measures, which are often implemented in the curves, are road safety barriers. Road safety barriers are part of the road restraint systems (RRSs) intended to prevent the vehicle from leaving the roadway [78,79]. The main contribution of a safety barrier is that when the driver loses control of the vehicle and leaves the roadway, the barrier can save his life by stopping the vehicle or redirecting it back to the road. It is self-proved that their presence in a roadway environment is very important for the purpose of safe driving.

2.5. Importance of Accessing and Managing Data

If most of the above conditions were met, the infrastructure condition would be ideal, functional, environmentally aesthetic, and most importantly fully-understandable by drivers, resulting in a safe and sustainable driving performance. In many aspects of road safety, including infrastructure management, crash data are essential because they form the basis for risk assessment and the application of efficient safety measures. The availability and dependability of crash data, however, frequently present serious difficulties. Hence, ensuring road crash inventory data might be a challenge for those wishing to invest in reliable road safety research. A typical procedure of how data should be handled is given in Figure 5.

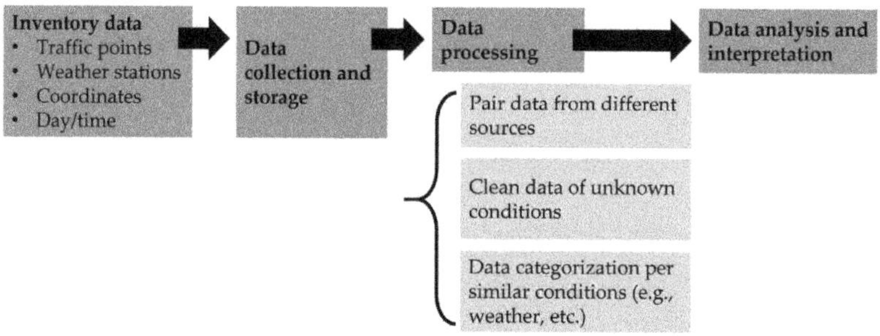

Figure 5. Conceptualization of effective road crash data management.

Of course, it is critical to look beyond obvious causation variables when studying crash data, since crashes can arise from a confluence of contributing events (e.g., traffic conflict points, weather status, poor pavement condition, bad road design, etc.). Again, multiple sources of data might be needed that could play a crucial role in the assessment of investments, especially in situations where funds are limited. To this end, it is crucial to show the efficacy of road safety initiatives in order to allocate scarce funds effectively and to identify and mitigate any potential increases in crash risk. Preserving the integrity of data is significant in order to avoid misreading or underestimating the number of casualties, which could jeopardize attempts to address road safety as a pressing public health issue [80].

Nowadays, with several advancements in data collecting techniques, the efficiency and accuracy of data analysis is much more improved, thereby yielding reliable results. Errors can be minimized and data gathering and interpretation procedures can be more easily homogenized. Rational and standardized data processing can definitively increase the usefulness for developing road safety policies and making well-informed decisions.

3. Data and Discussion

3.1. Overview

Following the scope of this study, an analysis of official statistical data from road crashes in Greece is made for the period 2016–2022. First, an overview of total road crashes and fatal crashes is given in Tables 1 and 2 per road category according to the process followed by the police, during the crash reporting and recording procedures. The percentages shown in Table 2 correspond to a specific road category and year; for instance, in 2016, the rate of fatal crashes on freeways was 38/284, or 13.4%.

Table 1. Official records of total and fatal crashes [27].

Road Category	Total Number of Crashes							*Number of Fatal Crashes*						
	2016	2017	2018	2019	2020	2021	2022	*2016*	*2017*	*2018*	*2019*	*2020*	*2021*	*2022*
Freeways	284	313	354	279	210	281	312	*38*	*47*	*58*	*47*	*31*	*35*	*42*
Intercity highways	1173	994	942	800	581	723	793	*232*	*213*	*142*	*151*	*106*	*119*	*134*
Local roads	1183	1092	1138	1072	709	1043	1114	*192*	*166*	*155*	*166*	*128*	*167*	*159*
Municipal roads	8632	8404	8252	8496	7512	8344	8184	*300*	*239*	*273*	*271*	*268*	*253*	*268*
Other	46	45	51	65	71	63	84	*10*	*14*	*17*	*21*	*19*	*10*	*16*

Table 2. Percentage of fatal crashes per year and road category.

Road Category	2016	2017	2018	2019	2020	2021	2022
Freeways	13.4%	15.0%	16.4%	16.9%	14.8%	12.5%	13.5%
Intercity highways	19.8%	21.4%	15.1%	18.9%	18.2%	16.5%	16.9%
Local roads	16.2%	15.2%	13.6%	15.5%	18.1%	16.0%	14.3%
Municipal roads	3.5%	2.8%	3.3%	3.2%	3.6%	3.0%	3.3%
Other	21.7%	31.1%	33.3%	32.3%	26.8%	15.9%	19.1%

As can be seen, crashes are more frequent on municipal roads, which implies urban areas; a fully anticipated result because of the higher number of conflict points within an urban road network. However, fatal crashes on municipal roads are of marginal magnitude compared to the rates shown in Table 2 for the rest of the road categories. This is mainly because of the higher speeds that vehicles can reach while travelling outside an urban area, thereby resulting in crashes with much more intensity. Focusing, on the first three road types, data are also given in Figure 6 (total crashes on the left axis, fatal crashes on the right axis), from which it can be stated that the rate of fatal crashes on freeways is nearly stable over the investigation period, whereas on intercity highways and local roads, there is a tendency for a progressive decrease in the percentage of fatal crashes.

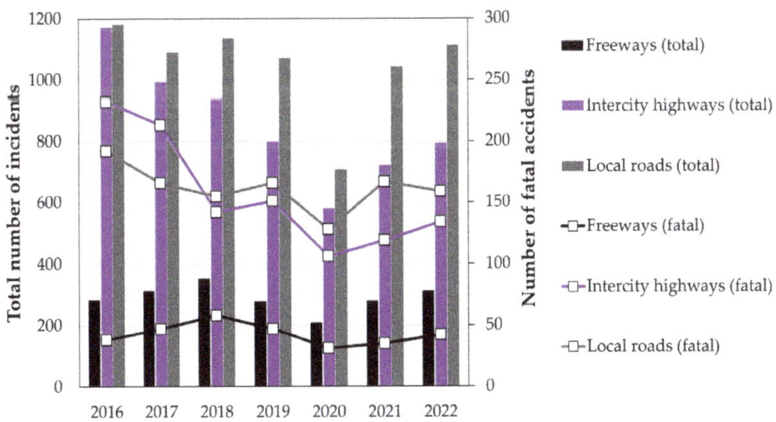

Figure 6. Evolution of total and fatal crashes on non-urban roadways (bars for the left vertical axis and lines for the right vertical axis).

The lowest number of total crashes as well the lowest rate of fatal crashes was observed in 2020, probably because of the COVID-19 pandemic period. Nevertheless, from Figure 6, it can be argued that the road status upgrade because of the construction of core freeways in Greece during the last two decades led to the descending trend in the rate of fatal crashes on the rest of the regional road network. Nevertheless, a potential future focus for road redesign procedures should be put on that type of roadway, as in absolute numbers, non-freeway crashes are of a considerable amount.

3.2. *Surface Status*

Following the description of the impact of pavement surface characteristics, the total number of incidents is now distributed under various road surface conditions, including normal (dry) surfaces, wet surfaces, surfaces contaminated with oil or dust, icy surfaces, and snowy surfaces. It is noted that no roughness or other distress data are provided in the datasets of [27]. Given this, results related to the road's surface are given in Tables 3 and 4.

Table 3. Distribution of annual total and fatal crashes per year and road surface condition [27].

Road Surface Status	Total Number of Crashes							*Number of Fatal Crashes*						
	2016	2017	2018	2019	2020	2021	2022	*2016*	*2017*	*2018*	*2019*	*2020*	*2021*	*2022*
Normal (dry)	10,416	10,020	9825	9740	8442	9708	9809	*685*	*599*	*569*	*586*	*505*	*515*	*568*
Wet	810	721	826	847	555	648	601	*79*	*73*	*71*	*63*	*41*	*65*	*45*
Contaminated	28	27	33	29	24	28	23	*1*	*0*	*1*	*0*	*1*	*0*	*0*
Icy	32	32	16	27	11	9	11	*3*	*5*	*3*	*2*	*2*	*0*	*1*
Snowy	18	39	25	33	23	38	24	*2*	*1*	*0*	*1*	*0*	*4*	*1*
Other	14	9	12	36	28	23	19	*2*	*1*	*1*	*4*	*3*	*0*	*4*

Table 4. Percentage of fatal crashes per year and road surface condition.

Road Surface Status	2016	2017	2018	2019	2020	2021	2022
Normal (dry)	6.6%	6.0%	5.8%	6.0%	6.0%	5.3%	5.8%
Wet	9.8%	10.1%	8.6%	7.4%	7.4%	10.0%	7.5%
Contaminated	3.6%	0.0%	3.0%	0.0%	4.2%	0.0%	0.0%
Icy	9.4%	15.6%	18.8%	7.4%	18.2%	0.0%	9.1%
Snowy	11.1%	2.6%	0.0%	3.0%	0.0%	10.5%	4.2%
Other	14.3%	11.1%	8.3%	11.1%	10.7%	0.0%	21.1%

Excluding the normal (dry) surface and the "other" surface, an increased rate of crashes is observed on wet road surfaces. Grouping the categories "wet, contaminated, icy, and snowy" as abnormal surface conditions, it appears that nearly 90% of the total crashes occur on wet surfaces (Table 3, left part), whereas 10% occur on contaminated, icy, or snowy surfaces. This rate seems to be constant over the years. Noticeably, nearly 7–10% of "wet" crashes are fatal (Table 4). Focusing on the impact of wet surfaces (recall Figure 1), the amount of water that is present on the road's surface is critical for the vulnerability of a particular subsection. Figure 7 demonstrates the comparison between crashes because of a short rain shower versus crashes because of rainfall. Note that the annual sum shown in Figure 7 is lower than the "wet" crashes of Table 3, since Table 3 is related to crashes that may have occurred on a wet surface, including those after the rainfall event, whereas Figure 7 includes only those occurred during the climatic events.

From Figure 7, it can be observed that a surface with wet dust is much more dangerous for crash generation than a fully wet surface. Drivers might overlook the previously mentioned condition of "summer ice" described in the literature. This is the phenomenon when a road surface becomes very slippery because pre-existing contaminants, dust, or other loose debris on the road surface are mixed with a small amount of water because of short rain showers, thereby resulting in a surface with poor skid resistance and a reduced braking capability. Drivers may erroneously maintain their speed, and as a consequence, they can be involved in road incidents much more easily.

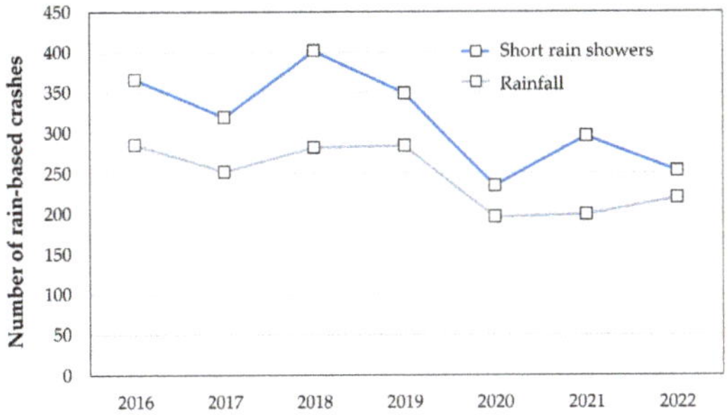

Figure 7. Crashes because of climatic events including short rain showers and rainfall.

Road authorities may also frequently overlook this factor, which can be eliminated by properly cleaning the surface at periodical intervals, especially after a prolonged dry period, by using routine maintenance methods (e.g., water blasting) in advance. Basically, in Greece some road authorities occasionally incorporate surface cleaning into their maintenance schedules mainly in the case of motorways. This is deemed much more necessary, especially in the case of tunnels, where the road surface cannot be completely dried and cleaned after a rainy event. However, for the rest of the roadway network, this still remains an issue because of the criticality of crashes during rainfall. It is noted that nearly 7% to 11% of the rain-based crashes in Greek territory are fatal according to [27].

3.3. The Role of Lighting and Signage

As previously explained, lighting conditions are related to the visibility of drivers, whose directional and maneuvering capabilities can lead to a considerable number of crashes. Data about crashes because of visibility conditions are given in Tables 5 and 6. Because of different data availability patterns for this condition status, the number of deaths was given in [27] instead of the number of fatal crashes shown in the previous subsections. So, a slightly different concept is presented in Table 6 that gives the "crash repetitions" per one death. A nearly constant frequency of 15–20 crashes is observed for cases of sufficient physical lighting (i.e., daytime and dusk). Sufficient lighting at nighttime can have an equivalent performance in terms of fatal crash frequency with those during daytime.

On the contrary, a considerable difference is observed during nighttime under no lighting conditions, where nearly one death is recorded every four crashes. Noticeably, the inadequacy or the malfunction of lighting results into a considerable rate of fatal crashes, i.e., one death is recorded every 5–10 crashes. This can provide a significant remark for the related authorities and decision-makers about the impact of good lighting conditions. Low or inadequate lightning, as well as the complete absence of lighting, does not satisfy drivers' needs and may alert for investments and greener solutions to ensure lighting through smart and cost-effective illumination systems, like the one suggested by Figure 4. As a result of good lighting, visibility can be improved and road collisions might be reduced.

Table 5. Distribution of annual total crashes per year and visibility conditions [27].

Visibility Status	Total Number of Crashes						
	2016	2017	2018	2019	2020	2021	2022
Daytime	7341	7136	7058	7019	5991	6907	6778
Dusk	614	566	579	532	508	546	553
Nighttime, sufficient lighting	2553	2420	2379	2418	2027	2421	2548
Nighttime, insufficient lighting	411	387	392	404	283	335	285
Nighttime, out of order lighting	63	42	50	45	37	30	40
Nighttime, no lighting	336	297	279	294	237	254	299

Table 6. Crash occurrence per deaths (i.e., one death per *X* crashes or else fatal crash frequency).

Visibility Status	Number *X* of Crashes per One Death						
	2016	2017	2018	2019	2020	2021	2022
Daytime	16	17	17	18	18	20	18
Dusk	17	16	18	17	17	16	15
Nighttime, sufficient lighting	15	17	19	19	17	17	19
Nighttime, insufficient lighting	6	7	7	8	9	10	7
Nighttime, out of order lighting	5	4	6	5	5	8	8
Nighttime, no lighting	4	4	4	4	4	4	5

Proper design of signage can be extremely useful for drivers, especially for those belonging to vulnerable categories (i.e., younger or elderly drivers), or for those who are new drivers in a particular area, and thus not fully familiarized. Moreover, a roadway path should be self-explaining for all drivers irrespective of their age or driving experience. Absence of signage or mal-design of road signs raises the potential for crashes, especially within urban areas because of the increased number of conflict points [81]. The interaction with pedestrians is also another aspect to consider in terms of strictly selecting and placing related signage.

In [27], information about signage comes from a category named "traffic regulation: markings and signaling". Subcategories include police enforcement, visible or invisible traffic signals, visible or invisible signs for stoppage, signage for sharp horizontal and vertical curves, etc. Based on the subcategories that account for a considerable number of crashes, such that meaningful conclusions can be made, those appearing in Figure 8 prove that most crashes occur in areas with visible traffic signals or visible stoppage signs. This fact implies that road crashes are more likely to occur in urban areas with many intersections, something which confirms the statistics given in Table 1.

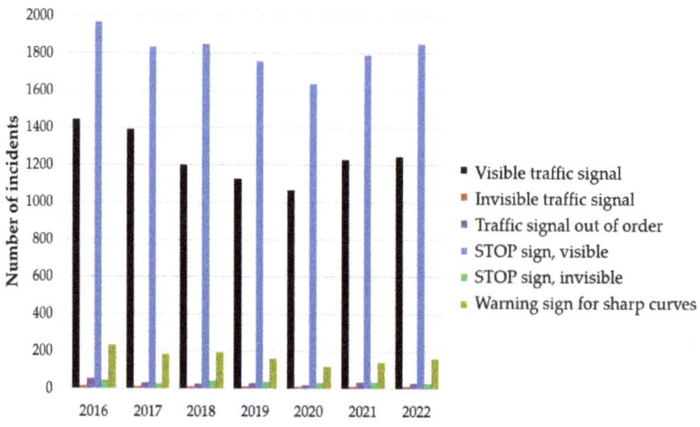

Figure 8. Crashes because of traffic regulatory measures (signals, signage, etc.).

According to Figure 8, a first reaction creates controversy considering that signing is visible. Upon closer inspection, the results could be impacted by the fact that most incidents occur on municipal roads (recall Table 1), where a number of factors, including signage, could contribute to the incident. Despite their visibility, the high number of incidents may suggest that suitable and effective signage design and placement need to be reconsidered, and more research and implementation should be carried out in this area.

Finally, identifying black spots in an urban area could help locate the exact points where re-assessment of the existing signage efficacy is needed. Moreover, the execution of black spot analysis can help identify a weak road or a traffic design that requires redesign assessment [82]. For the sake of completeness, it is noted that apart from infrastructure-based interventions, some additional behavioral driving patterns have been proposed to improve drivers' safety, like the speed limit of 30 km/h in urban areas [83].

3.4. Geometric Characteristics of Road Design

Inevitably, road geometrics are the key factor for designing safe infrastructure. As explained earlier, several design features have been found to correlate to some extent with safe driving conditions and accident ratio, including among others curve radii, number of lanes, superelevation, existence of emergency lane, slope, speed limit, etc. [49,84]. For the present study, the contribution of each sole geometric category is shown in Figure 9. The available statistics provide evidence that the majority of incidents occurs in the case of straight alignment, followed by at-grade intersections.

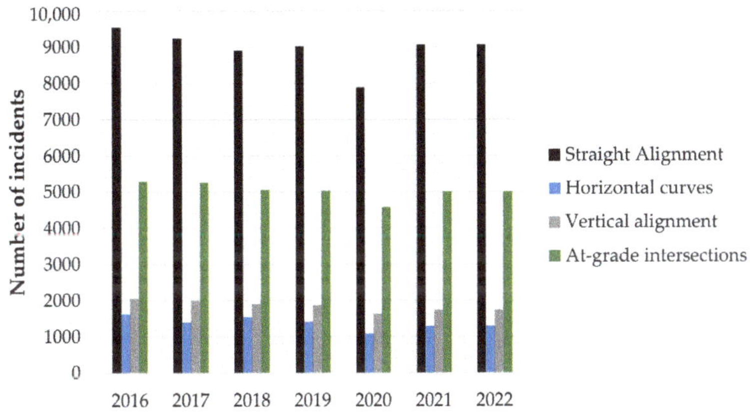

Figure 9. Crashes per different characteristics of road geometric design.

An issue with this discrimination is that it difficult to organize network-level data much more precisely, such that different combinations of vertical and horizontal alignment could form additional subcategories. In other words, the bars in Figure 9 provide duplicate information, considering that an individual crash on both horizontal and vertical curves has been counted double. This is why tables depicting total and fatal crashes cannot be shown at this stage alike the previous analysis steps, because they might provide some duplicate information and could mislead the reader. Furthermore, the combination of a location at an intersection and in the area of a horizontal curve could also account for a considerable number of crashes. This means that many more additional categories could be formed. Nevertheless, in order to extract some information about the criticality of road design aspects, crash data per geometric design category are given as a percentage out of the total number of crashes per year in Table 7.

Table 7. Percentage of crashes per year depending on the geometric design feature.

Geometric Characteristic	2016	2017	2018	2019	2020	2021	2022
Straight Alignment	84%	85%	83%	84%	87%	87%	86%
Horizontal curves	14%	13%	14%	13%	12%	12%	12%
Vertical curves	18%	18%	18%	17%	18%	17%	17%
At-grade intersections	47%	48%	47%	47%	50%	48%	48%

From Table 7, the following remarks can be made:

- The overwhelming majority of road incidents take place on straight alignments, probably because drivers overestimate their capabilities and increase their moving speed. This is why many international design standards for roads alert for a maximum length of straight alignment, e.g., [85], to avoid leaving drivers to navigate there for a long time. Although there might be cases when the previous remark is violated, there are solutions to limit or at least control the increased crash rate on straight alignments. These could include refining signage and road markings to help drivers adapt to the current road environment, and much more law enforcement and police patrols that could help drivers adopt a different behavioral perspective while driving. Moreover, reconsideration of speed limits could help drivers better understand the environmental component of safe driving, because of the lower pollutant emission and fuel consumption that could save money. It is noted that the rate of fatal crashes on straight alignments falls in the range of 4–6%;
- Nearly one out of two incidents occur as a result of at-grade intersections. Hopefully, only 3% of them are fatal accidents, and this rate appears to be constant over the investigation period. Profoundly, a considerable rate of these crashes belongs to the subcategory of straight alignment too. A limitation of the available data is the inability to discriminate what portion of the intersection-related crashes is in urban areas or not. Nevertheless, since intersections are characterized at an international level as black spots for incidents, they still pose a significant risk to the public, which often leads to serious accidents between road crossings, or even road and rail crossings. This should alert the related authorities about the need to upgrade the existing vehicle interaction system at intersections, by integrating modern design perspectives (e.g., roundabouts, etc.) and novel technologies. These could include smart lighting, proper road markings, and improved guide signs to provide a better directional management of the vehicle position, especially for vulnerable road users trying to interact with surrounding vehicles and/or pedestrians as well;
- The horizontal and vertical profile of a roadway appears to cause the lowest percentage of the total crashes (recall Table 7). However, nearly 10–15% of them are fatal. This proves the criticality of road accidents on curves, thereby emphasizing the necessity for a continuous and well-planned upgrade of the existing road network on curves. Of course, a complete road redesign in such cases might be impossible because of budget constraints, terrain configuration, limitations due to expropriations, and relief constraints that could require costly technical works, like bridges or tunnels, etc. Nevertheless, more minor, yet strategic interventions could mitigate the road crash potential, like (i) the use of HFST courses to ensure skid resistant road surfaces near curves, or areas with high longitudinal slopes, (ii) the improvement or complete replacement of signage and road markings, (iii) the reconsideration of speed limits, as well as (iv) the proper lighting of critical roadway segments to ensure visibility at nighttime.

4. Conclusions

4.1. Main Remarks

A multitude of research and experience has shown that a scientific, rigorous, consistent, and proactive approach to highway planning and design may considerably improve overall driver safety. Given the expense of engineering studies and limited funding, simulation studies, ad hoc safety evaluations, on-site inspections, and reviews of available historical data are among the major practices to follow in order to propose engineering remedies and improve the crash susceptibility of high-risk road subsections. This study used network-level historical records of crashes in the Greek territory over a seven-year period, and proposed four core and contributing pillars on which the focus for future road redesign planning could be put. These were all related to the road infrastructure status and included:

- the need to preserve adequate performance status of the road surface, which indicates the necessity for skid resistance measures at periodical intervals in order to plan any kind of restoring interventions;
- the need to ameliorate the signage and lighting status on the entire road network with an emphasis on high-risk locations;
- the need to improve the driving conditions near at-grade intersections, by improving signage, road markings, and lighting, and examining the potential for transforming core intersections into roundabouts; and
- the ability to intervene, to some extent (if possible), given several techno-economical constraints, at specific road segments because of current questionable design features. It is noted that the purpose of this study was not to propose specific segments that are subject to improvement. Moreover, the implementation scale is a multi-parametric task that requires a multitude of information with road design and geolocation data that were absent from the online dataset of crash records. This should be acknowledged as a study limitation.

The ultimate goal of these intervention pillars is to stress the importance of a good infrastructure status irrespective of the behavioral component of driver safety; or in other words, the target is a well-maintained road surface and a self-explaining road design that can jointly ensure improved directional capabilities of all drivers, thereby resulting in fewer crashes, or more accurately, fewer fatal crashes.

4.2. Challenges and Prospects

Actions for at least one of these pillars should aim at enhancing the safety and functionality of roadways. Towards this, the following challenges are mentioned when moving from a strategic planning to an implementation scale:

- data should be oriented from project-level records and reporting to identify a list of the most hazardous sites;
- crash data should be collected and reported through a standardized procedure including as much and as accurate information as possible, by experienced personnel;
- a prioritization process should follow for the scheduling of the most cost-effective redesign projects.

Overall, once crash data are to be used for research purposes by engineers and for selecting intervention actions by policy-makers and road stakeholders, a robust procedure to collect, manage, analyze, and interpret data is important. It is to be noted that although the trigger for this research was data coming from a single country, the methodological approach presented should be replicated elsewhere, by redefining those pillars (or even identifying new ones) that are representative and adaptive to the conditions met elsewhere, including specifications for road design and maintenance.

Aspects of future research developments and prospects could include more detailed analysis of crash records homogenized for specific road types and design data (e.g., study of crashes on horizontal curves, vertical curves, straight alignments, etc.), so that more informative conclusions can be drawn, and more targeted recommendations can be made, especially for the pillar of road redesign, which is the most difficult to achieve and implement. More in-depth cross-referenced data might be needed to perform advanced statistical analysis and verify the sole and joint impact of the contributing factors to road incidents.

Author Contributions: Conceptualization, K.G. and M.P.; Methodology, K.G. and M.P.; Analysis, K.G. and M.P.; Writing—review and editing, K.G. and M.P. All authors have read and agreed to the published version of the manuscript.

Funding: This research received no external funding.

Data Availability Statement: Raw data considered for this paper is publicly available at the Hellenic Statistical Authority (ELSTAT) website, "https://www.statistics.gr/en/statistics/-/publication/SDT04/2000 (accessed on 6 June 2024)".

Conflicts of Interest: The authors declare no conflicts of interest.

References

1. Kalita, K.; Maurya, A.K. Probabilistic geometric design of highways: A review. *Transp. Res. Procedia* **2020**, *48*, 1244–1253. [CrossRef]
2. Huang, Z.; Chen, F.; Xu, R. Research on Highway Landscape Design Based on Driver's Visual Characteristics. *IOP Conf. Ser. Earth Environ. Sci.* **2019**, *330*, 022127. [CrossRef]
3. World Health Organization (WHO). *Global Status Report on Road Safety*; WHO: Geneva, Switzerland, 2018.
4. Nikolaou, D.; Folla, K.; Yannis, G. Impact of Socioeconomic and Transport Indicators on Road Safety during the Crisis Period in Europe. *Int. J. Inj. Contr. Saf. Promot.* **2021**, *28*, 479–485. [CrossRef] [PubMed]
5. European Transport Safety Council (ETSC). *15th Annual Road Safety Performance Index (PIN) Report*; ETSC: Brussels, Belgium, 2021.
6. European Commission. Available online: https://transport.ec.europa.eu/2021-road-safety-statistics-what-behind-figures_en (accessed on 18 July 2024).
7. Gkyrtis, K. Theoretical Considerations from the Modelling of the Interaction between Road Design and Fuel Consumption on Urban and Suburban Roadways. *Modelling* **2024**, *5*, 737–751. [CrossRef]
8. European Automobile Manufacturers Association (EAMA). *Paving the Way to Carbon-Neutral Transport: 10-Point Plan to Help Implement the European Green Deal*; ACEA: Brussels, Belgium, 2020.
9. Nasir, M.K.; Md Noor, R.; Kalam, M.A.; Masum, B.M. Reduction of Fuel Consumption and Exhaust Pollutant Using Intelligent Transport Systems. *Sci. World J.* **2014**, *2014*, 836375. [CrossRef] [PubMed]
10. Liu, J.; Feng, L.; Li, Z. The Optimal Road Grade Design for Minimizing Ground Vehicle Energy Consumption. *Energies* **2017**, *10*, 700. [CrossRef]
11. Plati, C.; Gkyrtis, K.; Loizos, A. A Practice-Based Approach to Diagnose Pavement Roughness Problems. *Int. J. Civ. Eng.* **2024**, *22*, 453–465. [CrossRef]
12. Gkyrtis, K.; Loizos, A.; Plati, C. A mechanistic framework for field response assessment of asphalt pavements. *Int. J. Pavement Res. Technol.* **2021**, *14*, 174–185. [CrossRef]
13. Marecos, V.; Fontul, S.; Antunes, M.L.; Solla, M. Evaluation of a highway pavement using non-destructive tests: Falling Weight Deflectometer and Ground Penetrating Radar. *Constr. Build. Mater.* **2017**, *154*, 1164–1172. [CrossRef]
14. Loizos, A.; Gkyrtis, K.; Plati, C. Modelling Asphalt Pavement Responses Based on Field and Laboratory Data. In *Accelerated Pavement Testing to Transport Infrastructure Innovation*; Lecture Notes in Civil, Engineering; Chabot, A., Hornych, P., Harvey, J., Loria-Salazar, L., Eds.; Springer: Cham, Switzerland, 2020; Volume 96, pp. 438–447.
15. Loizos, A.; Spiliopoulos, K.; Cliatt, B.; Gkyrtis, K. Structural pavement responses using nonlinear finite element analysis of unbound materials. In Proceedings of the 10th International Conference on Bearing Capacity of Roads, Railways and Airfields (BCRRA), Athens, Greece, 28–30 June 2017; pp. 1343–1350.
16. Georgouli, K.; Pomoni, M.; Cliatt, B.; Loizos, A. A simplified approach for the estimation of HMA dynamic modulus for in service pavements. In Proceedings of the 6th International Conference on Bituminous Mixtures and Pavements (ICONFBMP), Thessaloniki, Greece, 10–12 June 2015; Taylor and Francis: Abingdon, UK, 2015; pp. 661–670.
17. Ferne, B.; Nunn, M. The European Approach to Long Lasting Asphalt Pavements: A state-of-the-art review by ELLPAG. In Proceedings of the Paper presentation in the International Conference on Perpetual Pavements, Columbus, OH, USA, 13–15 September 2006.
18. Gkyrtis, K.; Pomoni, M. An Overview of the Recyclability of Alternative Materials for Building Surface Courses at Pavement Structures. *Buildings* **2024**, *14*, 1571. [CrossRef]
19. National Highway Traffic Safety Administration. *Fatality Analysis Reporting System*; National Highway Traffic Safety Administration: Washington, DC, USA, 2012.
20. Connelly, L.B.; Supangan, R. The economic costs of road traffic crashes: Australia, states and territories. *Accid. Anal. Prev.* **2006**, *38*, 1087–1093. [CrossRef] [PubMed]
21. Fiolić, M.; Babić, D.; Babić, D.; Tomasović, S. Effect of road markings and road signs quality on driving behaviour, driver's gaze patterns and driver's cognitive load at night-time. *Transp. Res. Part F Traffic Psychol. Behav.* **2023**, *99*, 306–318. [CrossRef]
22. Council of the European Union; European Parliament. Directive 2008/96/EC of the European Parliament and the Council of 19 November 2008 on Road Infrastructure Safety Management. *Off. J. Eur. Union* **2008**, *319*, 59–67.
23. Colonna, P.; Intini, P.; Berloco, N.; Fedele, V.; Masi, G.; Ranieri, V. An Integrated Design Framework for Safety Interventions on Existing Urban Roads—Development and Case Study Application. *Safety* **2019**, *5*, 13. [CrossRef]
24. Crispino, M.; Camozzi, K.; Ketabdari, M.; Antoniazzi, A.; Toraldo, E. Safety Impact Prediction of Redesigning National Roads Crossing Residential Areas: An Italian Case Study. *Appl. Sci.* **2024**, *14*, 4984. [CrossRef]
25. Bella, F. Driving simulator for speed research on two-lane rural roads. *Accid Anal Prev.* **2008**, *40*, 1078–1087. [CrossRef]
26. Gemou, M. Transferability of driver speed and lateral deviation measurable performance from semi-dynamic driving simulator to real traffic conditions. *Eur. Transp. Res. Rev.* **2013**, *5*, 217–233. [CrossRef]
27. Hellenic Statistical Authority (ELSTAT). Road Crash Data. Available online: https://www.statistics.gr/en/statistics/-/publication/SDT04/2000 (accessed on 8 July 2024).
28. Intini, P.; Berloco, N.; Ranieri, V.; Colonna, P. Geometric and operational features of horizontal curves with specific regard to skidding proneness. *Infrastructures* **2020**, *5*, 3. [CrossRef]

29. Pomoni, M.; Plati, C.; Kane, M.; Loizos, A. Polishing behaviour of asphalt surface course containing recycled materials. *Int. J. Transp. Sci. Technol.* **2022**, *11*, 711–725. [CrossRef]
30. Flintsch, G.W.; McGhee, K.K.; de Leon Izeppi, E.; Najafi, S. The Little Book of Tire Pavement Friction. In *Pavement Surface Properties Consortium*; Version 1.0; 2012. Available online: https://wegarten.com/media/ngcs_tire_pavement_friction_ang.pdf (accessed on 8 July 2024).
31. Wilson, D.J. An Analysis of the Seasonal and Short-Term Variation of Road Pavement Skid Resistance. Ph.D. Thesis, The University of Auckland, Auckland, New Zealand, 2006.
32. Pomoni, M. Exploring Smart Tires as a Tool to Assist Safe Driving and Monitor Tire–Road Friction. *Vehicles* **2022**, *4*, 744–765. [CrossRef]
33. Federal Highway Administration (FHWA). Available online: https://highways.dot.gov/safety (accessed on 28 July 2024).
34. Lyon, C.; Persaud, B.; Merritt, D. Quantifying the safety effects of pavement friction improvements—Results from a large-scale study. *Int. J. Pavement Eng.* **2018**, *19*, 145–152. [CrossRef]
35. Geedipally, S.R.; Pratt, M.P.; Lord, D. Effects of geometry and pavement friction on horizontal curve crash frequency. *J. Transp. Saf. Secur.* **2017**, *11*, 167–188. [CrossRef]
36. Lyon, C.; Persaud, B.; Merritt, D. Developing crash modification factors and functions for high friction surface treatments on curves and ramps—An empirical Bayes before-after study. *Transp. Res. Rec.* **2020**, *2674*, 505–514. [CrossRef]
37. Cafiso, S.; Montella, A.; D'Agostino, C.; Mauriello, F.; Galante, F. Crash modification functions for pavement surface condition and geometric design indicators. *Accid. Anal. Prev.* **2021**, *149*, 105887. [CrossRef] [PubMed]
38. Elvik, R.; Høye, A.; Vaa, T.; Sørensen, M. *The Handbook of Road Safety Measures*, 2nd ed.; Emerald Group Publishing Limited: Bingley, UK, 2009.
39. Gkyrtis, K.; Loizos, A.; Plati, C. Integrating Pavement Sensing Data for Pavement Condition Evaluation. *Sensors* **2021**, *21*, 3104. [CrossRef] [PubMed]
40. Karballaeezadeh, N.; Mohammadzadeh, D.S.; Moazemi, D.; Band, S.S.; Mosavi, A.; Reuter, U. Smart Structural Health Monitoring of Flexible Pavements Using Machine Learning Methods. *Coatings* **2020**, *10*, 1100. [CrossRef]
41. Park, K.; Thomas, N.E.; Lee, K.W. Applicability of the International Roughness Index as a Predictor of Asphalt Pavement Condition. *J. Transp. Eng.* **2007**, *133*, 706–709. [CrossRef]
42. Chan, C.Y.; Huang, B.; Yan, X.; Richards, S.H. Investigating effects of asphalt pavement conditions on traffic accidents in Tennessee based on the pavement management system (PMS). *J. Adv. Transp.* **2010**, *44*, 150–161. [CrossRef]
43. Nemtsov, I.; Jafari, A.; Persaud, B.; Lindley, I. Safety effects of pavement maintenance treatments for two-lane rural roads: Insights for pavement management. In Proceedings of the 98th Annual Meeting of the TRB, Washington, DC, USA, 13–17 January 2019.
44. Montella, A.; Imbriani, L.L. Safety performance functions incorporating design consistency variables. *Accid. Anal. Prev.* **2015**, *74*, 133–144. [CrossRef]
45. D'Agostino, C.; Cafiso, S.; Kiec, M. Comparison of Bayesian techniques for the before–after evaluation of the safety effectiveness of short 2 + 1 road sections. *Accid. Anal. Prev.* **2019**, *127*, 163–171. [CrossRef]
46. Llopis-Castello, D.; Findley, D.J.; García, A. Comparison of the highway safety manual predictive method with safety performance functions based on geometric design consistency. *J. Transp. Saf. Secur.* **2021**, *13*, 1365–1386. [CrossRef]
47. Brenac, R. Safety at curves and road geometry standards in some European countries. *Transp. Res. Rec.* **1996**, *1523*, 99–106. [CrossRef]
48. Eboli, L.; Forciniti, C. The Severity of Traffic Crashes in Italy: An Explorative Analysis among Different Driving Circumstances. *Sustainability* **2020**, *12*, 856. [CrossRef]
49. Jima, D.; Sipos, T. The Impact of Road Geometric Formation on Traffic Crash and Its Severity Level. *Sustainability* **2022**, *14*, 8475. [CrossRef]
50. Stamatiadis, N.; Psarianos, B.; Apostoleris, K.; Taliouras, F.; Montella, A.; Garofoli, G. A case for differentiating design consistency evaluation between day and night. *Transp. Res. Procedia* **2020**, *45*, 643–650. [CrossRef]
51. PIARC. *Road Safety Manual, A Guide for Practitioners*; World Road Association: Paris, France, 2019.
52. Ma, Q.; Zhang, S.; Zhou, Q. Development of a conflict-free unsignalized intersection organization method for multiple connected and autonomous vehicles. *PLoS ONE* **2021**, *16*, e0249170. [CrossRef]
53. Tay, R.; Rifaat, S.M. Factors contributing to the severity of intersection crashes. *J. Adv. Transp.* **2017**, *41*, 245–265. [CrossRef]
54. Elvik, R. Speed and road safety: Synthesis of evidence from evaluation studies. *Transp. Res. Rec.* **2005**, *1908*, 59–69. [CrossRef]
55. Kronprasert, N.; Sutheerakul, C.; Satiennam, T.; Luathep, P. Intersection Safety Assessment Using Video-Based Traffic Conflict Analysis: The Case Study of Thailand. *Sustainability* **2021**, *13*, 12722. [CrossRef]
56. Tay, R. A random parameters probit model of urban and rural intersection crashes. *Accid. Anal. Prev.* **2015**, *84*, 38–40. [CrossRef]
57. Gkyrtis, K.; Kokkalis, A. An Overview of the Efficiency of Roundabouts: Design Aspects and Contribution toward Safer Vehicle Movement. *Vehicles* **2024**, *6*, 433–449. [CrossRef]
58. Retting, R.A.; Mandavilli, S.; McCartt, A.T.; Russell, E.R. Roundabouts, Traffic Flow and Public Opinion. *Traffic Eng. Control* **2006**, *47*, 268–272.
59. Damaskou, E.; Kehagia, F. Quality of service (QOS) of Urban roundabouts: A literature review. *Int. J. Transp. Syst.* **2017**, *2*, 37–45.
60. Distefano, N.; Leonardi, S.; Pulvirenti, G. Factors with the greatest influence on drivers' judgment of roundabouts safety. An analysis based on web survey in Italy. *IATSS Res.* **2018**, *42*, 265–273. [CrossRef]

61. Wang, C.; Wang, Y.; Peeta, S. Cooperative Roundabout Control Strategy for Connected and Autonomous Vehicles. *Appl. Sci.* **2022**, *12*, 12678. [CrossRef]
62. Burdett, B.; Alsghan, I.; Chiu, L.H.; Bill, A.R.; Noyce, D.A. Analysis of Rear-End Collisions at Roundabout Approaches. *Transp. Res. Rec.* **2016**, *2585*, 29–38. [CrossRef]
63. De Brabander, B.; Nuyts, E.; Vereeck, L. Road safety effects of roundabouts in Flanders. *J. Saf. Res.* **2005**, *36*, 289–296. [CrossRef]
64. Leich, A.; Fuchs, J.; Srinivas, G.; Niemeijer, J.; Wagner, P. Traffic Safety at German Roundabouts—A Replication Study. *Safety* **2022**, *8*, 50. [CrossRef]
65. Polders, E.; Daniels, S.; Casters, W.; Brijs, T. Identifying Crash Patterns on Roundabouts. *Traffic Inj. Prev.* **2015**, *16*, 202–207. [CrossRef]
66. Johnson, M.T. Effects of Phi and View Angle Geometric Principles on Safety of Multi-Lane Roundabouts. *Transp. Res. Rec.* **2023**, *2677*, 362–371. [CrossRef]
67. Nalin, A.; Simone, A.; Lantieri, C.; Rosatella, U.; Dondi, G.; Vignali, V. Indexing the Maintenance Priority of Road Safety Barriers in Urban and Peri-Urban Contexts: Application of a Ranking Methodology in Bologna, Italy. *Infrastructures* **2023**, *8*, 181. [CrossRef]
68. Mofolasayo, A. Towards 'Vision-Zero' in Road Traffic Fatalities: The Need for Reasonable Degrees of Automation to Complement Human Efforts in Driving Operation. *Systems* **2024**, *12*, 40. [CrossRef]
69. Cho, M.; Park, J.; Kim, S.; Lee, Y. Estimation of Driving Direction of Traffic Accident Vehicles for Improving Traffic Safety. *Appl. Sci.* **2023**, *13*, 7710. [CrossRef]
70. Choi, W.C.; Chong, K.S. Analysis of Road Sign-Related Factors Affecting Driving Safety with Respect to City Size. *Appl. Sci.* **2022**, *12*, 10163. [CrossRef]
71. Choi, W.; Sung, H.; Chong, K. Impact of Illuminated Road Signs on Driver's Perception. *Sustainability* **2023**, *15*, 12582. [CrossRef]
72. Mustapha, A.; Abdul-Rani, A.M.; Saad, N.; Mustapha, M. Ergonomic principles of road signs comprehension: A literature review. *Transp. Res. Part F Psychol. Behav.* **2024**, *101*, 279–305. [CrossRef]
73. Chang, K.; Ramirez, M.V.; Dyre, B.; Mohamed, M.; Abdel-Rahim, A. Effects of longitudinal pavement edgeline condition on driver lane deviation. *Accid. Anal. Prev.* **2019**, *128*, 87–93. [CrossRef]
74. Park, E.S.; Carlson, P.J.; Porter, R.J.; Andersen, C.K. Safety effects of wider edge lines on rural, two-lane highways. *Accid. Anal. Prev.* **2012**, *48*, 317–325. [CrossRef]
75. CIE Central Bureau. *Road Transport Lighting for Developing Countries*; CIE Central Bureau: Vienna, Austria, 2007.
76. Wanvik, P.O. Effects of Road Lighting on Motorways. *Traffic Inj. Prev.* **2009**, *10*, 279–289. [CrossRef]
77. Xu, N.; Xu, Y.; Yan, Y.; Guo, Z.; Wang, B.; Zhou, X. Evaluating Road Lighting Quality Using High-Resolution JL1-3B Nighttime Light Remote Sensing Data: A Case Study in Nanjing, China. *Remote Sens.* **2022**, *14*, 4497. [CrossRef]
78. Roque, C.; Cardoso, J.L. Observations on the relationship between European standards for safety barrier impact severity and the degree of injury sustained. *IATSS Res.* **2013**, *37*, 21–29. [CrossRef]
79. Ferko, M.; Babić, D.; Babić, D.; Pirdavani, A.; Ševrović, M.; Jakovljević, M.; Luburić, G. Influence of Road Safety Barriers on the Severity of Motorcyclist Injuries in Horizontal Curves. *Sustainability* **2022**, *14*, 14790. [CrossRef]
80. PIARC. Analysis and Use of Data to Improve Safety. Available online: https://roadsafety.piarc.org/en/road-safety-management-safety-data/data-analysis (accessed on 2 August 2024).
81. Desai, M.; Chowdhury, A. Eye-Tracking Analysis of Proposed Signage Design to Prevent Accidents Caused by the Abrupt Appearance of Dividers on Indian Roads. *Designs* **2024**, *8*, 18. [CrossRef]
82. Karamanlis, I.; Nikiforiadis, A.; Botzoris, G.; Kokkalis, A.; Basbas, S. Towards Sustainable Transportation: The Role of Black Spot Analysis in Improving Road Safety. *Sustainability* **2023**, *15*, 14478. [CrossRef]
83. Yannis, G.; Michelaraki, E. Review of City-Wide 30 km/h Speed Limit Benefits in Europe. *Sustainability* **2024**, *16*, 4382. [CrossRef]
84. Karimi, A.; Kashi, E. Investigating the effect of geometric parameters influencing safety promotion and accident reduction (Case study: Bojnurd-Golestan National Park road). *Cogent Eng.* **2018**, *5*, 1525812. [CrossRef]
85. Pei, Y.-L.; He, Y.-M.; Ran, B.; Kang, J.; Song, Y.-T. Horizontal Alignment Security Design Theory and Application of Superhighways. *Sustainability* **2020**, *12*, 2222. [CrossRef]

designs

MDPI

Article

Study on the Mechanical Properties and Calculation Method of the Bearing Capacity of Concrete-Filled Steel Pipes under Axial Pressure Load

Xin Liu [1,2], Jisheng Hu [1] and Yuzhou Zheng [3,*]

1 Guangzhou Second Municipal Engineering Co., Ltd., Guangzhou 510030, China
2 Guangzhou Construction Group Co., Ltd., Guangzhou 511356, China
3 College of Civil Engineering, Nanjing Tech University, Nanjing 211816, China
* Correspondence: zyz2024@njtech.edu.cn

Abstract: Circular steel pipe concrete can give full play to the combination of steel pipes and concrete, resulting in an improvement in the steel pipe's concrete bearing capacity and ductility. In this study, the axial compression load capacities of nine steel pipe concrete columns, including one traditional steel pipe concrete column and eight steel pipe self-stressed concrete columns, were analyzed using an axial pressure test. The damage patterns and stress–strain curves of all the specimens under axial compression load were analyzed, and a comparison analysis was made between the test results of the different specimens. The test results show that the longitudinal expansion displacement of concrete increases with the increase in the expansion agent content. The greater the self-stress, the higher the bearing capacity of steel-tube concrete columns under axial compressive load within a certain range of the expansion agent, indicating that self-stress can increase the bearing capacity of steel-tube concrete columns under axial compressive load, but the effect of the magnitude of the self stress on the damage pattern of the specimens is limited. The damage patterns of all the specimens were bulging in the center and concave at both ends. In addition, the existing theoretical calculation method of the bearing capacity of steel pipe concrete columns is modified, and a theoretical calculation method applicable to steel pipe self-stressed concrete columns is proposed to simplify the calculation method of the bearing capacity of steel pipe self-stressed concrete columns, which provides a basis for decision-making in practical engineering.

Keywords: self-stress concrete; bearing capacity; steel tube; axial compression

1. Introduction

Concrete-filled steel tube (CFST) components have the advantages of good levels of plasticity and toughness. Due to the high brittleness of concrete, it is rarely directly used in structures. After the concrete has been restrained by a steel tube, the toughness of concrete-filled steel tube members is improved. During the service period of members, CFST structures offer high plastic deformation and good ductility before failure. In practical engineering applications of concrete-filled steel pipe columns, as the concrete shrinks, the compactness of the concrete-filled steel pipe components is reduced, resulting in a reduction in the cooperative working ability of the steel pipe and concrete. Therefore, a short column of steel tube self-stress concrete is proposed to reduce the internal shrinkage of concrete and ensure optimal joint bearing of the external load between steel tubes and micro-expansion concrete. Micro-expansion concrete can make the concrete expand by adding an expansion agent into the concrete, produce expansion stress under the constraint of the steel pipe, and make steel pipes and concrete work together better. The main functions of adding an expansion agent are: (1) to compensate for the inherent shrinkage of concrete and effectively reduce the gap between the steel pipe and concrete; and (2) to use the expansion property to improve the interfacial bond strength between the steel pipe and concrete [1–6].

Therefore, it is indispensable to study micro-expansion concrete-filled steel tubular columns. Knowles [7,8], Tomii [9], Luksha [10], Schneide [11], and other scholars have conducted experimental investigations on the elements influencing the static performance of CFST columns under axial compression. Specimens with a small slenderness ratio have a better restraint effect on concrete. Giakoumelis and Lam [12], Sakino [13], Elobody [14], Gupta et al. [15], Liang and Fragomeni [16], Abed [17], and other scholars have conducted axial compression experiments on CFST columns, which have showed that the calculation formula of the bearing capacity of high-strength CFST columns is obtained by modifying the expression of ordinary CFST columns. The restraint effect of steel pipes on ordinary concrete is preferable to that of steel pipes on high-strength concrete. Huang et al. [18] investigated the expansion performance of six steel tubes after adding self-stressing concrete, and conducted an axial compression test on 18 self-stressing CFST columns. The results show that the carrying capacity of self-stressing concrete-filled steel tubes is approximately 20% higher than that of traditional CFSTs. According to the studies conducted by some researchers [19,20], the carrying capacity of steel pipe concrete mixed with micro-expansion is better than that of traditional CFSTs, and when it comes to the value of the self-stress, it is not a case of the larger the better. There is an optimal value of self-stress for improving the carrying capacity of CFSTs.

Based on the above summary [21–23], there have been some achievements in the study of steel tube self-stressing concrete columns under static load, but there is still no definite conclusion on whether the influence of self-stress on traditional steel tube concrete columns increases linearly, and there is no fast calculation formula for the bearing capacity of steel tube micro-expansion concrete columns. The primary objectives of this study are to investigate the rationality of micro-expanded concrete application in steel tubes to address the debonding issue resulting from the shrinkage and creep of normal concrete in steel tubes. In this regard, in this study we conducted an axial compression experiment on micro-expanded concrete-filled steel tubes, and the influence of an expansion agent on the failure mode, load-displacement curve, and load-strain curve are carefully analyzed. Furthermore, based on solid experimental observations, a high-precision formula for predicting the axial bearing capacity of micro-expansion concrete-filled steel tube columns is proposed, remedying the lack of a fast calculation method for the design of micro-expansion concrete column-filled steel tubes.

2. Test Process

2.1. Specimen Design

The expansion rates of 8% and 12% were used to test the expansion rate of micro-expansion concrete-filled steel tubes (MCFSTs). Then, 9 specimens with a section size of 159 mm (diameter) × 400 mm (length) and 168 mm (diameter) × 400 mm (length) micro-expansion concrete-filled steel tube columns were experimented to explore the effect of micro-expansion concrete on the working mechanism of MCFST columns. Test sample CS1 was the test reference, and test specimens CS2 and CS3 were studied in relation to the distinct effects of 8% and 12% samples on MCFST columns. Samples CS4, CS5, CS6, and CS7 were studied in relation to the effect of steel pipe diameter and wall thickness on the axial compression performance of CFST columns, and specimens CS8 and CS9 were studied in relation to the influence of end constraints on the axial compression performance. There are three *L/D* ratios for steel tube micro-expansion concrete columns, and the *L/D* ratios of 159 mm, 168 mm, and 219 mm are less than 2.52, 2.38, and 1.82, respectively. The parameters of the sample are shown in Table 1. Diameter is expressed as D, concrete strength is expressed as f_c, volume expansion rate is expressed as V, and the thickness of the steel pipe is expressed as t in Table 1.

Table 1. Parameters of the specimens.

ID	D (mm)	Length (mm)	t (mm)	f_c (MPa)	V	Remarks
CS1	168	400	4	80	0	
CS2	168	400	4	80	8%	
CS3	168	400	4	80	12%	
CS4	168	400	3	80	12%	
CS5	159	400	3	80	12%	
CS6	159	400	4	80	12%	
CS7	159	400	5	80	12%	
CS8	219	400	3	80	12%	
CS9	219	400	3	80	12%	end restraint

2.2. Material Properties

The proportion of the raw materials' need for micro-expansion concrete is shown in Table 2. Three micro-expansion concrete cube samples with the sizes 150 mm × 150 mm × 150 mm were reserved. The strength measured by the cube sample was 85 MPa. The Young's Modulus (E) of the steel tube was 208 GPa, and the uniaxial tensile strength was 230 MPa.

Table 2. Mix design of micro-expansion concrete.

ID	Cement (kg)	Ganister Sand (kg)	Slag (kg)	Zeeospheres (kg)	Expansion Agent (kg)	Sand (kg)	Big Cobble (kg)	Small Cobble (kg)	Water (kg)	Water-Reducing Admixture (kg)	f_c (MPa)
M_0	16.5	0.89	1.08	1.91	/	22.7	23.93	10.25	6.19	0.12	83.6
M_8	12.6	0.75	0.92	1.62	1.38	19.3	20.30	8.70	4.03	0.07	85.2
M_{12}	11.9	0.75	0.92	1.62	2.08	19.3	20.36	8.70	4.03	0.07	86.2

2.3. Specimen Preparation

The steel tubes were soldered with a 4-mm thick steel metal on their undersides, and then transported to the concrete pouring site. Prior to pouring, strain gauges were affixed and a grinding machine was utilized. The field workers sanded the connections of the strain gauges, cleansed them with alcohol, and finally applied a layer of AB glue for fixation. It should be noted and highly appreciated that the micro-expansive concrete mixture and laboratory for the concrete-pouring process were both provided by Jiangsu Subote New Materials Company Limited. Figure 1 illustrates the production process of the tested specimens. Firstly, the process began with the careful pouring of the concrete from the top of the steel tube, ensuring even distribution and filling. Next, the freshly poured concrete was subjected to vibration on a specialized vibrating table, a crucial step for removing air bubbles and enhancing the density of the material. Then, skilled workers meticulously smoothed the concrete, working it along with the upper surface of the steel pipe using a trowel to achieve a uniform and level finish before proceeding to the testing phase. Finally, the curing conditions for the micro-expanded concrete were identical to those of regular concrete, with a temperature of 20 °C and a relative humidity of 95%. The age of the concrete was 28 days.

(a) concrete mixing (b) concrete mixing (c) indoor maintenance

Figure 1. The production process of all the specimens.

2.4. Test Setup and Loading Method

Two experiments were carried out to gauge the longitudinal expansion rate of concrete under steel pipe restraint after adding the expansion agent into the concrete. Firstly, a test to assess the longitudinal expansion rate was carried out, and the displacement of the vertical expansion was recorded, mainly by a laser displacement meter. This is shown in Figure 2.

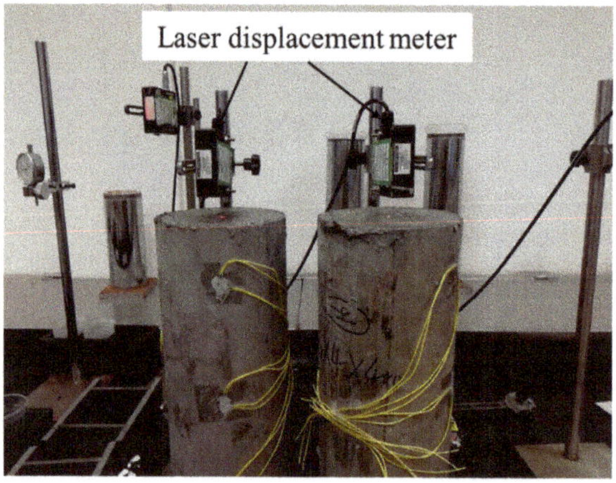

Figure 2. Volume expansion experiment equipment.

In order to explore the working mechanism of micro-expansion concrete-filled steel pipes, axial compression experiments of nine samples were carried out. The experiment process was as follows:

(1) Prior to the axial compression test, the MCFST members were placed in such a way that the steel tubes and the micro-expanded concrete were in the same straight line, so that the steel tubes and the micro-expanded concrete could share the load and avoid eccentric compression of the members. Similarly, the steel pipe members filled with micro-expanded concrete were leveled where they were placed.

(2) The center of the short micro-expansion CFST column was aligned with the loading center of the axial compression test apparatus to ensure that the member was subjected to axial compression rather than eccentric compression, and a displacement meter was placed between the test devices to measure the displacement, as shown in Figures 3 and 4.

(3) The steel pipe micro-expansion concrete member was placed on the testing machine, and the strain gauge was ground with a ten-color wire, and then connected to the static compound resistance strain gauge of Donghua Company for debugging. After debugging, the specimen was preloaded in the elastic range to ensure that the strain gauges worked properly. The strain gauge model was 120-5AA, which belonged to the terminal and did not require welding. The strain gauge above and below the specimen was mainly used to measure the changing trend of the end part of the specimen under load, while the strain gauge in the middle was mainly used to measure the changing trend of the middle part of the specimen under load. The upper and lower strain gauges were 50 mm away from the end, and the middle strain gauge divided the specimen into two equal parts. The preloading time was about 2 min. The preloading load should not exceed 30% of the theoretical calculated value of the load capacity of the specimen.

(4) When the loading reached the yield point of the specimen, the displacement loading was adopted, and the experiment was finished when the displacement reached 100 mm.

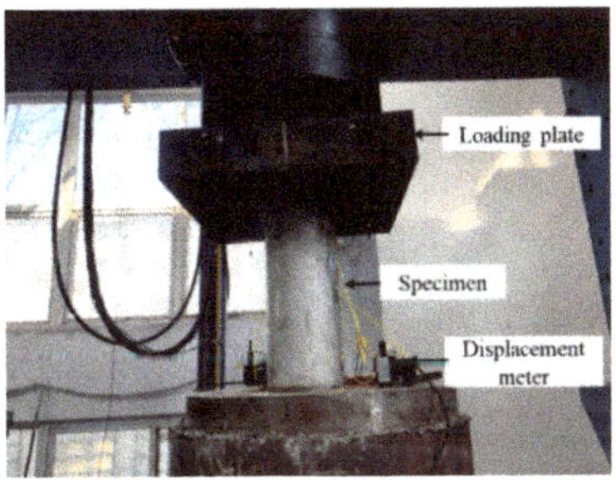

Figure 3. Loading equipment.

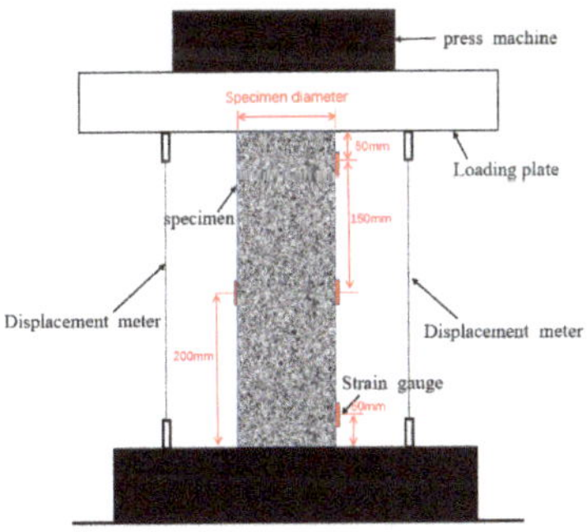

Figure 4. Schematic diagram of loading equipment.

3. Experiment Results

Through the axial compression test of the steel tube micro-expansion concrete short columns in this section, the load–displacement curve, destruction mode, and load–strain curve of the specimens under axial compression load were obtained. The failure mode and failure mechanism of the specimens are described. The load–displacement curve and load–strain curve of the samples were also analyzed.

3.1. Failure Mode and Mechanism Analysis

1. Longitudinal expansion experiment

The longitudinal expansion rate of concrete mixed with an expansion agent was obtained according to the method provided in the national code *Technical Code for Application of Concrete Admixtures* (*GB501119-2013* [24]). For the test equipment, a laser displacement meter was used. Under the condition of lateral restraint, the longitudinal expansion rate of concrete in this study was calculated according to the following formula:

$$\varepsilon_c = \frac{L_c - L_0}{L} \times 100\% \tag{1}$$

where:

ε_c—Longitudinal free expansion rate of concrete after curing;
L—Base length of concrete (400 mm);
L_0—The initial length of concrete, 400 mm (the initial shrinkage of the concrete is not considered in this paper);
L_c—Displacement meter reading of concrete after curing (mm).

In order to investigate the free expansion rate for different expansion agent contents, Figure 5 exhibits the free expansion rate–age curves. In the first phase, the expansion rate of micro-expansion concrete increases rapidly, and then the expansion results gradually reduce due to the shrinkage of micro-expansion concrete. When the concrete shrinkage guide is expanding, a descending section of the curve appears, and expansion and contraction are balanced. There are three distinct phases in the concrete after the admixture of expansion agents: the expand phase, shrink phase, and unchanged phase. This is due to the fact that before the unchanged phase, the expansion capacity produced by concrete expansion is greater than the shrink capacity, and the expansion capacity gradually diminishes during the expansion process. When the expansion capacity is smaller than the shrink capacity, the contraction appearance starts to emerge, and finally the equilibrium phase is reached.

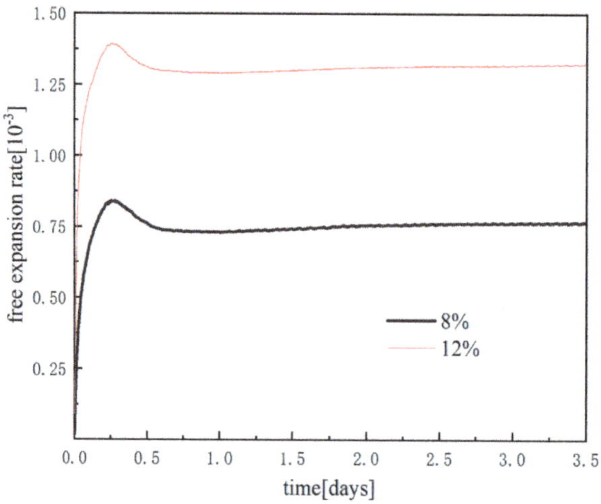

Figure 5. Free expansion rate-age curve.

2. Axial compression experiment results

The failure modes of the specimens are shown in Table 3. It can be seen from Table 3 that all the specimens showed the failure mode of bulging in the middle and concave at both ends. The failure modes of micro-expansion concrete-filled steel tube short columns are essentially the same as those of ordinary CFST short columns. In the initial stage of loading, the short column of steel tube micro-expansion concrete is in an elastic state, the round steel tube and micro-expansion concrete bear external axial force, respectively, and the Poisson's ratio of micro-expansion concrete is less than that of outer steel tube. Therefore, under the action of an external load, the steel pipe and the micro-expanded concrete are not subjected to mutual compression. At this stage, there is no mutual binding between the steel pipes and micro-expansion concrete. There is also no obvious deformation. With the continuous increase in the load, both steel pipes and concrete transition from the elastic phase to the yield phase, and the longitudinal strain becomes larger and larger. Micro-cracks begin to appear and develop in micro-expansion concrete. The Poisson's ratio of micro-expansion concrete gradually goes over that of steel tubes. At this stage, steel pipe concrete elements have entered the plastic phase. After the steel tube completely yields and the concrete has been crushed, the carrying capacity of the member decreases rapidly. When the lateral deformation coefficient of the micro-expansion concrete is greater than that of the outer steel tube, circumferential stress is generated between the steel tube and the micro-expansion concrete, and the steel tube begins to experience radial forces.

Table 3. Failure mode of all the specimens.

Specimens	Final Failure Mode	Description of Failure Mode
CS1		(1) No expansion agent (2) The yield bearing capacity is 2399 kN (3) The specimen shows the failure mode of bulging in the middle and concave at both ends (4) The displacement of the specimen is loaded to 130 mm, which is more serious than other specimens.
CS2		(1) 8% expansion agent (2) The yield bearing capacity is 2612 kN (3) The specimen shows the failure mode of bulging in the middle and concave at both ends (4) The lower concave deformation is larger than the upper one.

Table 3. *Cont.*

Specimens	Final Failure Mode	Description of Failure Mode
CS3		(1) 12% expansion agent (2) The yield bearing capacity is 2758 kN (3) The specimen shows the failure mode of bulging in the middle and concave at both ends (4) Obvious shear slip lines.
CS4		(1) 12% expansion agent (2) The yield bearing capacity is 2661 kN (3) The specimen shows the failure mode of bulging in the middle and concave at both ends (4) Significant shear slip lines.
CS5		(1) 12% expansion agent (2) The yield bearing capacity is 2275 kN (3) The specimen shows the failure mode of bulging in the middle and concave at both ends (4) No significant shear slip lines.
CS6		(1) 12% expansion agent (2) The yield bearing capacity is 2422 kN (3) The specimen shows the failure mode of bulging in the middle and concave at both ends (4) No significant shear slip lines.

Table 3. *Cont.*

Specimens	Final Failure Mode	Description of Failure Mode
CS7		(1) 12% expansion agent (2) The yield bearing capacity is 2700 kN (3) The specimen showed the failure mode of bulging in the middle and concave at both ends (4) No significant shear slip lines.
CS8		(1) 12% expansion agent (2) The yield bearing capacity is 4345 kN (3) The specimen showed the failure mode of bulging in the middle and concave at both ends.
CS9		(1) 12% expansion agent (2) The yield bearing capacity is 4134 kN (3) The specimen showed the failure mode of bulging in the middle and concave at both ends.

According to the Von Mises yield criterion, when the circumferential stress of the steel tube increases gradually, the corresponding longitudinal stresses have decreased and the forces between the steel pipe and the micro-expanded concrete have been redistributed [20]. Micro-expansion concrete has a higher compressive strength due to the constraint of the steel pipe. The steel pipe changes from mainly bearing the axial load to mainly bearing the circumferential stress caused by micro-expansion deformation [20]. However, the yield of the steel pipe does not mean that the whole specimen has lost its carrying capacity. After the yield of the steel tube, the internal forces resulting from the interaction between the steel pipe and the micro-expanded concrete have been redistributed, which strengthens the deformation coordination ability between them. At the same time, the Poisson's ratio of concrete in the plastic stage is much larger than that of the steel tube. After the concrete is crushed, the bearing capacity of the member may continue to rise due to the hooping effect of the steel pipe on it and the further joint work of the steel pipe and concrete. Therefore, the carrying capacity is further improved after the steel tube yields.

3.2. Load–Displacement Curve

In order to study the change in displacement of all the specimens under axial compressive loading, the displacement of the specimen was obtained from two pre-placed displacement meters to obtain the load displacement curve. The load–displacement curve of the axial compression test of the micro-expansion concrete-filled steel tube short column is shown in Figures 6–9. From Figure 6, it can be seen that at the initial stage of loading, specimens C1, C2, and C3 are in the elastic stage, and the load displacement curve increases linearly. With the continuous increase in load, the short column begins to enter the elastic-plastic stage, and the curve slope gradually slows down, indicating that the stiffness of the short column begins to change. At this time, the three-dimensional restraint effect of the micro-expansion concrete receiving the steel tube is more obvious. Under the condition of the steel tube restraint effect, the compressive strength of the micro-expansion concrete increases. Then, when the steel tube begins to buckle, the restraint effect of the steel tube on the micro-expansion concrete decreases, and the bearing capacity begins to decrease in varying degrees. However, due to the hoop effect of the steel tube, the specimen does not appear to show any obvious damage. After the bearing capacity decreases to a certain value, the load keeps increasing slowly and the displacement increases. It can be seen from the load–displacement curve in Figure 6 that the yield strength of the MCFST column is greater than that of ordinary CFST. The bearing capacity of the ordinary CFST column is 2399 kN, and the bearing capacities of the MCFST columns with C2 and C3 are 2612 kN and 2758 kN, respectively. The yield loading of the MCFST pier of C2 and C3 is 8.9% and 14.9% greater than the ordinary CFST column. Obviously, the micro-expansion concrete can notably increase the bearing capacity of the MCFST column. This is mainly due to the fact that expansion agents in concrete play an important role in compensating for the shrinkage defects of ordinary concrete. After curing, the micro-expansion concrete will produce self-stress under the double influence of concrete micro-expansion and steel pipe restraint.

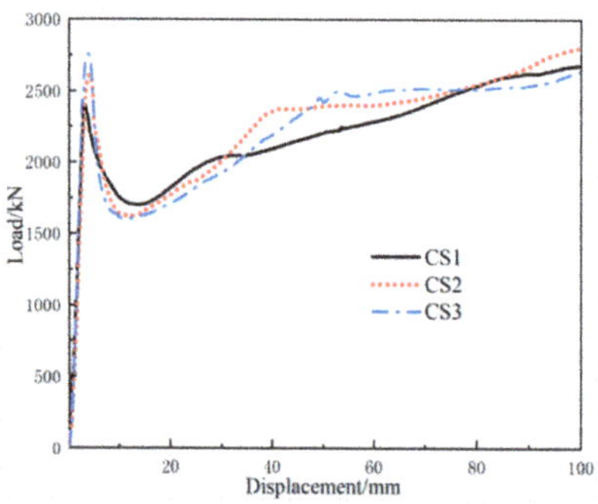

Figure 6. Load–displacement curves with different expansion rates.

As can be seen from the load-displacement curves of different thicknesses in Figure 7, the wall thicknesses of CS5, CS6, and CS7 are 3 mm, 4 mm, and 5 mm, respectively, and the corresponding bearing capacities are 2275 kN, 2422 kN, and 2700 kN, respectively. When the wall thickness increases from 3 mm to 4 mm, the bearing capacity increases by 6.5%, and when the wall thickness increases from 3 mm to 5 mm, the bearing capacity increases by 18.7%. The results show that the bearing capacity of steel tube micro-expansion concrete short columns will increase with the increase in the steel tube wall thickness. Because the load–displacement curve shows the elastic modulus of the specimen in the elastic stage, so we can see the bearing capacity at yield and the ductility after yield. Therefore, it can be seen from Figure 7 that the stiffness of different thicknesses in the elastic stage is essentially the same, but when the elastic stage transitions to the elastic-plastic stage, the stiffness with high wall thickness is significantly greater than that with small wall thickness. The yield bearing capacity is significantly improved, and the level of ductility is good. As can be seen from the load–displacement curves of different diameters in Figure 8, the outer diameters of the steel tubes of CS5, CS4, and CS8 are 159 mm, 168 mm, and 219 mm, respectively, and the corresponding bearing capacities are 2275 kN, 2661 kN, and 4345 kN, respectively. When the outer diameter of the steel tube increases from 159 mm to 168 mm, the bearing capacity increases by 16.9%, and when the outer diameter of the steel tube increases from 159 mm to 219 mm, the bearing capacity increases by 90.9%. The results show that the bearing capacity of steel tube micro-expansion concrete short columns will increase with the increase in the outer diameter of the steel tube. It can be seen from the load–displacement curves of different end constraints in Figure 9 that the load–displacement curves of CS9 and CS8 are the load–displacement curves of end unconstrained and end constrained, respectively, and the corresponding bearing capacities are 4134 kN and 4345 kN, respectively. When the end of the steel tube is unconstrained relative to the end constraint, the bearing capacity is increased by 5.1%, indicating that the end constraint will reduce the bearing capacity of the test specimen.

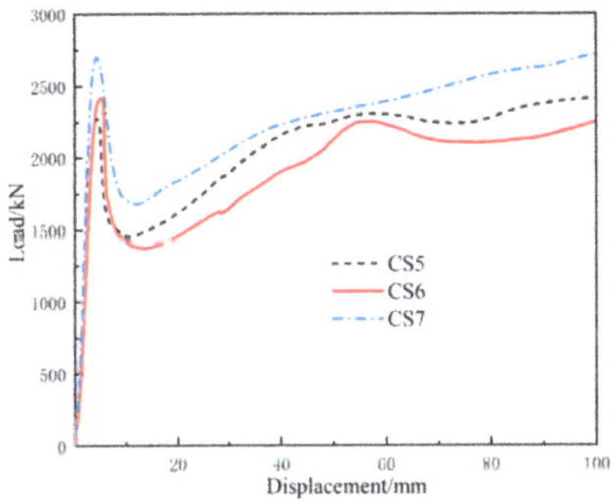

Figure 7. Load–displacement curves of different thicknesses.

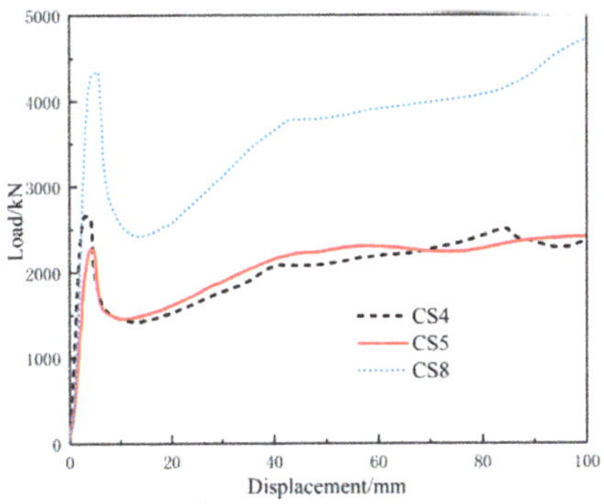

Figure 8. Load–displacement curves with different diameters.

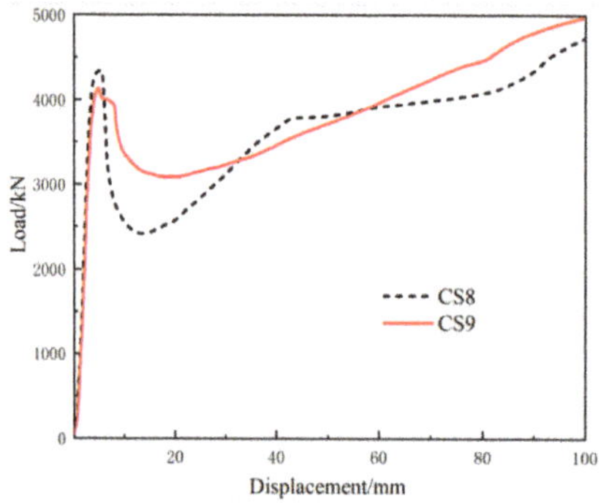

Figure 9. Load–displacement curves with and without end constraints.

3.3. Load–Strain Curve

In order to study the deformation of MCFST and CFST under axial compressive loading, the longitudinal strain and circumferential strain of the middle section are used to represent the longitudinal strain and circumferential strain of the whole specimen. According to the test results, the load-strain (N-ε) of all the test specimens is obtained. The load-strain of all the specimens is shown in Figures 10–13. It can be seen from Figure 10 that the strain of the micro-expansion concrete-filled steel tube under the same load is greater than that of steel tube ordinary concrete from the initial stage of loading to yield. This is mainly because the stiffness of the whole specimen is reduced after adding the expansion agent into the concrete. After yielding, for the ordinary concrete-filled steel tube, the ultimate load is reached soon after yielding and enters the descending section quickly. For the micro-expansion concrete-filled steel tube, after yielding, the specimen reaches the ultimate load after experiencing considerable deformation, and then enters the descending section. This is mainly due to the self-stress generated when the steel tube constrains the micro-expansion concrete, which strengthens the cooperative working ability of the steel tube and the micro-expansion concrete after the specimen is subject to buckling. When the ultimate load is reached, the tensile zone and compression zone reach the peak at the same time. This shows that the cooperation between the steel tube and concrete is good. Throughout the whole loading process, under the same load, the strain in the compression zone is greater than that in the tension zone. This means that the specimen will reach the yield stage first in the compression zone. From Figure 10, it can be seen that the initial stiffness of the specimen is reduced for specimens with an expansion agent, but the stiffness will essentially remain unchanged with the increase in the content of the expansion agent. With the increase in the content of the micro-expansion agent, the bearing capacity of the specimen gradually decreases. The ductility of the specimens is also enhanced. As can be seen from Figures 11 and 12, with the increase in the steel tube wall thickness and the steel tube diameters, the bearing capacity of the test piece gradually increases. As shown in Figure 13, the bearing capacity with end restraint is lower than that without end restraint. This is because the same expanding agent may have different mechanisms of action in different constraint states. The self-stress produced by the end restraint specimen is large, but the bearing capacity is reduced. Therefore, the content of the concrete expansion agent will have the most appropriate proportion.

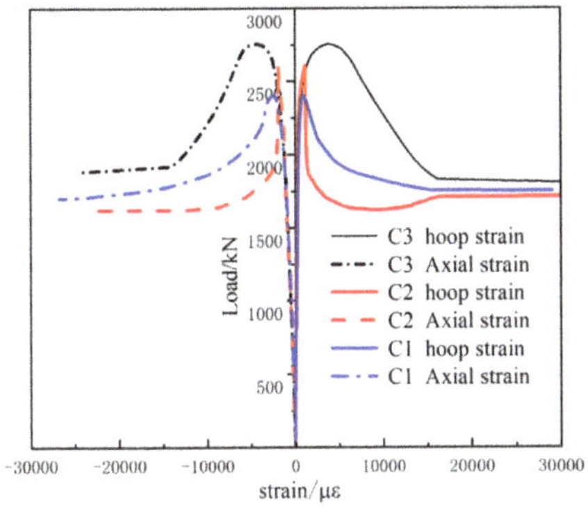

Figure 10. Load strain curves with different expansion rates.

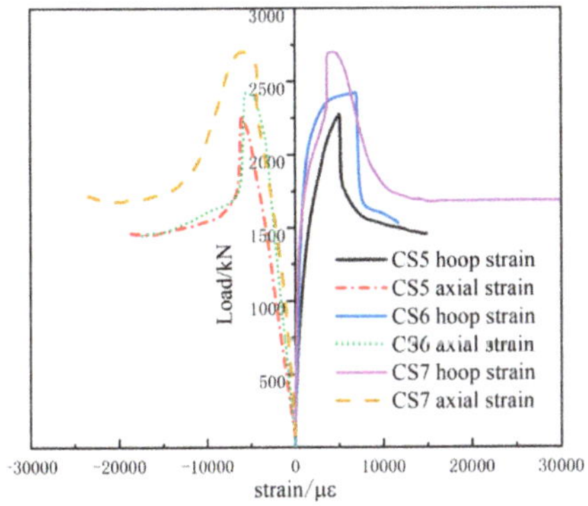

Figure 11. Load strain curves of different thicknesses.

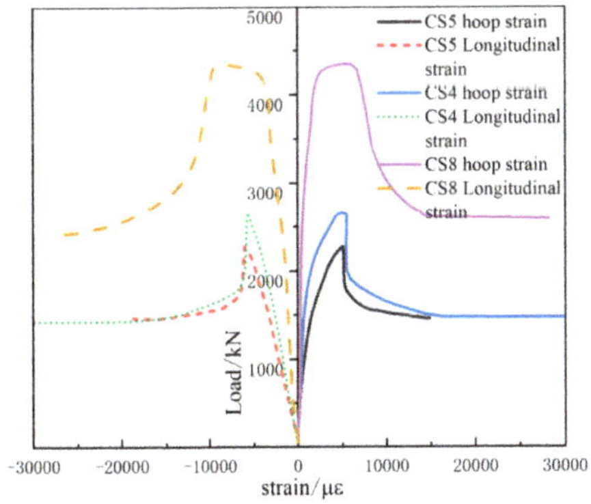

Figure 12. Load strain curves with different diameters.

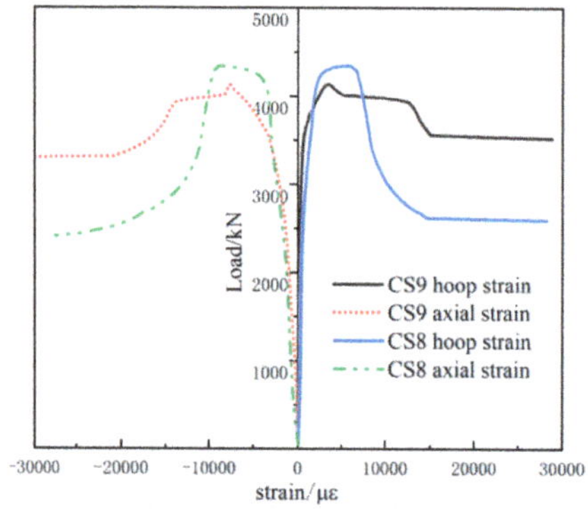

Figure 13. End constrained and unconstrained load strain curves.

In the past decades, China has made great breakthroughs and progress in the research and application of concrete-filled steel tube structures, and the research on the theoretical calculation formula of CFST structures is also relatively mature. However, there is no specific code for calculating the axial compressive capacity of concrete-filled steel tube columns. Therefore, this study modifies the practical calculation method of the axial pressure of CFST columns based on the existing experiment results, while referring to the calculation formula of the carrying capacity of ordinary concrete-filled steel tube columns in the existing specifications. The aim was to predict the axial bearing capacity of short columns of MCFST, there are three main methods for calculating the axial compressive carrying capacity of ordinary CFST: the unified theory; the superposition theory; and the limit equilibrium theory. Zhong [25] put forward the unified theory and unified design method. The main conclusion of the unified theory of concrete-filled steel tubes is that the two materials of steel tube and concrete are combined to form a composite material by using the deformation coordination condition, and the mechanical properties of concrete-filled steel tube structures change with the change in the geometric parameters and section form. However, the calculation is unified. The unified formula is used to calculate the bearing capacity of concrete-filled steel tube structures under axial compression load:

$$N_u = A_{sc} \times f_{sc} \quad (2)$$

$$\begin{gathered} f_{sc} = \left(1.212 + B\theta + C\theta^2\right) f_{ck} \\ \theta = \alpha f / f_c \end{gathered} \quad (3)$$

where:

N_0—Design value of the axial compression bearing capacity of a concrete-filled steel tube;
A_{sc}—The cross-sectional area of the concrete-filled steel tube composite structure is equal to the sum of the steel tube area and the concrete area;
f_{sc}—Compressive strength value of CFST;
θ—Confinement coefficient of specimen;
α—The steel content of the specimen is the ratio of the cross-sectional area of the steel tube to the cross-sectional area of the concrete;
f, f_c—The design values of the compressive strength of steel and concrete, respectively;
B,C—The influence coefficient of the confinement coefficient (depending on different section shapes); value according to Table 4.

Table 4. The influence coefficient of B and C.

Section Shapes		B	C
solid	circle and dodecagon	$0.176f/213+0.974$	$-0.104f_c/14.4+0.031$
	regular octagon	$0.140f/213+0.778$	$-0.070f_c/14.4+0.026$
	square	$0.131f/213+0.723$	$-0.070f_c/14.4+0.026$
hollow	circle and dodecagon	$0.106f/213+0.584$	$-0.037f_c/14.4+0.011$
	regular octagon	$0.056f/213+0.311$	$-0.011f_c/14.4+0.004$
	square	$0.039f/213+0.217$	$-0.006f_c/14.4+0.002$

Han [26] proposed the theory of the superposition of the bearing capacity of both steel pipes and concrete. The superposition theory is mainly composed of the bearing capacity of steel tubes and concrete when they are stressed separately. As with the carrying capacity of CFST composite structures, in the existing specifications, the round steel tube partially considers the influence of the three-dimensional restraint of concrete by the steel tube, and the compressive strength of core concrete will be improved. There are also other codes that do not take into account the three-dimensional keep within bounds effect of steel tubes on concrete, while the three-dimensional restraining effect of steel pipes on concrete is seldom considered for square steel pipe concrete. The following bearing capacity formula is used for cases where the restraining effect is not considered:

$$N_u = \eta_s f_y A_s + \eta_c f_{ck} A_c \tag{4}$$

where:

η_s—Variation coefficient of steel tube bearing capacity;
η_c—Variation coefficient of bearing capacity of concrete.

If the three-dimensional restraint effect of the steel tube on concrete is not considered, it is taken as $\eta_s = 1, \eta_c = 1$. If the three-dimensional action of the steel tube on concrete is considered, the bearing capacity coefficient of the steel tube remains unchanged as $\eta_s = 1$, and the bearing capacity coefficient of concrete is expressed as follows:

$$\eta_c = 1 + (\sqrt{4-3\beta^2} - 1)\alpha f_y / f_{ck} \tag{5}$$

where: $\beta = 0.25 + 0.32\alpha$; α is the steel content of the specimen; and the value is the ratio of the cross-sectional area of the steel tube to the concrete area.

Cai [27] put forward the limit equilibrium theory of concrete-filled steel tube composite structures. The main conclusion of this theory is that the bearing capacity of concrete-filled steel tube structures is less affected by the deformation mode and loading mode. The bearing capacity is calculated by using the equilibrium condition in the limit state, and the bearing capacity obtained by experimental research and theoretical calculation is based on the hoop coefficient θ. There are different calculation methods:

$$\begin{aligned} N_u &= A_c f_c(1+2\theta); & \theta \le 1.235 \\ N_u &= A_c f_c(1+\sqrt{\theta}+1.1\theta); & \theta \ge 1.235 \end{aligned} \tag{6}$$

where:

$\theta = \frac{A_s \times f_y}{A_c \times f_c}$—Formula for calculating the hoop coefficient;
f_y—Steel tube yield strength;
f_c—Compressive strength of concrete.

Eurocode EC4-2004 [28]: The calculation formula of the axial compression bearing capacity of concrete-filled steel tube with different sections is proposed. When the concrete-filled steel tube column meets both the slenderness ratio $\overline{\lambda} \le 0.5$ and eccentricity $e/d \le 0.1$, the three-dimensional restraint effect of concrete should be considered. The calculation formula of the axial compression bearing capacity is as follows:

$$N \leq N_u \tag{7}$$

$$N_u = \frac{f_y}{\gamma_s} A_s + \frac{f_c'}{\gamma_c} A_c \tag{8}$$

where:

N—Axial compression bearing capacity design value;
N_u—Axial load bearing capacity limit.

When considering the hoop effect of the steel tube on the concrete, the calculation formula of the axial compressive bearing capacity of the member is:

$$N_u = \eta_s \frac{f_y}{\gamma_s} A_s + (1 + \eta_c \times \frac{t}{D} \frac{f_y}{f_c'}) \frac{f_c'}{\gamma_c} \times A_c \tag{9}$$

$$\eta_c = 4.9 - 18.5\overline{\lambda} + 17\overline{\lambda}^2 \geq 0 \tag{10}$$

$$\eta_s = 0.25(3 + 2\overline{\lambda}) \leq 1 \tag{11}$$

where:

A_s, A_c—The steel tube section area and concrete section area;
f_y, f_c'—The yield strength of the steel pipe and compressive strength of the concrete;
γ_s, γ_c—The partial coefficient of steel (generally 1.0), partial coefficient of concrete (generally 1.5);
η_s, η_c—Coefficient.

When the slenderness ratio of the concrete-filled steel tube column is not considered, the calculation formula of the axial compression bearing capacity can be simplified as follows:

$$N_u = 0.75 \frac{f_y}{\gamma_s} A_s + (1 + 4.9 \times \frac{t}{D} \frac{f_y}{f_c'}) \times f_c' A_c \tag{12}$$

There have been many studies on the bearing capacity formula of concrete-filled steel tubes, as shown in Table 5, which presents the typical codes for the calculation of the axial compressive ultimate bearing capacity of CFST at home and abroad.

Table 5. Typical codes for the calculation of the axial compressive ultimate bearing capacity of CFST columns at home and abroad.

Design Code	Calculation Formula	Remarks
ACI [29]/AS [30]	$N_{ACI/AS} = 0.85 f_c A_c + f_y A_s$	
AIJ [31]	$N_{AIJ} = f_c A_c + (1+\eta) f_y A_s$	$\eta = 0$ for square steel tube; $\eta = 0.27$ for circular steel tube
Eurocode 4 [28]	$N_u = 0.75 \frac{f_y}{\gamma_s} A_s + (1 + 4.9 \times \frac{t}{D} \frac{f_y}{f_c'}) \times f_c' A_c$	Simplified formula without regard to the slenderness ratio
CECS28:2012 [32]	$0.5 < \theta \leq [\theta]$, $N_{CECs} = 0.9 A_c f_{ck}(1 + \alpha_0 \theta)$ $2.5 > \theta > [\theta]$, $N_{CECS} = 0.9 A_c f_{ck}(1 + \sqrt{\theta} + \theta)$	α_0 is coefficient; $[\theta]$ is limit value of hoop coefficient
GB50936 [33]	$N_{GB} = A_{sc} f_{sc}$	$f_{sc} = (1.212 + B\theta + C\theta^2) f_{ck}$ $A_{sc} = A_s + A_c$; B, C as shown Table 4

Note: t is the wall thickness of the concrete-filled steel tube, D is the diameter of the concrete-filled steel tube, f_y is the yield strength of the steel tube, f_{cd} is the compressive strength of the concrete prism, γ_s, γ_c is the coefficient of steel (generally 1.0) and concrete (generally 1.5), respectively, and $= f_y A_s / f_c A_c$ is the hoop coefficient of the MCFST.

3.4. Empirical Formula Based on Experimental Results

Under the action of axial compression load, the bearing capacity of MCFST columns is higher than that of ordinary concrete-filled steel tubes. The main reason for this is that the three-dimensional restraint effect of steel tubes on micro-expansion concrete is strengthened due to the expansion of the micro-expansion concrete. Due to the use of micro-expansion concrete, there will be large errors in the existing calculation formula of the axial compression bearing capacity of concrete-filled steel tube columns.

The calculation and test results of different standard formulas are shown in Table 6. It can be seen that the error between the calculation formula of the bearing capacity of concrete-filled steel tube columns in the European Code (Eurocode 4) and the test results in this study is the smallest, with an error of 9.8%. The increase in the axial compression bearing capacity of steel tube micro-expansion concrete columns is due to the increase in the passive confining pressure of steel tubes on concrete due to the expansion of concrete. Therefore, referring to the bearing capacity calculation formula of Eurocode 4, this paper introduces the strength improvement coefficient considering the three-dimensional restraint of steel tubes to concrete, and puts forward an intuitive and convenient axial compression bearing capacity calculation formula of steel tube micro-expansion concrete column for engineering applications.

$$N_{\mathrm{u}} = 0.75\frac{f_{\mathrm{y}}}{\gamma_{\mathrm{s}}}A_{\mathrm{s}} + (1 + 4.9 \times \frac{\mathrm{t}}{D}\frac{f_{y}}{f_{c}'}) \times \beta f_{c}' A_{c} \tag{13}$$

Table 6. Calculation results of different specifications of specimen CS2.

Design Code	Calculation Formula	Error Rate
ACI [29]/AS [30]	1935.9 kN	25.9%
AIJ [31]	2322.9 kN	11.1%
Eurocode 4 [28]	2656 kN	9.8%
CECS28:2012 [32]	θ does not fit the formula	Inapplicable
GB50936 [33]	1646.2 kN	36.9%

These include the improvement coefficient β of strength under the constraint of the steel tube after adding the expansion agent into the concrete; the meaning of the other symbols is consistent with the previous text. Substitute the test data into the formula N_u, and the corresponding coefficient of MCFST is shown in Table 7.

Table 7. The coefficient β of all the specimens.

Specimen	Coefficient β
CS2	1.20
CS3	1.28
CS4	1.28
CS5	1.48
CS6	1.24
CS7	1.32
CS8	1.27

The steel pipe diameter and steel content of each specimen are different, but the coefficient is very close. The coefficient β = 1.2 can be obtained by linear regression of the coefficient, that is, the practical bearing capacity calculation formula based on the test results is as follows:

$$N_{\mathrm{u}} = 0.75\frac{f_{\mathrm{y}}}{\gamma_{\mathrm{s}}}A_{\mathrm{s}} + (1 + 4.9 \times \frac{\mathrm{t}}{D}\frac{f_{y}}{f_{c}'}) \times 1.2 f_{c}' A_{c} \tag{14}$$

4. Conclusions

In order to explore the mechanical properties of micro-expansion concrete-filled steel tube columns under axial compression, axial compression tests were carried out on 9 specimens. The experimental results show that:

1. The experiment results of the longitudinal volume expansion rate show that the displacement of concrete expansion increases with the increase in the expansion agent content.
2. The damage pattern of all the specimens was bulging and curved in the center and concave at the ends. With the increase in the expansion agent content, the bearing capacity of the specimen increased. The yield strengths of micro-expansion concrete-filled steel tube columns added with 8% and 12% expansion agent were 8.9% and 14.9% higher than the concrete-filled steel tube short column. This indicates that the addition of an expansion agent to concrete has a positive effect on its axial compressive load-bearing capacity.
3. When the expansion agent exceeded a certain size, the bearing capacity decreased. This indicates that there is an optimum content of expansion agent incorporated into concrete. The next step is to prepare an experimental study to ascertain the specific ratio of the optimum content.
4. A practical bearing capacity calculation formula based on the test results is proposed, which provides a reference for practical engineering applications. The next step is to carry out the bearing capacity test on more specimens to optimize the formula, so that the formula can be used in a wider range of applications.
5. While this study confirmed the reliability of micro-expansion concrete in filling steel tube members through experimental methods, the testing conditions limited the ability to conduct experiments on full-scale specimens with high slenderness ratios. Future research efforts could focus on establishing simulation methods for micro-expansion concrete and utilizing validated numerical simulation techniques to investigate the response of full-scale specimens. This approach would enhance the understanding of micro-expansion concrete performance in practical applications and improve the reliability and applicability of the research findings.

Author Contributions: X.L.: methodology, original draft preparation, project administration, software; J.H.: investigation, review and editing; Y.Z.: project administration, writing—review and editing, supervision, methodology, data curation, validation. All authors have read and agreed to the published version of the manuscript.

Funding: This research was funded by the Guangzhou Construction Group Technology Plan Project (Grant no. [2024]-KJ012).

Data Availability Statement: The raw data supporting the conclusions of this article will be made available by the authors on request.

Conflicts of Interest: Author Xin Liu was employed by the company Guangzhou Second Municipal Engineering Co., Ltd. and Guangzhou Construction Group Co., Ltd. and Jisheng Hu was employed by the company Guangzhou Second Municipal Engineering Co., Ltd. The remaining authors declare that the research was conducted in the absence of any commercial or financial relationships that could be construed as a potential conflict of interest.

References

1. Güneyisi, E.M.; Nour, A.I. Axial compression capacity of circular CFST columns transversely strengthened by FRP. *Eng. Struct.* **2019**, *191*, 417–431. [CrossRef]
2. Yan, J.-B.; Dong, X.; Wang, T. Axial compressive behaviours of square CFST stub columns at low temperatures. *J. Constr. Steel Res.* **2020**, *164*, 105812. [CrossRef]
3. He, L.; Zhao, Y.; Lin, S. Experimental study on axially compressed circular CFST columns with improved confinement effect. *J. Constr. Steel Res.* **2018**, *140*, 74–81. [CrossRef]
4. Dong, H.; Li, Y.; Cao, W.; Qiao, Q.; Li, R. Uniaxial compression performance of rectangular CFST columns with different internal construction characteristics. *Eng. Struct.* **2018**, *176*, 763–775. [CrossRef]

5. Huang, Y.-H.; Liu, A.-R.; Fu, J.-Y.; Pi, Y.-L. Experimental investigation of the flexural behavior of CFST trusses with interfacial imperfection. *J. Constr. Steel Res.* **2017**, *137*, 52–65. [CrossRef]
6. Ouyang, Y.; Kwan, A. Finite element analysis of square concrete-filled steel tube (CFST) columns under axial compressive load. *Eng. Struct.* **2018**, *156*, 443–459. [CrossRef]
7. Knowles, R.B.; Park, R. Strength of concrete filled steel tubular columns. *J. Struct. Div. ASCE* **1969**, *95*, 2565–2587. [CrossRef]
8. Knowles, R.B.; Park, R. Axial load design for concrete filled steel tubes. *J. Struct. Div. ASCE* **1970**, *96*, 2125–2153. [CrossRef]
9. Tomii, M. Experimental studies on concrete filled steel tubular stub columns under concentric loading. In Proceedings of the International Colloquium on Stability of Structures Under Static and Dynamic Loads, SSRC/ASCE, Washington, DC, USA, 17–19 May 1977.
10. Luksha, L.K.; Nesterovich, A.P. Strength testing of large-diameter concrete filled steel tubular members. In *Proceedings of the Third International Conference on Steel-Concrete Composite Structures, Fukuoka, Japan, 26–29 September 1991*; Wakabayashi, M., Ed.; Association for International Cooperation and Research in Steel-Concrete Composite Structures: Fukuoka, Japan, 1991; pp. 67–72.
11. Srinivasan, C.N.; Schneider, S.P. Axially loaded concrete-filled steel tubes. *J. Struct. Eng.* **1988**, *125*, 1202–1206. [CrossRef]
12. Giakoumelis, G.; Lam, D. Axial capacity of circular concrete-filled tube columns. *J. Constr. Steel Res.* **2004**, *60*, 1049–1068. [CrossRef]
13. Sakino, K.; Nakahara, H.; Morino, S.; Nishiyama, I. Behavior of centrally loaded concrete-filled steel-tube short columns. *J. Struct. Eng.* **2004**, *130*, 180–188. [CrossRef]
14. Ellobody, E.; Young, B.; Lam, D. Behaviour of normal and high strength concrete-filled compact steel tube circular stub columns. *J. Constr. Steel Res.* **2006**, *62*, 706–715. [CrossRef]
15. Gupta, P.; Sarda, S.; Kumar, M. Experimental and computational study of concrete filled steel tubular columns under axial loads. *J. Constr. Steel Res.* **2007**, *63*, 182–193. [CrossRef]
16. Liang, Q.Q.; Fragomeni, S. Nonlinear analysis of circular concrete-filled steel tubular short columns under axial loading. *J. Constr. Steel Res.* **2009**, *65*, 2186–2196. [CrossRef]
17. Abed, F.; AlHamaydeh, M.; Abdalla, S. Experimental and numerical investigations of the compressive behavior of concrete filled steel tubes (CFSTs). *J. Constr. Steel Res.* **2013**, *80*, 429–439. [CrossRef]
18. Huang, C.K.; Xu, L.; Liu, Y. Experimental study on axial compressive properties of self-compacting and self-stressing concrete filled steel tube short columns. *J. Dalian Univ. Technol.* **2006**, 77–82. (In Chinese)
19. Lu, Y.Y.; Yan, B.; Li, N.; Li, S. Compressive Behavior of Steel Tube Columns Filled with Steel Fiber Reinforced Self-stressing and Self-compacting Concrete. *J. Civ. Eng. Manag.* **2017**, *34*, 1–4.
20. Xu, L.H.; Wu, M.; Zhou, P.H. Experimental study on axial compression bearing capacity of self-stressing and self-compacting high strength concrete filled steel tubular short columns. *Eng. Mech.* **2017**, 93–100. (In Chinese)
21. Armaghani, D.J.; Safari, V.; Fahimifar, A.; Amin, M.F.M.; Monjezi, M.; Mohammadi, M.A. Uniaxial compressive strength prediction through a new technique based on gene expression programming. *Neural Comput. Appl.* **2018**, *30*, 3523–3532. [CrossRef]
22. Sarir, P.; Chen, J.; Asteris, P.G.; Armaghani, D.J.; Tahir, M.M. Developing GEP tree-based, neuro-swarm, and whale optimization models for evaluation of bearing capacity of concrete-filled steel tube columns. *Eng. Comput.* **2021**, *37*, 1–19. [CrossRef]
23. Sarir, P.; Shen, S.-L.; Wang, Z.-F.; Chen, J.; Horpibulsuk, S.; Pham, B.T. Optimum model for bearing capacity of concrete-steel columns with AI technology via incorporating the algorithms of IWO and ABC. *Eng. Comput.* **2021**, *37*, 797–807. [CrossRef]
24. *GB 50119-2013*; Code for Utility Technical of Concrete Admixture. China Standard Press: Beijing, China, 2013.
25. Zhong, S.T. *Unified Theory of Concrete Filled Steel Tubular: Research and Application*; Tsinghua University Press: Beijing, China, 2006.
26. Han, L.H. *Concrete Filled Steel Tube Structure-Theory and Practice*; Science Press: Beijing, China, 1990.
27. Cai, S.H. *Modern Concrete Filled Steel Tube Structure*; People's Communications Press: Beijing, China, 2007.
28. *EN 1994-1-1*; Eurocode4: Design of Composite Steel and Concrete Structures—Part 1-1: General Rules and Rules for Buildings. European Committee for Standardization: Brussels, Belgium, 2004.
29. *ACI-318R*; Building Code Requirements for Structural Concrete and Commentary. American Concrete Institute: Farmington Hills, MI, USA, 2005.
30. *AS3600*; Australian Standards. Concrete Structures. Standards Association of Australia: Sydney, Australia, 2001.
31. AIJ. *Standards for Structural Calculation of Steel Reinforced Concrete Structures*, 5th ed.; Architectural Institute of Japan: Tokyo, Japan, 2001. (In Japanese)
32. *CECS 28:2012*; Technical Specification for Concrete Filled Steel Tube. China Planning Press: Beijing, China, 2012.
33. *GB 50936-2014*; Code for Design of Concrete Filled Steel Tube Structures. China Construction Industry Press: Beijing, China, 2014.

designs

Article

Seismic Design of Steel Frames with Protected Connections

Luigi Palizzolo [1,*], Santo Vazzano [2] and Salvatore Benfratello [1]

[1] Department of Engineering, University of Palermo, Viale delle Scienze, 90128 Palermo, Italy; salvatore.benfratello@unipa.it
[2] Italian National Fire and Rescue Service, Department of Firefighters, Public Rescue and Civil Defence, Italian Ministry of Interior, 00184 Rome, Italy; santo.vazzano@vigilfuoco.it
* Correspondence: luigi.palizzolo@unipa.it

Abstract: The present paper is devoted to the seismic design of steel frames constituted by multistep I-shaped cross-section beam elements. The proposed design problem formulation is aimed at protecting the connections among beams and columns. In particular, reference is made to beams welded at their ends to appropriate steel plates connected by bolts to the columns. Therefore, the protection against brittle failure of the beam end sections is ensured by appropriate constraints of the optimal design problem. A useful comparison is made between the adoption of the so-called Reduced Beam Sections (RBS) and the use of multistep beam elements. In particular, the RBS approach here considered is the well-known dogbone technique consisting of reducing the width of the beam cross-sections in correspondence with suitably located beam portions, while the typical multistep beam element is constituted by a factory-made I-shaped uniform piecewise profile. To perform the necessary comparison, reference is made to a three-story, two-span plane steel frame constituted by elastic, perfectly plastic material and subjected to static and seismic loads. The load conditions and the relevant combinations have been imposed in compliance with the Italian structural code. The frame is first studied as constituted by European standard steel profiles on sale, and the related design is obtained using the optimization tool contained in SAP2000 software. A linear dynamic analysis is performed to determine the response of the frame. Later, the same frame, either equipped with dogbone and constituted by multistep beam elements, subjected to serviceability load conditions, is studied in terms of inter-story drifts and beam deflections. The geometry of the multistep beam elements is obtained by the solution to the proposed optimization problem. Furthermore, a nonlinear static analysis is performed to evaluate the capacity curves of the same frames. The results obtained for the frames equipped with the described different devices, compared with those related to the original frame, provide very interesting information on the sensitivity of the seismic response of the structure, showing the full reliability of the multistep beam element approach.

Keywords: seismic design; steel frames; dogbone; multistep beam elements; moment resisting connections

1. Introduction

The wide selection of frame structures to be adopted in both civil and industrial buildings allows the design of structural elements of different typologies, constituted by standard or innovative materials, considering the use of suitable specific devices to improve the local or global behavior, achieving a safer structure with higher performance.

The most commonly adopted structural typologies are the trusses and the flexural frames, whose selection strongly depends on the structural geometry and on the acting loadings. It is worth noting that, often in frame structures, the two typologies are both present.

From the material point of view, the designer has a wide range of choices. In the last century, reinforced concrete has been widely adopted due to its intrinsic simple production and implementation. This wide adoption brings many drawbacks, mainly related to actual environmental sustainability standards and to its durability related to weathering. Indeed, as it is well known, ordinary Portland cement in concrete construction has led to significant environmental concerns, like depletion of natural resources and CO_2 emissions in its production process. Furthermore, practical experience with concrete structures widely shows they suffered material decay during their service life due to natural actions, especially related to steel oxidation. Several studies have been proposed to enhance the environmental performance of concrete structures, as well as their durability (see, e.g., [1]).

In recent years, interesting research has been devoted to new conglomerates with mechanical performances analogous to those of Ordinary Portland Cement (OPC) concrete, but they are characterized by strong environmental sustainability and show a very important role in the circular economy of construction. The most studied new conglomerate is based on geopolymers [2–4], which are materials developed from alumino-silicate source materials (usually fly ash or slag) with alkaline activators and natural materials such as Kaolinite clay. As reported in many research papers, the mechanical characteristics of such materials can be improved by a suitable adoption of artificial or natural fibers [5–9]. Another important construction material not commonly adopted is wood, which shows high environmental sustainability aspects but whose use is limited to very simple structures.

The second most common structural material widely adopted worldwide is steel. It shows high mechanical performances and limited drawbacks in terms of environmental sustainability; it fulfills the circular economy criteria and guarantees long durability (see, e.g., [10]). Finally, the actual steel production in highly qualified factories ensures high ductility characteristics that are in perfect agreement with the requirements of international standards, especially in the case of structures under dynamic loadings. Furthermore, steel structures provide cost-effective constructions, increased usable floor space, aesthetic pleasing, future adaptability, ease of design, and green production.

In the present paper reference is made to steel frame structures which have been the topic of a huge amount of research devoted both to the analysis [11–17] and to the optimal design problems [18–24]. In particular, the focus is on moment resisting steel frame equipped with devices able to improve the seismic behavior, with a specific attention paid to the Reduced Beam Sections (RBS) connections and to the adoption of multistep beam elements.

The benefit of equipping the frame with such devices is to obtain a structure that fulfills the requirements of the reference standard, taking simultaneously advantage of their intrinsic mechanical characteristics. Briefly, the standard requirements [25,26] can be summarized as follows: (a) to possess suitable stiffness and resistance both in serviceability limit state (SLS) and in ultimate limit state (ULS) conditions; (b) to exhibit appropriate plastic dissipation under high-intensity seismic loadings; (c) to avoid brittle ruptures, especially in correspondence of the welded connections.

Therefore, an important role is played by the connections among structural elements, which must guarantee specific performances in terms of safety, stiffness, and resistance and deeply influence the global behavior of the structure. Usually, these connections are made by welding and/or bolting the ends of the involved elements. It is worth noting that some recent seismic events, for example, the Northridge earthquake of 1994, evidenced the vulnerability of the welding connections [27–29], showing that the failure of the connections was due to the brittle behavior of the welding.

In practical engineering, the main solution utilized to face this problem is the realization of RBS connections. One of the most popular and appropriate RBS connections is represented by the dogbone technique [30–40], which mainly consists of a suitable cut of the flanges of the I-shape profile adopted for the beam aimed to determine a prefixed reduction of the resistance of the specific portion of the element where it is operated. The shape of the cut, symmetrically made with respect to the web of the profile, could be linear as well as parabolic. This technique concentrates the localization of the plastic hinge in a specific portion, but, for obvious reasons, it implies a non-negligible local flexural stiffness decrement, which affects even the overall stiffness of the frame. Other solutions were proposed by different authors [41–60], ranging from the use of special treatments on the beam elements to the adoption of special devices where a resistance reduction is imposed, identifying a special section devoted to the localization of plastic deformations.

In the framework of the RBS connections, some of the authors proposed a new special device called the Limited Resistance Plastic Device (LRPD), which is characterized by the advantages of the RBS connections and enhanced stiffness features. The main novelty of LRPD, whose positioning is typically at the end of beams and/or at the column base, is that its specific design allows the onset of plastic strain, always guaranteeing the unaltered bending stiffness of the overall involved structural element.

The first model of LRPD was referred to as Limited Resistance rigid Perfectly plastic flexural Hinges (LRPH) [61–63]; in subsequent studies, the design problem has been extended to consider both normal and tangential stresses acting simultaneously, taking advantage of smart beam elements to calculate the mechanical response of the beam elements [64]. Further studies have been devoted to the validation of the LRPD by numerical FE analyses and by appropriate experimental tests, and suitable applications have been devoted to the design of frames with assigned capacity curves [65]. Furthermore, the authors proposed many improvements in the formulation of the LRPD, comparing the results with those obtained by different RBS connections, considering the local buckling, and protecting the welded section from brittle rupture (see, e.g., [66]).

The actual version, referred to as LRPD, is a beam element with multistep geometry, constituted by three subsequent I-shaped portions whose width, depth, web, and flange thicknesses are suitably designed.

In this last version of LRPD, the inner portion with reduced dimensions shows limited strength and allows the onset of plastic strain, while the outer portions are equal to each other and show web and flange thicknesses greater to guarantee the unaltered bending stiffness of the overall involved element. As it can be easily deduced, LRPD design requires the solution to appropriate optimal design problems and their production must be factory performed by welding steel plates of assigned thickness.

It must be remarked that the use of LRPD requires specific factory manufacturing to realize the whole structural element constituted by the central portion (standard profile) and by the end portions (LRPD).

It follows that it can be very useful to extend the fundamental idea at the basis of LRPD to the entire structural element by defining a multistep beam constituted by welded I-shaped cross-sections suitably designed to fulfill all the above-described standard requirements. These beams are constituted by a given number of consecutive portions, each with a constant section. Generally speaking, the cross-sections of different portions show different geometric characteristics (in terms of width, depth, web, and flange thickness) that must be suitably designed to represent the main design variables. The length of each portion, as well as its position along the structural element, can represent further design variables.

The aim of the present paper is to propose a new special formulation of the optimal design problem of multistep beam elements, mainly devoted to the comparison between the structural performances of frames equipped with dogbones and with this new kind of structural beam. The presence of dogbones defines a structural element subdivided into five consecutive portions where the first, the third, and the fifth ones possess the same cross-section of the original element, while the second and fourth ones are equal to each other, are positioned symmetrically with respect to the middle cross-section of the reference beam element, possess an assigned length, and show a reduced cross-section variable along the beam axis due to the analogous cut of the flanges. Consequently, the multistep beam is designed to consist of five consecutive portions whose length is equal to the corresponding portion of the beam equipped with the dogbones. The first, third, and fifth portions possess the same constant cross-section with flange thickness obtained by the solution to the proposed optimal design problem. The second and the fourth portions (equal to each other) possess reduced geometric characteristics (in terms of flange width and flange thickness), which are always obtained by the solution to the optimal design problem. Finally, it is assumed that the web thickness of all the portions of the multistep beam is always the same and that the flanges of all portions possess a unique common medium plane.

To compare the different approaches described above, reference is made to a standard steel frame under both static and seismic loadings. The reference frame is first designed following the standard requirements [26], adopting European standard I-shaped steel profiles on sale, and its structural response is determined. The obtained reference frame is successively equipped with dogbones designed following the reference standard [25,67] utilizing two different dogbone sizing. The two obtained frames are studied, and their structural response is also determined. Similarly, just like was performed for the dogbone frames, two multistep beam frames are designed by solving the proposed optimal design problem, imposing that each cross-section of the weak portions of each beam of the typical frame exhibits the same flexural strength as the corresponding dogbone, while the strong portion cross-sections are designed in such a way that the overall element fulfills the above-described safety, strength, and stiffness requirements. Therefore, the proposed optimal design problem formulation for multistep beam element frames represents an interesting enhancement of the classical design problem formulation for steel structures, allowing a wider degree of freedom in the cross-section sizing and in the definition of their strength and stiffness features.

The proposed numerical application shows the great affordability of the proposed approach representing a first step for further deeper developments devoted to the overall optimal design of a frame equipped with multistep beam elements.

2. Geometry and Mechanical Properties of the Different Approaches

In this section, a description of the geometrical and mechanical behavior of the dogbone and of the proposed multistep beam element is provided. In both cases, it is possible to realize a specific portion of the beam element with reduced yield resistance, where the onset of plastic deformations is expected, but only the multistep beam element ensures appropriate characteristics of resistance and stiffness of the connections between beams and columns, as it will be proved in the following.

In Figure 1, the main geometric characteristics of the dogbone are reported.

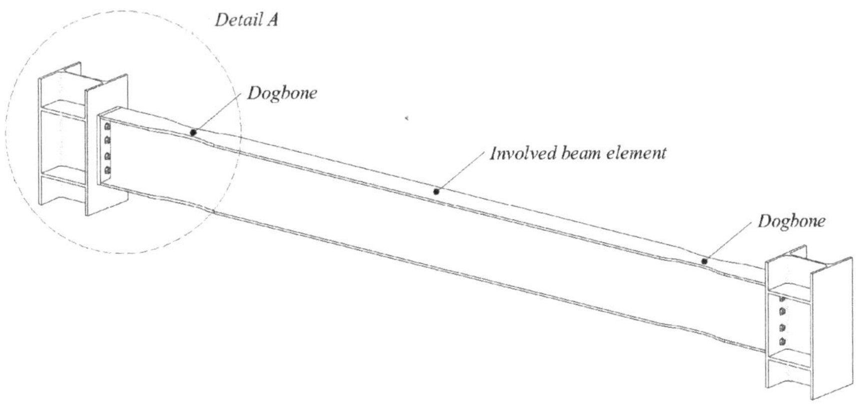

(a)

Figure 1. *Cont.*

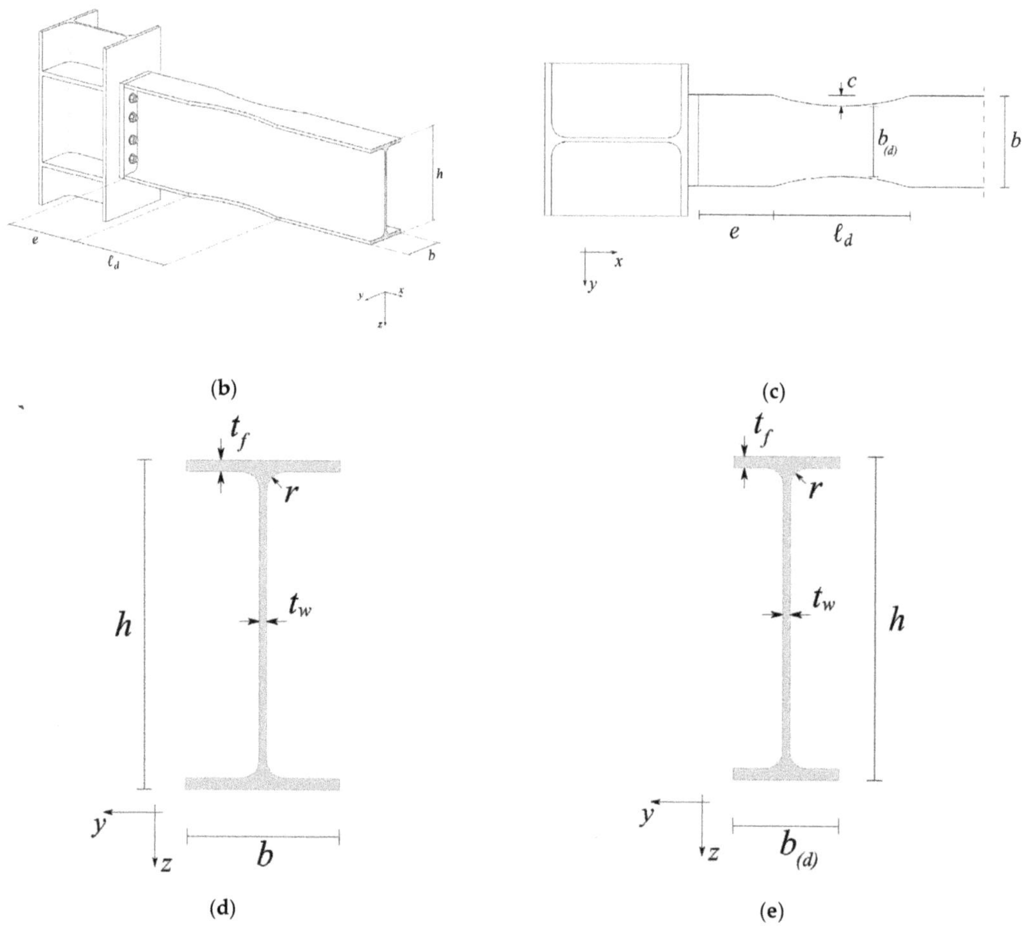

Figure 1. Dogbone geometry; (**a**) 3D sketch of the beam element equipped with dogbones; (**b**) Detail A with main geometrical features; (**c**) typical cut geometry; (**d**) typical geometry of the reference beam cross-section; (**e**) typical geometry of the dogbone middle cross-section.

In Figure 1a, a 3D representation of a typical beam element with the dogbone insertions is sketched. In Figure 1b, a detail of the connection with dogbone is represented; here, b and h are the width and depth of the involved I-shaped beam element (reference beam cross-section reported in Figure 1d, the parameter e indicates the distance between the initial cut section and the relevant column where the beam end is connected and ℓ_d measures the length of the portion with reduced variable width. Among the different shapes available for cutting the flanges, the parabolic one is chosen in the present study. As reported in Figure 1c,e, the middle cut section (i.e., the weaker cross-section) shows the minimum base length $b_{(d)} = b - 2c$, where c is usually defined as percentage of the original beam width and it represents the main parameter for the dogbone sizing. It is worth noting that the parameters e and ℓ_d usually depend on the depth of the original beam width, as will be specified in the application stage. Due to the assigned geometry of the dogbone, for a given $b_{(d)}$, it is possible to determine the corresponding geometric and mechanical resistance parameters as well as the related elastic and plastic axial forces and bending moments of the weaker cross-section, i.e., the one located in the middle of the interested portion and reported in Figure 1e, as follows:

$$A_{(d)} = 2b_{(d)}t_f + t_w\left(h - 2t_f\right) + (4 - \pi)r^2 \tag{1}$$

$$\begin{aligned} I_{y(d)} = b_{(d)}\frac{t_f^3}{6} + b_{(d)}t_f\frac{\left(h-t_f\right)^2}{2} + t_w\frac{\left(h-2t_f\right)^3}{12} + \frac{r^4}{3} + r^2\left(h - 2t_f - r\right)^2 \\ -r^4\left(\frac{\pi}{4} - \frac{16}{9\pi}\right) - \pi r^2\left(\frac{h}{2} - t_f - r + \frac{4r}{3\pi}\right)^2 \end{aligned} \tag{2}$$

$$W_{Ey(d)} = \frac{2I_{y(d)}}{h} \tag{3}$$

$$\begin{aligned} W_{py(d)} = b_{(d)}t_f\left(h - t_f\right) + t_w\frac{\left(h-2t_f\right)^2}{4} + 2r^2\left(h - 2t_f - r\right) \\ -\frac{\pi r^2}{2}\left(h - 2t_f - 2r + \frac{8r}{3\pi}\right) \end{aligned} \tag{4}$$

$$N_{E(d)} = N_{p(d)} = A_{(d)}\sigma_0 \tag{5}$$

$$M_{Ey(d)} = W_{Ey(d)}\sigma_0 \tag{6}$$

$$M_{py(d)} = W_{py(d)}\sigma_0 \tag{7}$$

where, besides the already defined quantities, $A_{(d)}$ is the area of the cross-section; t_w, t_f, and r are web thickness, flange thickness, and welding radius of the involved cross-section beam element; $I_{y(d)}$ is the second-order moment of area of the cross-section with respect to the central axis y; $W_{Ey(d)}$ and $W_{py(d)}$ are the elastic and plastic resistance moduli of the cross-section, respectively; $N_{E(d)}$ and $N_{p(d)}$ are the limit elastic and plastic axial forces (equal to each other); and, finally, $M_{Ey(d)}$ and $M_{py(d)}$ are the limit elastic and plastic bending moments of the cross-section. It is assumed that the material constituting the beam element shows elastic, perfectly plastic behavior, and, as usual, the tensile yield stress has been indicated with σ_0.

While the dogbone technique consists of direct intervention and modification on the extreme portions of the original beams of the frame, the use of multistep beams consists of the complete replacement of the original beams with the factory-made I-shaped multistep ones. For the aim of the present study, the typical multistep beam here is considered to consist of five subsequent portions with uniform piecewise I-shaped cross-sections, as reported in Figure 2.

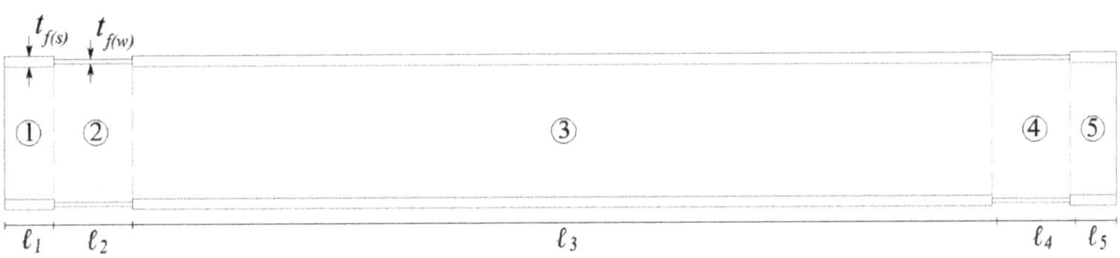

(a)

Figure 2. *Cont.*

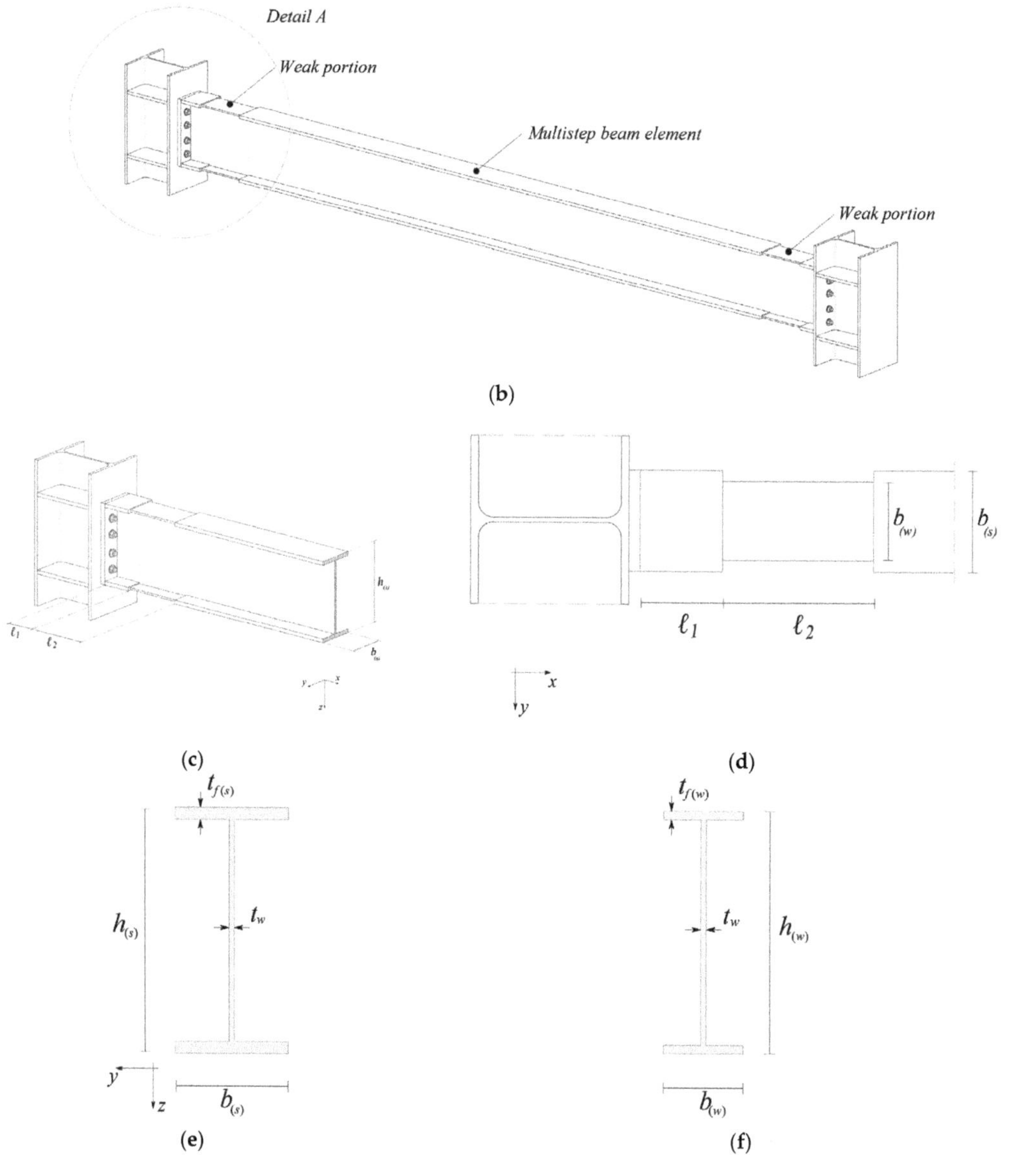

Figure 2. Multistep beam geometry: (**a**) typical multistep beam element constituted by five subsequent portions; (**b**) 3D sketch of the multistep beam; (**c**) Detail A with main geometrical features; (**d**) view of the beam extrados with main geometrical features; (**e**) typical geometry of the strong cross-section; (**f**) typical geometry of the weak cross-section.

Referring to an analogous scheme as for the beam element equipped with dogbones, portions 2 and 4 are constituted by equal cross-sections, as well as portions 1, 3, and 5 are constituted by equal cross-sections (but different from those of portions 2 and 4). Portions 2 and 4 represent the weak portions of the multistep beam, while portions 1, 3, and 5 represent the strong portions of the multistep beam. In the following, all the quantities related to the weak portions will be identified by the index (w), while the analogous quantities related to the strong portions will be identified by the index (s).

Furthermore, the length of the weak portions 2 and 4 is usually assumed the same as well as it happens for portions 1 and 5. To design these lengths, different approaches can be utilized. Referring to the weak portions, as already stated, the main objective is to create a zone where a fully plasticization can occur. As proved in [64], this requirement can be satisfied by choosing the weak portion length $\ell_2 = \ell_4$ not less than 0.5 h. On the other hand, referring to the strong portion length (1 and 5), the corresponding length can be defined, imposing both resistance and mechanical (stiffness) constraints in the relevant optimization problem (see, e.g., [61–63]). In the present paper, due to the main aim related to the comparison between dogbone and multistep beam, the following relations will be adopted as follows:

$$\ell_1 = \ell_5 = e \tag{8}$$

$$\ell_2 = \ell_4 = \ell_d \tag{9}$$

$$\ell_3 = \ell - 2e - 2\ell_d \tag{10}$$

Furthermore, the consecutive I-shape elements with different flange and web thicknesses are symmetrically disposed with respect to the medium cross plane x, y (Figure 2e,f). From a geometrical point of view, it is assumed that (a) the flanges of all the portions possess a unique common medium plane as well as the webs; (b) always referring to the dogbone scheme, the webs of all the cross-section portions possess the same thickness equal to that of the original cross-section beam element. From a mechanical point of view, in order to guarantee the desired behavior, the strong portions must possess flange thicknesses greater than the corresponding ones of the weak portions.

The geometric and mechanical resistance parameters, as well as the related elastic, are applied to all the quantities referred to in the relevant cross-section as follows:

$$A_{(w)} = 2b_{(w)}t_{f(w)} + t_w\left(h_{(w)} - 2t_{f(w)}\right) = b_{(w)}h_{(w)} - \left(b_{(w)} - t_w\right)\left(h_{(w)} - 2t_{f(w)}\right) \tag{11}$$

$$\begin{aligned} I_{y(w)} &= b_{(w)}\frac{t_{f(w)}^3}{6} + b_{(w)}t_{f(w)}\frac{\left(h_{(w)} - t_{f(w)}\right)^2}{2} + t_w\frac{\left(h_{(w)} - 2t_{f(w)}\right)^3}{12} \\ &= \frac{b_{(w)}h_{(w)}^3}{12} - \frac{\left(b_{(w)} - t_w\right)\left(h_{(w)} - 2t_{f(w)}\right)^3}{12} \end{aligned} \tag{12}$$

$$W_{Ey(w)} = \frac{2I_{y(w)}}{h} \tag{13}$$

$$\begin{aligned} W_{py(w)} &= b_{(w)}t_{f(w)}\left(h_{(w)} - t_{f(w)}\right) + t_w\frac{\left(h_{(w)} - 2t_{f(w)}\right)^2}{4} \\ &= \frac{b_{(w)}h_{(w)}^2}{4} - \frac{\left(b_{(w)} - t_w\right)\left(h_{(w)} - 2t_{f(w)}\right)^2}{4} \end{aligned} \tag{14}$$

$$N_{E(w)} = N_{p(w)} = A_{(w)}\sigma_0 \tag{15}$$

$$M_{Ey(w)} = W_{Ey(w)}\sigma_0 \tag{16}$$

$$M_{py(w)} = W_{py(w)}\sigma_0 \tag{17}$$

where, besides the already defined quantities, $A_{(w)}$ is the area of the cross-section, $b_{(w)}$, $h_{(w)}$ and $t_{f(w)}$ are width, depth and flange thickness of the cross-section, $I_{y(w)}$ is the second-order moment of area of the cross-section with respect to the central axis y, $W_{Ey(w)}$ and $W_{py(w)}$ are the elastic and plastic resistance moduli of the cross-section, respectively, $N_{E(w)}$

and $N_{p(w)}$ are the limit elastic and plastic axial forces (equal to each other) and, finally, $M_{Ey(w)}$ and $M_{py(w)}$ are the limit elastic and plastic bending moments of the cross-section.

Since the multistep beam is constituted by welded I-shaped cross-sections, in Equations (11)–(17) the welding size has been neglected. Analogous relations can be written for the cross-sections related to the strong portions, changing index (w) with index (s).

3. Design Criterion for the Multistep Beam

As previously stated, steel structures are often more advisable with respect to other typologies. Their full reliability is related to the capacity of fulfilling some appropriate fundamental requisites: they must possess adequate resistance and stiffness features both in serviceability and in ultimate limit conditions, must exhibit wide plastic dissipation capacity and the risk of brittle failure must be avoided in correspondence of the welded connections. It follows that the introduction of special devices in the structural elements (dogbone) can be very useful, as well as the design of special factory-made beam elements (multistep beams).

Before facing the design of the multistep beams, the following remarks can be useful. As clarified in the introduction, according to [67], where the allowed geometric characteristics and ranges are reported, the design of the dogbone does not require any specific computational procedure or numerical effort. Specifically, both the distance e between the end section of the beam and the initial section of the dogbone, and the length ℓ_d of the dogbone itself depend on the depth of the original cross-section, and the allowable ranges are $0.25\,\mathrm{h} \leq e \leq 0.5\,\mathrm{h}$ and $0.5\,\mathrm{h} \leq \ell_d \leq 0.75\,\mathrm{h}$.

An analysis of the structural effects due to the dogbones (see, e.g., [68]) immediately reveals that their application does not fulfill all the above-described requirements. Indeed, as it is also reported in the referenced standard [67], the dogbones implicate a reduction of the overall stiffness of the beam, and, further, their design does not take into account the stresses acting in the welded section to avoid dangerous brittle ruptures.

It is worth noting that brittle ruptures represent a very dangerous limit state identifying an overall structural behavior inappropriate for a steel structure, which must be characterized by a ductility behavior, as the international structural standards requires throughout the adoption of a suitable behavioral factor.

The design of the multistep beam, both regarding the geometry of the cross-section and the length of each portion, requires the solution of an optimal design (usually, the objective function is the minimum volume) with suitable constraints on stiffness, strength, and safety.

In the present paper, the multistep beam is designed by means of a simplified procedure, reported in the following, starting from some assigned dimensions to achieve a meaningful comparison with the dogbones.

First, the minimum volume is assumed to be the objective function of the multistep beam. The following geometrical characteristics are assumed:

- as the limit contour of the cross-section, the rectangle circumscribed to the section of the original beam is selected; clearly, this rectangle is characterized by a base b and a depth h equal to the width and to the depth of the original beam;
- the length of each portion is assumed as reported in Equations (8)–(10);
- all the cross-sections possess the same web thickness equal to that of the original beam, that is $t_{w(w)} = t_{w(s)} = t_w$, and the flanges of all the portions possess a unique common medium plane;
- the width of the cross-section of the strong portions is assumed to be equal to that of the cross-section of the original beam, that is $b_{(s)} = b$ while that of the weak portion $b_{(w)} \leq b$ is deduced by the solution to the optimization problem.

Therefore, the variables of the optimal design problem are the width $b_{(w)}$ and the thicknesses $t_{f(w)}$ and $t_{f(s)}$.

The constraints to be imposed are related to the following requirements:

- the limit plastic bending moment of the weak portions is equal to that of the dogbone;
- the cross-section of the weak portion has to belong to Class 1 of [25], avoiding any possible local buckling phenomenon;
- the limit elastic bending moment of the cross-section characterizing the strong portions must be such to prevent brittle failure;
- the bending stiffness of the overall multistep beam must not be lower than that of the original beam.

The fulfillment of the above-reported requirements leads to a multistep beam possessing adequate strength and ductility features, avoiding any reduction of the bending stiffness and limiting the stresses acting in the welded section.

It is worth noting that, with the above-described characteristics of the multistep beam, the use of the behavioral factor as reported in the structural standards is fully effective.

Therefore, the minimum volume design problem for the multistep beam is characterized by the following assigned quantities:

- $\ell_1 = \ell_5$, $\ell_2 = \ell_4$ and ℓ_3 (lengths of the different portions of the multistep beam);
- $b_1 = b_3 = b_5 = b$ (cross-section widths of the strong portions of the multistep beam, being b the width of the original beam element);
- $h_1 = h_3 = h_5 = h$ (cross-section depths of the strong portions of the multistep beam, being h the depth of the original beam element);
- $t_{w1} = t_{w2} = t_{w3} = t_{w4} = t_{w5} = t_w$ (web thickness of the different portions of the multistep beam, being t_w the web thickness of the original beam element);
- σ_0 (material yield stress);

It is also characterized by the following geometrical and technological constraints:

- $h_2 = h_4 = h_{(w)} = h - t_{f(s)} + t_{f(w)}$ (the cross-section flanges of subsequent beam element portions have the same medium plane);

by the following design variables:

- $b_2 = b_4 = b_{(w)}$ (width of the weak portion cross-sections of the multistep beam);
- $t_{f2} = t_{f4} = t_{f(w)}$ (flange thickness of the weak portion cross-sections of the multistep beam);
- $t_{f1} = t_{f3} = t_{f5} = t_{f(s)}$ (flange thickness of the strong portion cross-sections of the multistep beam);

and by the following objective function (volume of the multistep beam)

$$\begin{aligned} V = A_{(s)}(\ell_1 + \ell_3 \quad &+ \ell_5) + A_{(w)}(\ell_2 + \ell_4) \\ &= \left[bh - (b - t_w)\left(h - 2t_{f(s)}\right)\right](\ell_1 + \ell_3 + \ell_5) \\ &+ \left[b_{(w)}h_{(w)} - \left(b_{(w)} - t_w\right)\left(h_{(w)} - 2t_{f(w)}\right)\right](\ell_2 + \ell_4) \end{aligned} \tag{18}$$

Therefore, the above-described minimum volume design problem for the multistep beam can be written in the following form:

$$\min_{\left(b_{(w)}, t_{f(w)}, t_{f(s)}\right)} V \tag{19}$$

subjected to:

$$0 \leq b_{(w)} \leq b \tag{20}$$

$$0 \leq t_{f(w)} \leq t_f \tag{21}$$

$$t_f \leq t_{f(s)} \leq \frac{h}{2} \tag{22}$$

$$\frac{b_{(w)} - 3t_w}{18}\sqrt{\frac{\sigma_0}{235}} \leq t_{f(w)} \tag{23}$$

$$\begin{aligned} b_{(d)}t_f\left(h - t_f\right) + t_w\tfrac{\left(h-2t_f\right)^2}{4} + 2r^2\left(h - 2t_f - r\right) - \tfrac{\pi r^2}{2}\left(h - 2t_f - 2r + \tfrac{8r}{3\pi}\right) \\ \leq b_{(w)}t_{f(w)}\left(h - t_{f(s)}\right) + t_w\tfrac{\left(h-t_{f(s)}-t_{f(w)}\right)^2}{4} \end{aligned} \tag{24}$$

$$M_{max} \leq \left[\frac{bh^2}{6} - \frac{(b - t_w)\left(h - 2t_{f(s)}\right)^3}{6h}\right]\sigma_0 \tag{25}$$

$$\begin{aligned} 2\ell_1 + \ell_3 \leq \left[\tfrac{bh^3}{12} - \tfrac{(b-t_w)\left(h-2t_{f(s)}\right)^3}{12}\right] \\ \times \left\{ \tfrac{\ell}{I_{(o)}} - \tfrac{2\ell_2}{\left[\tfrac{b_{(w)}\left(h-t_{f(s)}+t_{f(w)}\right)^3}{12} - \tfrac{\left(b_{(w)}-t_w\right)\left(h-t_{f(s)}-t_{f(w)}\right)^3}{12}\right]} \right\} \end{aligned} \tag{26}$$

where Equation (20) provides the upper and lower bound on the width of the weak portion cross-section, Equations (21) and (22) provide the upper and lower bounds on the flange thicknesses of the weak and strong portion cross-sections, respectively, Equation (23) provide the constraint imposing that the weak portion cross-section belongs to the Class 1, Equation (24) ensures that the weak portion cross-section reaches the yield bending moment value as the related weaker dogbone cross-section, Equation (25) ensures that the strong portion cross-section, which realizes the welded connection, is safe against any brittle failure (being its limit elastic bending moment not lower than the maximum envelop bending moment M_{max} acting on the relevant section) and, finally, Equation (26) ensures that the global flexural stiffness of the multistep beam element is not lower than the corresponding one related to the original beam element, being $I_{(o)}$ the second-order moment of area of the original cross-section beam element.

Equations (20)–(22) represent trivial technological and mechanical constraints, Equation (23) is deduced by well-known formulae present in the international structural codes, Equation (24) relates the plastic resistance modulus of the middle dogbone cross-section (first member of the inequality) and the plastic resistance modulus of the weak portion cross-section (second member of the inequality), in Equation (25) the limit elastic bending moment of the strong portion cross-section, which realizes the welded connection, is expressed in terms of elastic resistance modulus. Finally, Equation (26) is deduced imposing that the global flexural stiffness of the multistep beam element is not lower than the corresponding one related to the original beam element, i.e., that the relative rotation of the end cross-sections of the multistep beam element is not greater than the analogous quantity related to the original beam element. With reference to Figure 3, the equality condition related to this last constraint reads as follows:

$$2\int_0^{\ell_1} \frac{1}{EI_{(s)}}dx + 2\int_0^{\ell_2} \frac{1}{EI_{(w)}}dx + \int_0^{\ell_3} \frac{1}{EI_{(s)}}dx = \int_0^{\ell} \frac{1}{EI_{(o)}}dx \tag{27}$$

that, with easy mathematics, can be written as follows:

$$2\ell_1 + \ell_3 = I_{(s)}\left(\frac{\ell}{I_{(o)}} - \frac{2\ell_2}{I_{(w)}}\right) \tag{28}$$

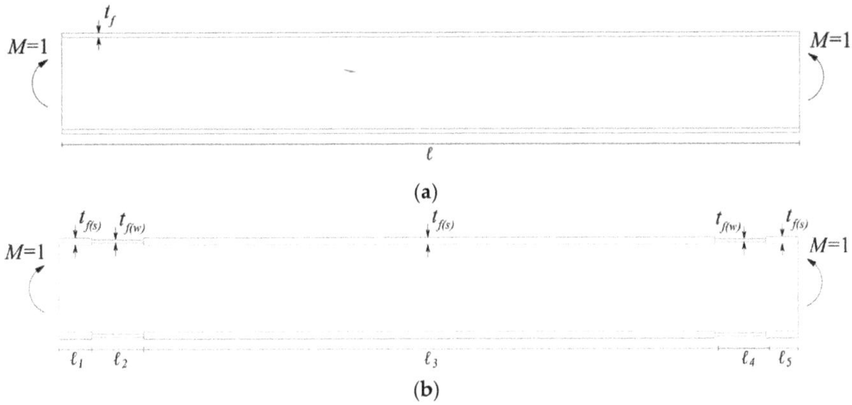

Figure 3. Load scheme related to the global stiffness constraint: (**a**) Reference beam element subjected to uniform unitary bending moment; (**b**) multistep beam subjected to uniform unitary bending moment.

4. Numerical Application

As a numerical application, a perfectly clamped three-story two-span plane steel frame is considered. In Figure 4 the axis scheme of the reference frame is sketched and in Table 1 the assigned geometry is detailed.

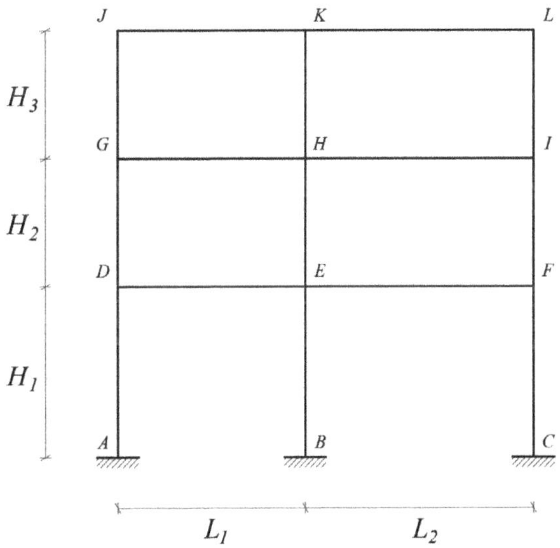

Figure 4. Axis scheme of the reference plane steel frame.

Table 1. Geometrical characteristics of the reference frame.

Frame Geometry [m]			
H_1	4.00	L_1	4.50
$H_2 = H_3$	3.00	L_2	5.50

The material is assumed to be S275 steel with an elastic-perfectly plastic behavior. The load condition, referring to the Italian structural code [26], is defined by the presence of the structure self-weight G_1, of the permanent loads G_2 and of the variable loads Q. It is assumed that these actions produce uniformly distributed loads on the beams with the values reported in Table 2.

Table 2. Values of the acting loads on the selected frame.

Loads [kN/m]	
$q(G_1)$	25.00
$q(G_2)$	15.00
$q(Q)$	20.00

For the aims of the present study and for the sake of simplicity, it is assumed that all the beams are subjected to the same uniformly distributed load combinations, whose values depend on appropriate load multipliers as indicated by the Italian code [26] and defined for the different considered limit states. The seismic actions are defined by referring to the Italian code for a building in Palermo, by assuming a class type B and a category type III. Therefore, the design of the frame is obtained considering all the load combinations as prescribed by the relevant code for serviceability and limit collapse conditions, as well as for static load and for seismic load conditions. As usual, the seismic response is obtained by the solution to a linear dynamic analysis. The structural software adopted for the numerical computations is SAP2000 v2017 [69].

The main aim of the numerical application is to verify the reliability of the proposed optimal procedure, to analyze for the given structure the effectiveness of the use of the dog-bone technique and of the multistep beams, and to compare the two different interventions in terms of structural safety and consistent behavior.

Therefore, at first, the structural design of the chosen frame is carried out by performing the auto-design procedure available in SAP2000 and making reference to European standard steel profiles on sale. The obtained results (optimal cross-sections of beams and columns) are reported in Table 3.

Table 3. Cross-sections obtained by the standard code design.

Columns						Beams			
A-D	HEA 320	*B-E*	HEA 550	*C-F*	HEA 400	*D-E*	IPE 330	*H-I*	IPE 360
D-G	HEA 320	*E-H*	HEA 550	*F-I*	HEA 400	*E-F*	IPE 360	*J-K*	IPE 360
G-J	HEA 320	*H-K*	HEA 550	*I-L*	HEA 400	*G-H*	IPE 330	*K-L*	IPE 450

In Figure 5 the exploitation percentage of beams and columns of the designed frame is represented in terms of resistance capacity; in Table 4, for each beam end, are reported: the ratio between limit plastic bending moment and maximum value of the envelope bending moment response and the ratio between limit elastic bending moment and maximum value of the envelope bending moment response.

Table 4. The ratio between plastic/elastic bending moment and maximum envelope acting bending moment in correspondence of the beam ends of the reference frame (*i* left end, *j* right end).

Beam	*D-E*		*E-F*		*G-H*		*H-I*		*J-K*		*K-L*	
End	*i*	*j*	*i*	*j*	*i*	*j*	*i*	*j*	*i*	*j*	*i*	*j*
Plastic/Envelope	1.05	1.01	1.01	1.01	1.11	1.04	1.01	1.04	1.90	1.43	1.56	1.91
Elastic/Envelope	0.93	0.90	0.89	0.89	0.98	0.92	0.90	0.93	1.69	1.27	1.38	1.68

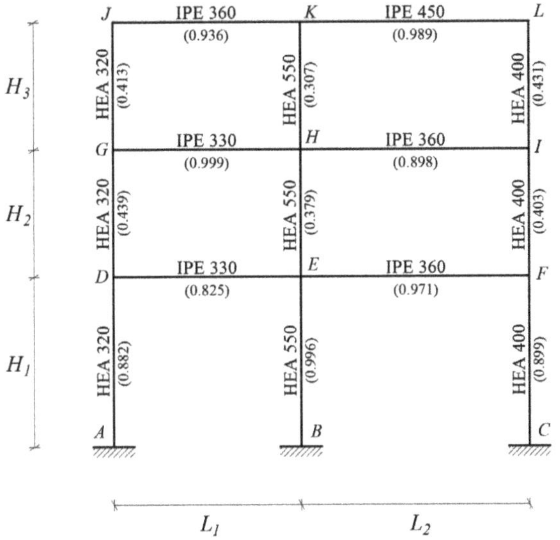

Figure 5. Exploitation percentage of beams and columns of the reference frame.

The examination of the results summarized in Table 4, with specific reference to the beam behavior, allows us to state that at the cross-section beam ends (where the connections between beams and columns are realized), the limit plastic safety condition is fulfilled, but in many cases (beams *D-E*, *E-F*, *G-H*, *H-I*) the limit elastic bending moment is lower than the acting one, i.e., some brittle damage can occur, and the safety is not ensured. This remark will be recalled to evaluate the goodness of the response of the frame equipped with dogbones and multistep beams.

Once the reference frame is designed, the dogbone intervention is defined.

Two different width reductions have been chosen for the dogbone intervention: $c = 20\%\ b$ (in the following indicated as DB60, $b_{(d)} = 0.6b$) and $c = 30\%\ b$ (in the following indicated as DB40, $b_{(d)} = 0.4b$). With reference to Figure 6 and Table 5, the geometry of the dogbone intervention is fully defined [67]. The dogbones have been considered to be realized at all the frame beam ends.

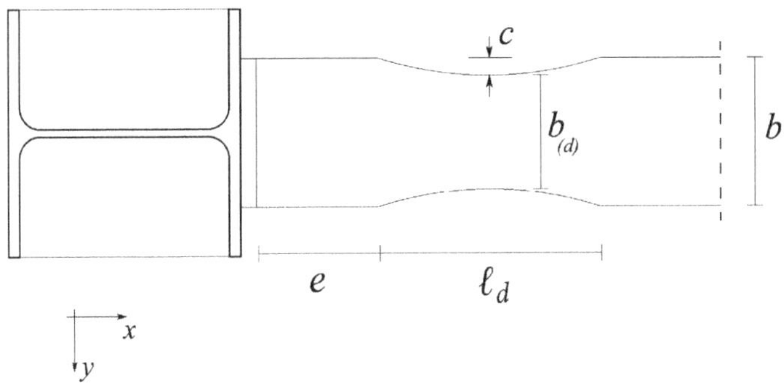

Figure 6. Geometric characteristics of the dogbone.

Table 5. Geometrical parameters characterizing the adopted dogbones (h is the depth of the reference cross-section).

DB60		DB40	
c	0.20 b	c	0.30 b
$b_{(d)}$	0.6 b	$b_{(d)}$	0.4 b
e	$h/4$	e	$h/2$
ℓ_d	$h/2$	ℓ_d	$3\,h/4$

Once the geometry of the dogbones is defined, the multistep beams can be designed. The optimal design is obtained by the solution for each beam element of the minimum volume problem (19). Assuming the length of the portions of the multistep beams as the corresponding ones of the beam equipped with dogbones, two different designs are determined, related to the two above-described different dogbone geometries considered: the first one (in the following referred to as MS60) is determined by imposing, besides the other common constraints, that the cross-sections of the weak portions of the multistep beams possess the same limit bending resistance as DB60; the second one (in the following referred to as MS40) is determined analogously but referring to the limit bending resistance of DB40. In Tables 6 and 7, the full geometry of the cross-sections of the obtained multistep beams is reported for MS60 and MS40, respectively.

Table 6. Geometric characteristics of the multistep beam element (MS60) in mm.

	Strong Portions				Weak Portions			
Beam	Width b	Depth h	Web Thickness t_w	Flange Thickness $t_{f(s)}$	Width $b_{(w)}$	Depth $h_{(w)}$	Web Thickness t_w	Flange Thickness $t_{f(w)}$
D-E	160	334	7.5	15.4	131	328	7.5	9.4
E-F	170	364	8.0	17.1	138	358	8.0	10.3
G-H	160	332	7.5	13.6	131	328	7.5	9.4
H-I	170	363	8.0	15.5	138	358	8.0	10.3
J-K	170	362	8.0	14.5	138	358	8.0	10.3
K-L	190	454	9.4	18.2	128	445	9.4	9.9

Table 7. Geometric characteristics of the multistep beam element (MS40) in mm.

	Strong Portions				Weak Portions			
Beam	Width b	Depth h	Web Thickness t_w	Flange Thickness $t_{f(s)}$	Width $b_{(w)}$	Depth $h_{(w)}$	Web Thickness t_w	Flange Thickness $t_{f(w)}$
D-E	160	334	7.5	15.4	109	326	7.5	7.8
E-F	170	364	8.0	17.1	114	356	8.0	8.6
G-H	160	333	7.5	14.1	109	326	7.5	7.8
H-I	170	363	8.0	15.5	114	356	8.0	8.6
J-K	170	363	8.0	15.8	114	356	8.0	8.6
K-L	190	452	9.4	16.8	154	447	9.4	11.9

The static and the seismic (linear dynamic) analyses of the frames related to DB60 and DB40 intervention, as well as those of the frames constituted by MS60 and MS40 multistep beams, are carried out, always by using the structural software SAP2000. It is worth noting that to perform these analyses each beam has been considered as constituted by five subsequent portions with suitably assigned geometry, as deduced by the previous computations.

The comparison between the two different approaches is made analyzing the responses of the frames in serviceability conditions (limit elastic regime), in limit collapse conditions (envelope of fundamental static and seismic combinations) and beyond the elastic limit (performing a push-over analysis).

Referring to serviceability conditions, the control is made on the relevant kinematical quantities: deflection of the middle point of the beams and inter-story drifts. In Table 8 the deflections of the middle point of all the beams for the studied frames (reference frame, DB60, DB40, MS60 and MS40 frames) is reported.

Table 8. Deflections of the beams [mm].

Beam	*D-E*	*E-F*	*G-H*	*H-I*	*J-K*	*K-L*
Reference	4.10	5.83	4.33	6.08	4.15	4.24
DB60	4.32	6.04	4.52	6.25	4.23	4.45
MS60	3.95	5.53	3.54	5.97	4.07	4.24
DB40	4.43	6.22	4.65	6.46	4.29	4.59
MS40	4.04	5.63	3.68	6.08	3.95	4.15

Even if all the results are fully acceptable in terms of structural code rule (the maximum deflection reaches about 1.2‰ of the related spam length), it is easy to observe that in all the cases, the deflections related to the dogbone frames are greater than the one of the reference frames; on the contrary, in all the cases the deflections related to the multistep beam frames are not greater than those of the reference frame.

In Figure 7, a diagram is sketched where the maximum frame inter-story drifts are reported for all the investigated frames as the percentage of the inter-story heights. Once again, all the evaluated drifts are consistent and respect the relevant structural code [26], which requires that drifts do not exceed 5‰ of the inter story height. Nevertheless, it is worth noting that just the multistep frames ensure less global deformability, better protecting the structure against secondary element damages.

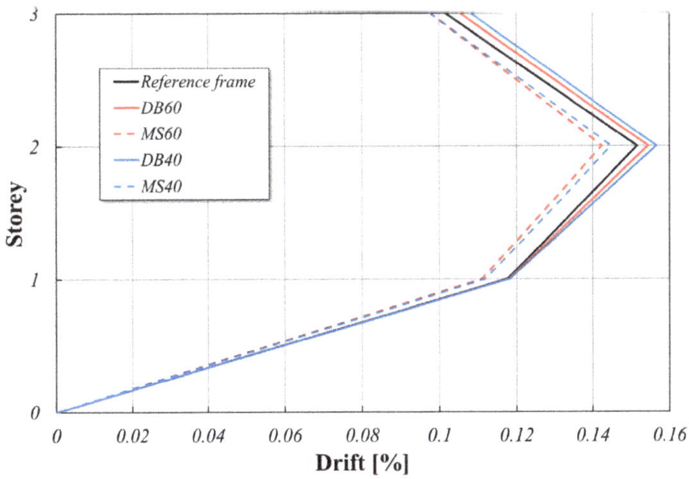

Figure 7. Inter-story drifts.

In limit collapse conditions, the control must be carried out by comparing the cross-section capacity with the structural demand resulting from the analysis. For the aim of the present study, reference has been made to the behavior of the end cross-sections of the beams, which represent the portions mainly involved by the presence of the dogbones and of the multistep beams.

Operating as already made for the reference frame, for each beam end of the dogbone frames and of the multistep beam frames, in Table 9 is reported the ratio between limit plastic bending moment and maximum value of the envelope bending moment response.

Table 9. The ratio between the plastic limit bending moment and maximum acting bending moment for the frames with dogbones and multistep beams (*i* is the left end, *j* is the right end).

Beam	*D-E*		*E-F*		*G-H*		*H-I*		*J-K*		K-L	
End	*i*	*j*	*i*	*j*	*i*	*j*	*i*	*j*	*i*	*j*	*i*	*j*
DB60	1.09	1.01	1.00	1.02	1.14	1.07	1.04	1.07	1.92	1.46	1.59	1.93
MS60	1.35	1.22	1.19	1.27	1.15	1.06	1.23	1.27	2.17	1.67	2.03	2.46
DB40	1.12	1.01	1.02	1.05	1.16	1.10	1.06	1.09	1.93	1.49	1.61	1.94
MS40	1.33	1.20	1.25	1.33	1.24	1.14	1.29	1.33	2.41	1.86	1.82	2.20

As expected, the safety is always ensured. But, as previously stated, this result can be accepted if the possibility of dangerous brittle failure is disregarded. Unfortunately, only the involved cross-sections constitute the welded connections, and, therefore, just for them, the risk of brittle failure is actual. To this aim, for each beam end of the dogbone frames and of the multistep beam frames, Table 10 reports the ratio between the limit elastic bending moment and the maximum value of the envelope bending moment response.

Table 10. The ratio between elastic limit bending moment and maximum envelope acting bending moment for the frames with dogbones and multistep beams (*i* left end, *j* right end).

Beam	*D-E*		*E-F*		*G-H*		*H-I*		*J-K*		K-L	
End	*i*	*j*	*i*	*j*	*i*	*j*	*i*	*j*	*i*	*j*	*i*	*j*
DB60	0.97	0.87	0.86	0.91	1.01	0.95	0.92	0.95	1.71	1.30	1.40	1.70
MS60	1.21	1.08	1.06	1.13	1.09	1.00	1.10	1.13	1.93	1.49	1.80	2.18
DB40	0.99	0.90	0.88	0.93	1.03	0.98	0.94	0.97	1.71	1.32	1.42	1.71
MS40	1.26	1.14	1.12	1.19	1.10	1.01	1.15	1.18	2.14	1.66	1.60	1.95

The examination of these results clearly highlights that just the multistep beam frames in all the cases satisfy the check, ensuring always a linear elastic behavior and, consequently, preventing any dangerous brittle failure.

Finally, to deduce some further information on the post elastic behavior of the investigated frames, a static non-linear analysis (push-over) has been performed for all the frames. The results, referred as capacity curves, are plotted in Figure 8, where the diagrams are sketched in terms of global base shear force vs displacement of the control point (chosen as the point at the top of the frame central column).

The following remarks can be derived:

- all the frames show an ideal consistent behavior beyond the elastic limit, exhibiting structural stability for increasing horizontal actions. In fact, this good behavior is really exhibited if no brittle failure occurs, i.e., if the cross-sections of the beam end behave elastically. As previously proved, such good behavior is just ensured for the multistep beam frames;

- the multistep beam frames and the dogbone frames are all characterized by a comparable lower limit for the intensity of the horizontal load with respect to the reference frame due to the weakening of the relevant beams; it is worth noting that in all the cases, the resistance capacity of the multistep beam frames is greater than the one of the dogbones frames;
- all the frames exhibit an initial elastic behavior with very close stiffness values; actually, in the initial stage of loading, the frame stiffness is mainly influenced by the column stiffnesses, which are the same for all the frames. Anyway, confirming the good consistency of the multistep beam frame approach, the initial stiffness of these frames is always greater than the corresponding one related to the dogbone frames, as reported in Table 11, where the difference, in terms of percentage, of elastic initial stiffness of the reference frame with respect to that of the frames with dogbones and of the frames constituted by multistep beam elements is given.

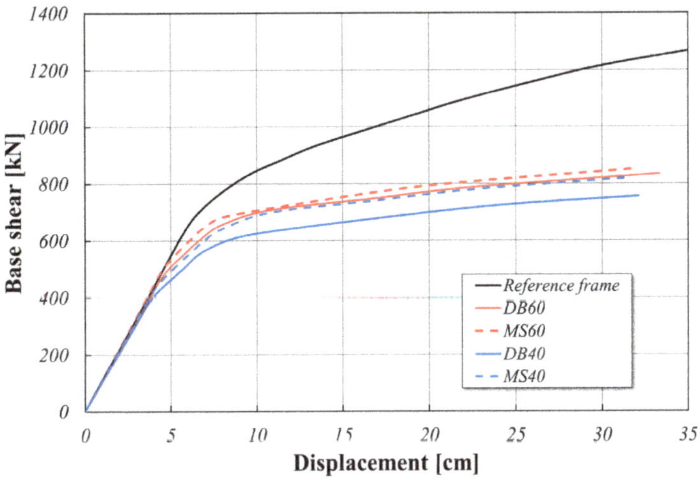

Figure 8. Push-over curves.

Table 11. Percentage difference of initial elastic stiffness of the original frame with respect to that of the frames with dogbones and of the frames constituted by multistep beam elements.

DB60	3.24%
MS60	−1.65%
DB40	5.71%
MS40	1.71%

It is worth noting that the negative value deduced for MS60 indicate that the related frame exhibits an initial stiffness greater than the one of the reference frame; this result must not surprise, indeed it depends on the circumstance that for this frame the demand in term of elastic resistance (related to the brittle failure prevention) mainly influence the optimization problem with respect to the stiffness requirements.

5. Conclusions

The present paper is devoted to the proposal of a simple optimal design problem formulation for steel frames constituted by multistep I-shaped cross-section beam elements aimed at protecting the connections among beams and columns. In order to investigate the reliability of the proposed formulation and on the effectiveness of the use of the multistep beam elements. A deep comparison has been carried out among the behavior of the standard code design of a chosen three-story, two-span plane steel frame of the same frame equipped with dogbones and of the frame constituted by multistep I-shaped cross-section beam elements.

The obtained results have been definitely encouraging; indeed, it has been proved that only the frame constituted by multistep I-shaped cross-section beam elements, designed by means of the proposed formulation, ensures the fulfillment of all the required good features for the structure in terms of resistance, deformability, ductility, and safety with respect to undesired brittle failure.

The proposed formulation, therefore, represents a consistent advancement of the concept of LRPD devices belonging to the broader field of RBS devices. At the same time, the reliability of the formulation and the perfect consistency of the obtained results represent an incentive to deepen the present study. Several future developments are expected, as the proposing of more general optimal design problem formulations devoted to the search of frame steel structures constituted by multistep beam elements in which all the geometric features can represent specific design variables, always respecting all the relevant behavioral constraints, the introduction of more rigorous elastic and yield resistance domains to evaluate the structure capacity, as well as the extension to three-dimensional structures.

Furthermore, appropriate experimental tests and the related due FEM analyses will be carried out in order to verify the consistency of the adopted computational model with the realization of the factory-made I-shaped uniform piecewise profiles also from a technological point of view.

Author Contributions: Conceptualization, L.P. and S.B.; methodology, L.P.; validation, S.V. and S.B.; writing—original draft preparation, L.P. and S.B.; writing—review and editing, L.P. and S.B. All authors have read and agreed to the published version of the manuscript.

Funding: The APC was funded by FFR 2024 funds from the University of Palermo.

Institutional Review Board Statement: Not applicable.

Informed Consent Statement: Not applicable.

Data Availability Statement: Data are contained within the article.

Conflicts of Interest: The authors declare no conflicts of interest.

References

1. Müller, H.S.; Haist, M.; Vogel, M. Assessment of the sustainability potential of concrete and concrete structures considering their environmental impact, performance and lifetime. *Constr. Build. Mater.* **2014**, *67*, 321–337. [CrossRef]
2. Davidovits, J. Geopolymers and geopolymeric materials. *J. Therm. Anal.* **1989**, *35*, 429–441. [CrossRef]
3. Arcones-Pascual, G.; Hernández-Olivares, F.; Sepulcre-Aguilar, A. Comparative properties of a lime mortar with different metakaolin and natron additions. *Constr. Build. Mater.* **2016**, *114*, 747–754. [CrossRef]
4. Zhang, P.; Zheng, Y.; Wang, K.; Zhang, J. A review on properties of fresh and hardened geopolymer mortar. *Compos. Part B Eng.* **2018**, *152*, 79–95. [CrossRef]
5. Lilargem Rocha, D.; Tambara Júnior, L.U.D.; Marvila, M.T.; Pereira, E.C.; Souza, D.; de Azevedo, A.R.G. A Review of the Use of Natural Fibers in Cement Composites: Concepts, Applications and Brazilian History. *Polymers* **2022**, *14*, 2043. [CrossRef]
6. de Lima, T.E.S.; de Azevedo, A.R.G.; Marvila, M.T.; Candido, V.S.; Fediuk, R.; Monteiro, S.N. Potential of Using Amazon Natural Fibers to Reinforce Cementitious Composites: A Review. *Polymers* **2022**, *14*, 647. [CrossRef]
7. Majumder, A.; Stochino, F.; Frattolillo, A.; Valdes, M.; Mancusi, G.; Martinelli, E. Jute fiber-reinforced mortars: Mechanical response and thermal performance. *J. Build. Eng.* **2023**, *66*, 105888. [CrossRef]
8. Greco, P.F.; Pepi, C.; Gioffré, M. A novel biocomposite material for sustainable constructions: Metakaolin lime mortar and Spanish broom fibers. *J. Build. Eng.* **2024**, *83*, 108425. [CrossRef]
9. Benfratello, S.; Cirello, A.; Palizzolo, L.; Sanfilippo, C.; Valenza, A. Experimental Analysis and Numerical Modelling of the Mechanical Behavior of a Sisal-Fiber-Reinforced Geopolymer. *Appl. Sci.* **2024**, *14*, 5216. [CrossRef]
10. Landolfo, R.; Cascini, L.; Portioli, F. Sustainability of steel structures: Towards an integrated approach to life-time engineering design. *Front. Archit. Civ. Eng. China* **2011**, *5*, 304–314. [CrossRef]
11. Gokhfeld, D.A.; Cherniavsky, D.F. *Limit Analysis of Structures at Thermal Cycling*; Springer: Dordrecht, The Netherland, 1980; ISBN 978-90-286-0455-1.

12. Marti, K. Limit load and shakedown analysis of plastic structures under stochastic uncertainty. *Comput. Methods Appl. Mech. Eng.* **2008**, *198*, 42–51. [CrossRef]
13. Chen, W.; Duan, L. *Plasticity, Limit Analysis, Stability and Structural Design: An Academic Life Journey from Theory to Practice*; World Scientific: Singapore, 2021; ISBN 978-981122974-9/978-981122973-2.
14. Huang, M.F.; Li, Q.; Chan, C.M.; Lou, W.J.; Kwok, K.C.S.; Li, G. Performance-based design optimization of tall concrete framed structures subject to wind excitations. *J. Wind. Eng. Ind. Aerodyn.* **2015**, *139*, 70–81. [CrossRef]
15. Benfratello, S.; Di Paola, M.; Palizzolo, L.; Tabbuso, P. Evaluation of the shakedown limit load multiplier for stochastic seismic actions. *Meccanica* **2017**, *52*, 2735–2750. [CrossRef]
16. Zhang, R.; Wang, W.; Alam, M.S. Seismic evaluation of friction spring-based self-centering braced frames based on life-cycle cost. *Earthq. Eng. Struct. Dyn.* **2022**, *51*, 3393–3415. [CrossRef]
17. Hu, S.; Zhu, S. Life-cycle benefits estimation for hybrid seismic-resistant self-centering braced frames. *Earthq. Eng. Struct. Dyn.* **2023**, *52*, 3097–3119. [CrossRef]
18. Banichuk, N.V. *Introduction to Optimization of Structures*; Springer: New York, NY, USA, 1990; ISBN 978-1-4612-7988-4. [CrossRef]
19. Se-Hyu, C.; Seung-Eock, K. Optimal design of steel frame using practical nonlinear inelastic analysis. *Eng. Struct.* **2002**, *24*, 1189–1201. [CrossRef]
20. Benfratello, S.; Giambanco, F.; Palizzolo, L.; Tabbuso, P. Structural design of frames able to prevent element buckling. In Proceedings of the 11th International Conference on Computational Structures Technology, CST 2012, Civil-Comp. Proceedings, Dubrovnik, Croatia, 4–7 September 2012; Volume 99, p. 102644, ISBN 978-190508854-6.
21. Degertekin, S.O. Improved harmony search algorithms for sizing optimization of truss structures. *Comp. Struct* **2012**, *92–93*, 229–241. [CrossRef]
22. Kaveh, H.; Nasrollahi, A. Performed-based seismic design of steel frames utilizing charged system search optimization. *Appl. Soft Comput.* **2014**, *22*, 213–221. [CrossRef]
23. Benfratello, S.; Palizzolo, L.; Tabbuso, P. Dynamic shakedown design of structures under repeated seismic loads. In *Research and Applications in Structural Engineering, Mechanics and Computation, Proceedings of the 5th International Conference on Structural Engineering, Mechanics and Computation, SEMC, Cape Town, South Africa, 2–4 September 2013*; Taylor & Francis Group: London, UK, 2013; pp. 241–246, Code 101169, ISBN 978-113800061-2. [CrossRef]
24. Zhang, R.; Hu, S. Optimal design of self-centering braced frames with limited self-centering braces. *J. Build. Eng.* **2024**, *88*, 109201. [CrossRef]
25. *EN 1993-1-8:2006*; Eurocode 3: Design of Steel Structures Part 1–8: Design of Joints. The European Union: Luxembourg, 2006.
26. Italian Ministry of Infrastructure and Transport. *National Standard NTC 2018*; Italian Ministry of Infrastructure and Transport: Rome, Italy, 2018.
27. Plumier, A. The dogbone: Back to the future. *Eng. J.* **1997**, *34*, 61–67. [CrossRef]
28. Miller, D.K. Lessons learned from the Northridge earthquake. *Eng. Struct.* **1998**, *20*, 249–260. [CrossRef]
29. Mahin, S.T. Lessons from damage to steel buildings during the Northridge earthquake. *Eng. Struct.* **1998**, *20*, 261–270. [CrossRef]
30. Iwankiw, N.R.; Carter, C.J. The dogbone: A new idea to chew on. *Mod. Steel Constr.* **1996**, *36*, 18–23.
31. Engelhardt, M.D.; Winneberger, T.; Zekany, A.J.; Potyray, T.J. Experimental investigation of dogbone moment connections. *Eng. J.* **1998**, *35*, 128–139. [CrossRef]
32. Mirghaderi, S.R.; Shahabeddin, T.; Imanpour, A. Seismic performance of the accordion-web RBS connection. *J. Constr. Steel Res.* **2010**, *66*, 277–288. [CrossRef]
33. Saleh, A.; Mirghaderi, S.R.; Zahrai, S.M. Cyclic testing of tubular web RBS connections in deep beams. *J. Constr. Steel Res.* **2016**, *117*, 214–226. [CrossRef]
34. Momenzadeh, S.; Kazemi, M.T.; Asl, M.H. Seismic performance of reduced web section moment connections. *Int. J. Steel Struct.* **2017**, *17*, 413–425. [CrossRef]
35. Horton, T.A.; Hajirasouliha, I.; Davison, B.; Ozdemir, Z. More efficient design of reduced beam sections (RBS) for maximum seismic performance. *J. Constr. Steel Res.* **2021**, *183*, 106728. [CrossRef]
36. Tabar, A.M.; Alonso-Rodriguez, A.; Tsavdaridis, K.D. Building retrofit with reduced web (RWS) and beam (RBS) section limited-ductility connections. *J. Constr. Steel Res.* **2022**, *197*, 107459. [CrossRef]
37. Shakeri, K.; Akrami, V.; Moradpour, S.; Khedmati, S. Post-earthquake Behavior of Steel Moment Resisting Frames with Connections Modified by Introducing Reduced Beam Section (RBS). *Int. J. Steel Struct.* **2024**, *24*, 462–476. [CrossRef]
38. Morshedi, M.A.; Dolatshahi, K.M.; Maleki, S. Double reduced beam section connection. *J. Constr. Steel Res.* **2017**, *138*, 283–297. [CrossRef]
39. Gerami, M.; Bahirai, M. Seismic Rehabilitation of Steel Frame Connections Through Asymmetrically Weakening the Beam. *Int. J. Steel Struct.* **2019**, *19*, 1209–1224.
40. Sofias, C.E.; Pachoumis, D.T. Assessment of reduced beam section (RBS) moment connections subjected to cyclic loading. *J. Constr. Steel Res.* **2020**, *171*, 106151. [CrossRef]
41. Tahamouli Roudsari, M.; Jamshidi, K.H.; Akbari, G.F.; Torkaman, M. An experimental and numerical investigation of reduced beam section connections with horizontal and vertical web stiffeners. *Asian J. Civ. Eng.* **2020**, *21*, 581–595. [CrossRef]

42. Plumier, A.; Doneux, C.; Castiglioni, C.; Brescianini, J.; Crespi, A.; Dell'Anna, S.; Lazzarotto, L.; Calado, L.; Ferreira Joao, J.; Feligioni, S.; et al. *Two Innovations for Earthquake Resistant Design—The INERD Project—Final Report*; Report EUR22044EN; Research Programme of the Research Fund for Coal and Steel; Steel RTD: Luxembourg, 2006.
43. Castiglioni, C.A.; Kanyilmaz, A.; Calado, L. Experimental analysis of seismic resistant composite steel frames with dissipative devices. *J. Constr. Steel Res.* **2012**, *76*, 1–12. [CrossRef]
44. Vayas, I.; Karydakis, P.; Dimakogianni, D.; Dougka, G.; Castiglioni, C.A.; Kanyilmaz, A.; Calado, L.; Proença, J.M.; Espinha, M.; Hoffmeister, B.; et al. *Dissipative Devices for Seismic Resistant Steel Frames—The FUSEIS Project, Final Report*; Research Programme of the Research Fund for Coal and Steel: Luxembourg, 2012.
45. Vayas, I.; Karydakis, P.; Dimakogianni, D.; Dougka, G.; Castiglioni, C.A.; Kanyilmaz, A.; Calado, L.; Proença, J.M.; Espinha, M.; Hoffmeister, B.; et al. *Dissipative Devices for Seismic Resistant Steel Frames—The FUSEIS Project, Design Guide*; Research Programme of the Research Fund for Coal and Steel: Luxembourg, 2012.
46. Calado, L.; Proenca, J.M.; Espinha, M.; Castiglioni, C.A. Hysteretic behaviour of dissipative bolted fuses for earthquake resistant steel frames. *J. Constr. Steel Res.* **2013**, *85*, 151–162. [CrossRef]
47. Calado, L.; Proenca, J.M.; Espinha, M.; Castiglioni, C.A. Hysteretic behavior of dissipative welded fuses for earthquake resistant composite steel and concrete frames. *Steel Compos. Struct.* **2013**, *14*, 547–569. [CrossRef]
48. Dougka, G.; Dimakogianni, D.; Vayas, I. Seismic behavior of frames with innovative energy dissipation systems (FUSEIS 1-1). *Earthq. Struct.* **2014**, *6*, 561–580. [CrossRef]
49. Dougka, G.; Dimakogianni, D.; Vayas, I. Innovative energy dissipation systems (FUSEIS 1-1)—Experimental analysis. *J. Constr. Steel Res.* **2014**, *96*, 69–80. [CrossRef]
50. Valente, M.; Castiglioni, C.A.; Kanyilmaz, A. Welded fuses for dissipative beam-to-column connections of composite steel frames: Numerical analyses. *J. Constr. Steel Res.* **2017**, *128*, 498–511. [CrossRef]
51. Valente, M.; Castiglioni, C.A.; Kanyilmaz, A. Numerical investigations of repairable dissipative bolted fuses for earthquake resistant composite steel frames. *Eng. Struct.* **2017**, *131*, 275–292. [CrossRef]
52. Calado, L.; Proença, J.M.; Sio, J.F.A. Composite Frames with Dissipative Beam Splices: Numerical Analyses and Design Guidelines. *Key Eng. Mater.* **2018**, *763*, 771–778. [CrossRef]
53. Kanyilmaz, A.; Muhaxheri, M.; Castiglioni, C.A. Influence of repairable bolted dissipative beam splices (structural fuses) on reducing the seismic vulnerability of steel-concrete composite frames. *Soil Dyn. Earthq. Eng.* **2019**, *119*, 281–298. [CrossRef]
54. Tsarpalis, P.; Bakalis, K.; Thanopoulos, P.; Vayas, I.; Vamvatsikos, D. Pre-normative assessment of behaviour factor for lateral load resisting system FUSEIS pin-link. *Bull. Earthq. Eng.* **2020**, *18*, 2681–2698. [CrossRef]
55. Avgerinou, S.; Vayas, I. High-strength steel on dissipative elements in seismic resistant systems: Tests and simulations. *J. Constr. Steel Res.* **2020**, *172*, 106173. [CrossRef]
56. Tsarpalis, P.; Vayas, I.; Thanopoulos, P.; Vamvatsikos, D. Rehabilitation of reinforced concrete building using the fuseis beam-link system. *Structures* **2021**, *34*, 3300–3314. [CrossRef]
57. Avgerinou, S.; Thanopoulos, P.; Hoffmeister, B.; Vayas, I. Seismic resistant buildings with dissipative elements made of high strength steel [Erdbebensichere Bauten mit dissipativen Elementen aus hochfestem Stahl]. *Stahlbau* **2022**, *91*, 326–337. [CrossRef]
58. Colajanni, P.; La Mendola, L.; Monaco, A.; Pagnotta, S. Seismic Performance of Earthquake-Resilient RC Frames Made with HSTC Beams and Friction Damper Devices. *J. Earthq. Eng.* **2022**, *26*, 7787–7813. [CrossRef]
59. Colajanni, P.; La Mendola, L.; Monaco, A.; Pagnotta, S. Design of RC joints equipped with hybrid trussed beams and friction dampers. *Eng. Struct.* **2021**, *227*, 111442. [CrossRef]
60. Colajanni, P.; Pagnotta, S. Friction-based beam-to-column connection for low-damage RC frames with hybrid trussed beams. *Steel Compos. Struct. Int. J.* **2022**, *45*, 231–248. [CrossRef]
61. Benfratello, S.; Palizzolo, L. Limited resistance rigid perfectly plastic hinges for steel frames. *Int. Rev. Civ. Eng.* **2017**, *8*, 286–298. [CrossRef]
62. Benfratello, S.; Cucchiara, C.; Palizzolo, L.; Tabbuso, P. Fixed strength and stiffness hinges for steel frames. In Proceedings of the 23rd Conference of the Italian Association of Theoretical and Applied Mechanics, AIMETA 2017, Salerno, Italy, 4–7 September 2017; Code 133541, ISBN 978-889424847-0.
63. Benfratello, S.; Palizzolo, L.; Tabbuso, P.; Vazzano, S. On the post elastic behavior of LRPH connections. *Int. Rev. Model. Simul.* **2019**, *12*, 341–353. [CrossRef]
64. Benfratello, S.; Palizzolo, L.; Tabbuso, P.; Vazzano, S. LRPH device optimization for axial and shear stresses. *Int. Rev. Model. Simul.* **2020**, *11*, 152–163. [CrossRef]
65. Benfratello, S.; Caddemi, S.; Palizzolo, L.; Pantò, B.; Rapicavoli, D.; Vazzano, S. Targeted steel frames by means of innovative moment resisting connections. *J. Constr. Steel Res.* **2021**, *183*, 106695. [CrossRef]
66. Benfratello, S.; Palizzolo, L.; Vazzano, S. A New Design Problem in the Formulation of a Special Moment Resisting Connection Device for Preventing Local Buckling. *J. Appl. Sci.* **2022**, *12*, 202. [CrossRef]
67. Federal Emergency Management Agency, FEMA-350. *Recommended Seismic Design Criteria for New Steel Moment-Frame Buildings*; Federal Emergency Management Agency: Washington, DC, USA, 2000.

68. Shen, J.; Kitjasateanphun, T.; Srivanich, W. Seismic performance of steel moment frames with reduced beam sections. *Eng. Struct.* **2000**, *22*, 968–983. [CrossRef]
69. CSI (Computers and Structures Inc.). *SAP2000 v22 Analysis Reference Manual*; CSI: Berkeley, CA, USA, 2020.

designs

MDPI

Article

Modeling and Comparison of Design Features of Pendulum and Radial Micro-Hydropower Plants Considering the Influence of Variable Design Parameters

Almira Zhilkashinova [1], Igor Ocheredko [1], Bagdat Azamatov [2], Mergen Nurbaev [1], Dmitry Dogadkin [2] and Madi Abilev [3,*]

[1] National Scientific Laboratory for Collective Use, Sarsen Amanzholov East Kazakhstan University, 34 Tridtsatoy Gvardeiskoy Divizii Str., Ust-Kamenogorsk 070002, Kazakhstan; azhilkashinova@vku.edu.kz (A.Z.); iocheredko@vku.edu.kz (I.O.); mnurbaev@vku.edu.kz (M.N.)

[2] Competence Center "Smart Engineering", Daulet Serikbayev East Kazakhstan Technical University, 19 Serikbayev Street, Ust-Kamenogorsk 070010, Kazakhstan; bazamatov@ektu.kz (B.A.); ddogadkin@ektu.kz (D.D.)

[3] Department of Analytical, Colloid Chemistry and Technology of Rare Elements, Al-Farabi Kazakh National University, 71 Al-Farabi Ave., Almaty 050040, Kazakhstan

* Correspondence: madi.abilev@kaznu.edu.kz

Abstract: This article provides a comparative analysis of pendulum and radial micro-hydropower plants. The novelty of this study lies in the comparative analysis of units that are fundamentally different in design to achieve the most rational option for low-speed rivers. It has been established that a pendulum micro-hydropower plant has a high torque with relatively small dimensions but operates cyclically. At a diameter of 1 m and a blade area of 0.3 m^2, the peak torque was 140 N·m. At the same time, the design is sensitive to the blade area and at 0.6 m^2 and a lever length of 1.5 m, the torque reached 430 N·m. A radial micro-hydropower plant has lower torque but operates constantly. At an area of 1.23 m^2 and a diameter of 1 m, the torque was 40.4 N·m. Accordingly, in terms of specific area with a diameter of 1 m, a pendulum micro-hydropower plant has up to 12 times more torque. It has been established that the pendulum hydropower plant best satisfies the requirements for converting a low river speed into high revolutions of a current generator.

Keywords: radial micro-hydropower plant; pendulum micro-hydropower plant; torque; blade area; low-speed rivers; computational modeling

1. Introduction

Micro- and mini-hydropower plants (HPPs) are cheap, environmentally friendly, and reliable sources of energy. This is one of the greenest ways to produce electricity today. Moreover, small hydropower plants have enormous untapped potential in most regions of the world. They can make a big contribution to meeting energy needs in the future. They are based on well-studied and proven technologies, but there is great potential for further optimization.

Currently, in the field of small damless hydropower plants, a search is underway for the most rational designs that provide maximum torque with minimum dimensions. Work is underway in the field of materials for mini-hydropower plants. Multi-component structures, according to Emmy Noether's theorem, will not have high efficiency, since they will have high mechanical losses due to the rotation of the mechanisms. These structures have complex symmetry. Such designs can only be considered for installation in high-pressure water flows with guide vanes. For slow-moving rivers, the simplest versions of hydropower plants with the axis of rotation located normal and vertical to the river flow are promising [1–9].

Damless hydropower plants, being mechanisms, are influenced by variable design parameters that affect torque. This suggests the possibility of finding the most rational design for use in shallow, slow-moving rivers. A feature of such structures, in contrast to structures installed in high-speed flows, should be the ability to convert the low speed of the river into high revolutions of the current generator [9–12]. At the same time, hydropower plants with a vertical axis of rotation should have the greatest variability, since the length of the lever is not limited by the depth of the river. The limitation on the depth of the river can be overcome by changing the geometry of the blade, reducing its height and increasing its width [13–15].

The development and optimization of HPP designs are made using modern tools and techniques in hydropower plant modeling. These tools help the researchers and developers to ensure that the plant operates at maximum efficiency by selecting the optimal location, turbine type, and size; to minimize the ecological footprint by analyzing the potential effects on local ecosystems and water resources; to conduct cost–benefit analyses to ensure the project is financially feasible and sustainable over the long term; and to identify and mitigate potential risks such as flooding, structural failures, and maintenance issues.

Hydropower plant design is a complex and multifaceted process that necessitates a thorough understanding of hydrology, structural engineering, environmental science, and economics. To achieve a successful and efficient hydropower plant design, engineers and researchers rely on a variety of advanced tools and techniques. These tools and techniques facilitate the simulation, analysis, and optimization of different aspects of the plant, from water flow to structural integrity and environmental impact. Modern tools and techniques in hydropower plant modeling include a variety of models and software. Among them, computational fluid dynamics, finite element analysis, hydrological modeling, optimization algorithms, and specialized simulation software are emphasized.

Computational fluid dynamics (CFD) is a crucial tool in hydropower plant modeling, offering a detailed simulation and analysis of fluid flow within hydraulic structures. By utilizing advanced numerical methods and algorithms, CFD provides in-depth insights into water behavior as it interacts with turbines, spillways, and other components. This allows engineers to optimize design parameters, enhance energy conversion efficiency, and minimize hydraulic losses. Furthermore, CFD helps identify potential issues such as cavitation and flow separation, ensuring the structural integrity and longevity of the plant. However, CFD has several disadvantages, such as high computational demand, complexity, data sensitivity, initial costs, simplifications, and assumptions [8,16–19].

Hydrological modeling is a critical tool in the design and optimization of hydropower plants, offering valuable insights into water flow, site selection, and environmental impacts. While it provides numerous benefits such as accurate water flow prediction and flood risk assessment, challenges like data requirements, model complexity, and inherent uncertainties must be carefully managed. Balancing these advantages and disadvantages is essential for effectively utilizing hydrological modeling in the development of efficient and sustainable hydropower systems [20–22].

Optimization algorithms are tools in hydropower plant modeling, having significant advantages in terms of efficiency, cost reduction, and complex problem-solving. They facilitate the identification of optimal solutions for various design and operational challenges, contributing to the development of more efficient and sustainable hydropower systems. However, the high computational demands, data dependence, and complexity of these algorithms present challenges in using these tools [23–25].

Specialized software tools like HEC-RAS (Hydrologic Engineering Center's river analysis system), SWMM (storm water management model), and HOMER (hybrid optimization of multiple energy resources) are used for various aspects of hydrological and energy modeling. They provide comprehensive platforms for simulating and analyzing complex hydropower systems, supporting detailed design and planning. By enabling the virtual testing of various scenarios, it helps optimize design and operational parameters, contributing to the development of efficient and reliable hydropower systems. However, the high initial cost, complexity, and dependence on accurate data pose challenges [26–29].

Finite element analysis (FEA) is a computational tool used in the design and assessment of hydropower plants. It involves breaking down complex structures into smaller, manageable elements and analyzing their behavior under various physical conditions. FEA is particularly valuable in evaluating the structural integrity and performance of dams, turbines, and powerhouses, ensuring that they can withstand different stresses and loads. FEA provides precise insights into the stress, strain, and deformation of complex structures, helping engineers identify potential weaknesses and optimize designs. Accurate prediction of how components will behave under various loads and conditions using FEA ensures that hydropower structures are safe and reliable, reducing the risk of failures and accidents. By enabling precise analysis of complex geometries and material behaviors, FEA helps engineers design robust and reliable hydropower systems. Its versatility and ability to solve complex problems make it an essential component in the development of efficient and durable hydropower infrastructure [30–32]. Due its ability to provide the most accurate calculation of volumetric mechanisms using hydrodynamic equations, FEA was chosen as a research tool in this study.

The goal is to study the design features of damless hydropower plants by resolving the contradiction between the low speed and depth of rivers on the one hand, and the need for high torque on the shaft from the turbine, by comparing the efficiency of a radial turbine and a pendulum hydropower plant on the other hand, and choosing the most promising design.

To eliminate the above gaps, it is necessary to solve specific problems:

(1) theoretical studies of the influence of variable design parameters on the efficiency of radial and pendulum hydropower plants;
(2) comparison of the efficiency of radial turbines and pendulum hydropower plants.

2. Materials and Methods

Radial and pendulum hydropower plants were studied through finite element analysis using continuum mechanics and the Navier-Stokes equations, k-ε, and turbulence models. The work was performed in the Solidworks 2022 software package. The working assumption is that due to the large lever in the pendulum HPP, it would be possible to compensate for the small depth of small rivers and obtain a high torque.

2.1. Operating Principle of a Pendulum Hydropower Plant

Under the influence of water flow, a force is exerted on the blade, which sets in motion a lever rigidly fixed to the axis (Figure 1). The lever rotates the gearbox, and a torque is transmitted to the current generator. The blade describes a trajectory from position 1 to 2 and back. At points 1 and 2, the blade, resting against the tilters, turns 90° and the cycle repeats.

The boundary conditions were as follows: river flow speed, 1 m/s; lever length (H), 1–1.5 m; blade size (t), 1–1.5 m × 0.3–0.5 m(c); blade angle (α), (45°, −45°); lever angle (β), (45°, −45°).

Figure 2 shows the calculation domain for the pendulum hydropower plant.

Figure 3 shows a 3D model of the pendular hydropower plant. Micro-HPP consists of a blade mounted on a movable lever; the movement of the lever is limited by stops set to ±45°. Stops are also installed on the blades; they limit movement to ±45°. The blade, under the influence of the water flow, transmits force to the lever, which in turn drives the current generator through a gearbox.

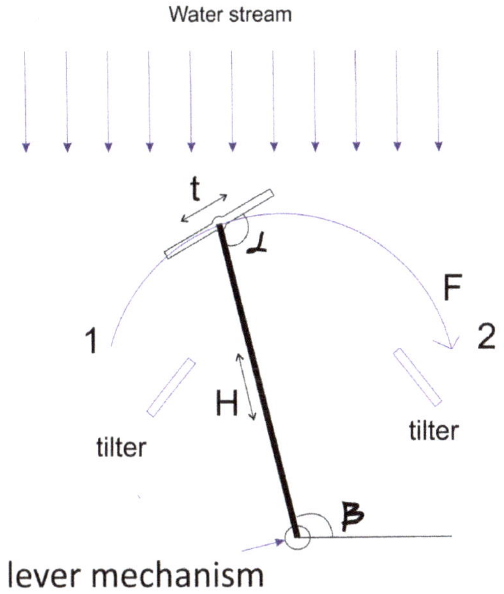

Figure 1. Scheme of operation of a pendular hydropower plant.

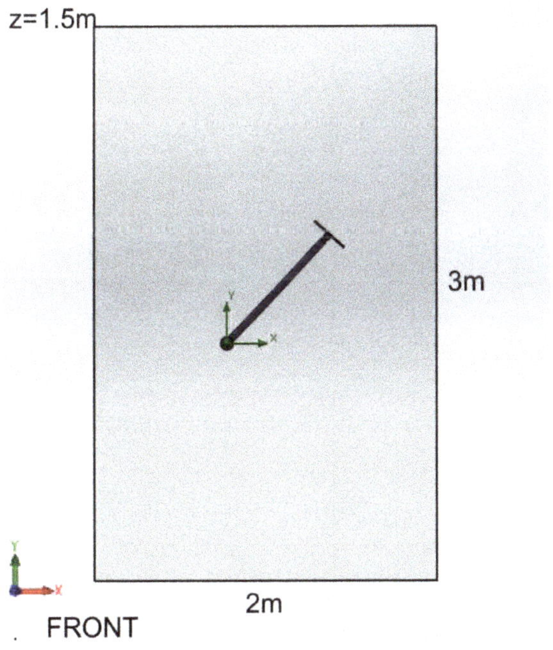

Figure 2. Size of the calculation domain of the pendular hydropower plant.

(a) (b)

Figure 3. 3D model of the pendulum hydropower plant: (**a**) general view; (**b**) top view.

2.2. Calculation of a Radial Hydropower Plant

The flow of water passing through a special guiding device acts on the rotor blades of the hydropower plant and rotates them. The rotation of the rotor axis is transmitted to the current generator. The process occurs continuously. The boundary conditions were as follows: river speed, 1 m/s and turbine diameter, 0.50–1.35 m. Figure 4 shows the calculation domain for the radial hydropower plant.

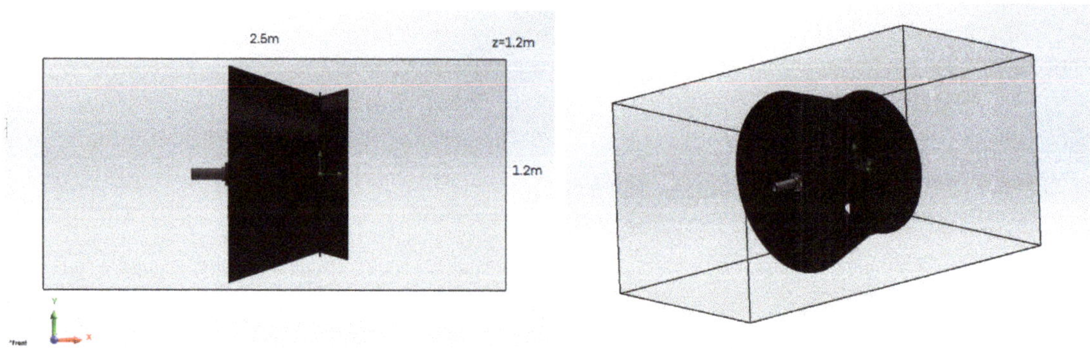

(a) (b)

Figure 4. Size of the calculated domain of the radial hydropower plant: (**a**) general view; (**b**) front view.

Figure 5 shows a general view of the impeller of a micro-HPP turbine. The diameter of the impeller was chosen taking into account the initial parameters: the river flow speed, 1 m/s; the depth of the Irtysh River at the installation site is approximately 4–6 m; and the required output power is at least 1 kW.

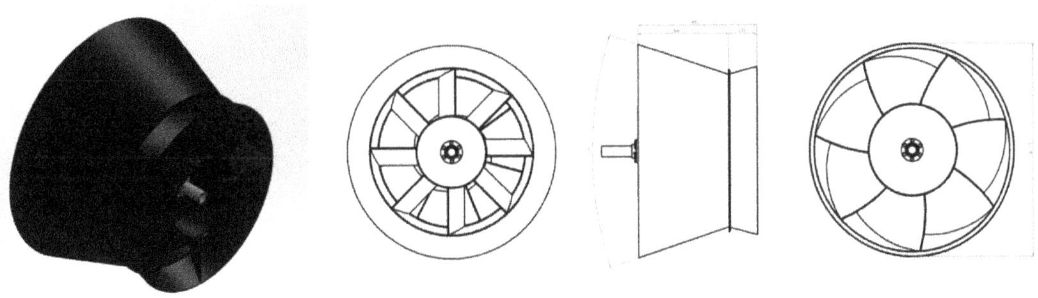

(a) (b)

Figure 5. General view of the turbine impeller: (**a**) general view; (**b**) scheme of the impeller.

The micro-HPP turbine is made of polyethylene. Polyethylene has a resistance to corrosion, acid resistance, is a lightweight material, and is easy to process using various processing methods. The body of the micro-HPP consists of two halves: a guide channel, which serves to direct the flow of water to the blades of a hydraulic turbine, and the main body with a turbine wheel. The flow of water, passing through the guide channel, rotates the rotor of the hydro-turbine, the shaft of which drives the current generator. The outer diameter of the wheel is 1131 mm.

Figures 6 and 7 show diagrams of the external and internal casings of the hydropower plant turbine with a diameter of 1.35 m.

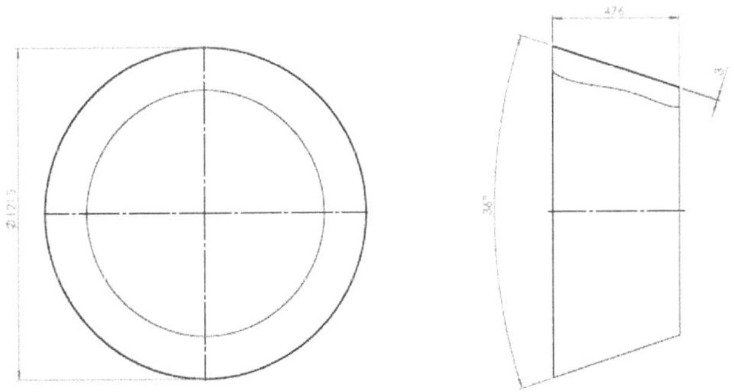

Figure 6. General view of the external casing of hydropower plant turbine.

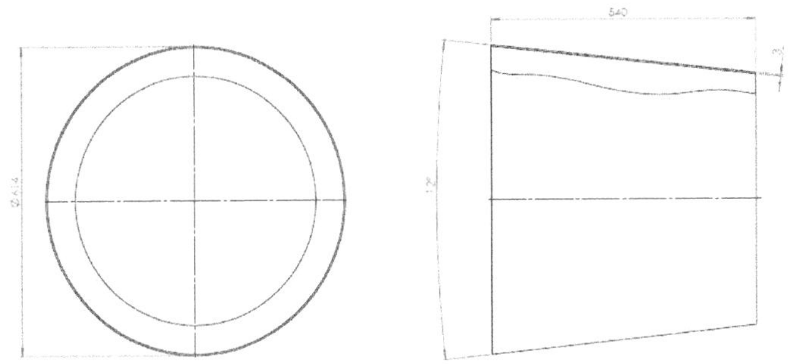

Figure 7. General view of the internal casing of hydropower plant turbine.

Figure 8 shows the working blades of hydropower plant and a diagram of the guide channel. The height of the working blade is 606 mm and the width is 503 mm. The turbine blades have a complex geometry.

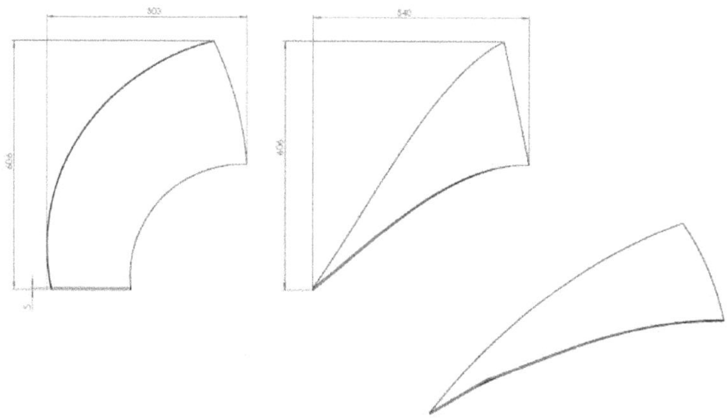

Figure 8. Working blades of hydropower plant.

3. Results and Discussion

3.1. Calculation Results for a Pendulum Hydropower Plant

Figure 9 shows the velocity field during swelling of the blade of a pendulum hydropower plant. The figure shows that in front of the blade, the speed drops, and behind the blade, the water speed is zero, which indicates that pressure is created on the blade. The color gradient represents different velocity magnitudes, with blue indicating lower velocities (0 m/s) and red indicating higher velocities (up to 1.08 m/s). The area immediately downstream of the blade shows a significant reduction in velocity, represented by the blue color. This indicates that the blade is effectively slowing down the water flow in this region. Moving away from the blade, the velocity increases, transitioning from blue to green, yellow, and eventually red. This gradient indicates the acceleration of water as it moves past the blade and resumes a more natural flow pattern. The highest velocities, depicted in red and orange, are concentrated around the edges of the blade and downstream, suggesting that water is forced to accelerate as it bypasses. This acceleration could be due to the narrowing of the flow path around the blade. The white arrow indicates the direction of flow, which appears to be deflected by the blade. This deflection creates a wake region behind the blade, characterized by lower velocities (blue area). The presence of such a wake is typical in fluid dynamics when a solid object obstructs the flow of a fluid. The blade causes a significant variation in the velocity field, which can have several implications. These include increased energy extraction efficiency in certain regions due to higher velocities and potential areas of turbulence and vortices in the wake region.

The regions with higher velocities (yellow to red) are likely the most efficient for energy extraction. Placing turbines or other energy-harvesting devices in these areas could maximize the efficiency of the hydropower plant. The wake region (blue area) behind the blade may experience lower pressures and potential flow recirculation. This should be considered in the structural design of the plant to ensure stability and durability. The flow deflection and wake formation could lead to turbulence and vortex shedding, which might affect the performance and lifespan of the hydropower equipment. These factors should be analyzed further to mitigate adverse effects.

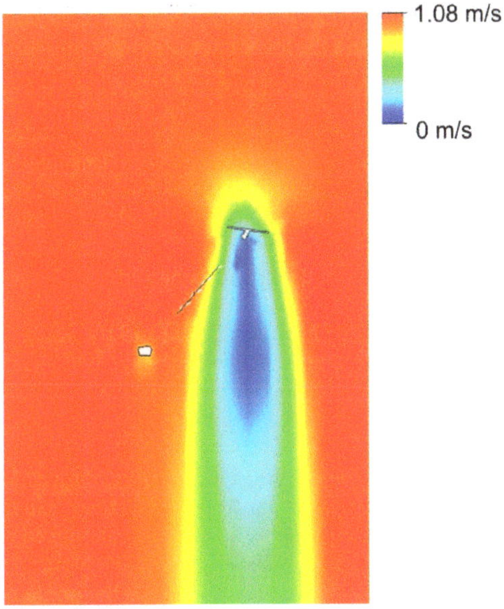

Figure 9. Velocity field during operation of a pendular hydropower plant.

Figure 10 shows the dependence of torque and hydrodynamic force on the working angle for 1-m and 1.5-m blades. Both torque and force show a linear increase with the angle from negative to positive values. The torque values are higher than the force values for the same working angle, indicating that the lever length contributes to the increase in torque. At a lever length of 1 m, the graphs coincide. At a lever length of 1.5 m, the torque value exceeds the force value. The working angle ranges from −40° to +40°. At −40°, the torque is at its minimum (~20 N·m), and it increases steadily as the angle moves towards +40°, where the torque reaches its maximum (~140 N·m).

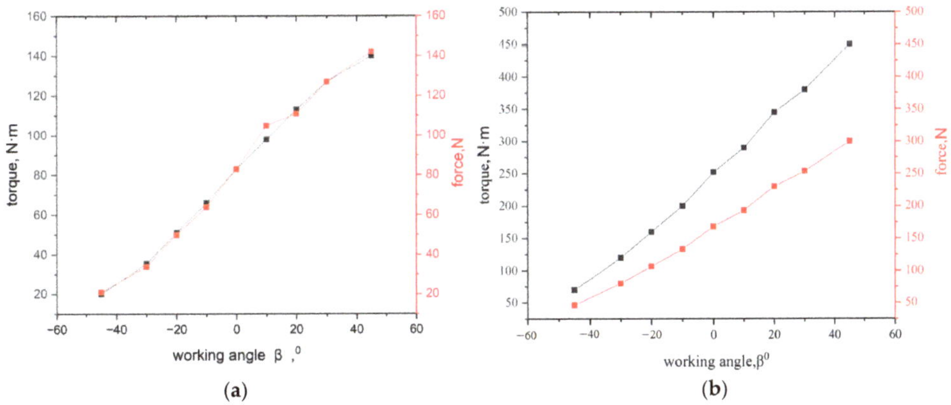

Figure 10. Dependence of torque and hydrodynamic force on the angle of inclination of the lever at different blade size: (**a**) 1 m; (**b**) 1.5 m.

The graph shows a fairly linear relationship, especially as it moves from −20° to +40°. This linearity suggests that the system responds predictably to changes in the working angle within this range. There is a slight non-linearity and asymmetrical response around the negative angle region, suggesting possible variations in mechanical efficiency or resistance.

The variation in torque at negative angles should be considered when designing control systems to ensure smooth operation. Ensuring the lever operates predominantly within the positive angle range could enhance overall performance and reliability.

Figure 10a illustrates the dependence of torque on the working angle of the lever in a hydropower system. The near-linear increase in torque with the angle indicates an efficient and predictable system, particularly in the range from −20° to +40°. This insight is valuable for optimizing the design and operation of the hydropower plant, ensuring maximum efficiency and performance. The slight asymmetry at negative angles should be addressed to maintain smooth and reliable operation.

Next, we doubled the area of the blade, the length of the lever by one and a half times, and made calculations at a river speed of the same 1 m/s.

The graph in Figure 10b illustrates the relationship between torque and the angle of inclination of a lever. The horizontal axis (x-axis) represents the working angle of the lever, denoted in degrees (°). The range of the angle is from −40° to 40°. The vertical axis (y-axis) represents the torque, measured in Newton meters (N·m). The torque values range from 0 to 450 N·m.

The graph shows a positive correlation between the working angle and the torque. As the working angle increases from −40° to 40°, the torque also increases. The relationship appears to be nonlinear, with the rate of increase in torque becoming more pronounced at higher angles. At a working angle of 40°, the torque reaches approximately 400 N·m.

Figure 10b suggests that the torque exerted by the lever increases as the angle of inclination increases. This could imply that the lever is more effective at generating torque at higher angles of inclination. This analysis could be useful in mechanical engineering and design of hydropower systems, where understanding the torque generated by levers at different angles is crucial for optimizing performance. Comparing Figure 10a,b, it can be concluded that the torque is most sensitive to the blade's surface area.

Increasing the length of the lever from 1 m to 1.5 m, while maintaining the blade area, resulted in an increase from 145 N·m to 177 N·m. At the same time, doubling the blade area to 0.6 m^2 led to an increase in torque to 430 N·m, with a lever length of 1.5 m (Table 1). Table 1 shows the dependence of torque on variable geometric parameters of lever length and blade area.

Table 1. Dependence of torque on lever length and blade area at peak torque.

Blade Area, m^2	Lever Length, m	Torque, N·m
0.3	0.5	70
	1	145
	1.5	177
	2	199
0.6	0.5	158
	1	257
	1.5	430
	2	528
0.9	0.5	-
	1	378
	1.5	620
	2	815
1.2	0.5	-
	1	505
	1.5	820
	2	1005

Table 1 shows that the maximum torque of 1000 N·m is achieved with a lever length of 2 m and a blade area of 1.2 m^2. The increase in torque in the peak state with an increase in the lever length from 1 m to 2 m, and an increase in the blade area from 0.3 m^2 to 1.2 m^2, was 145 N·m and 1005 N·m, respectively, and amounted to 860 N·m.

3.2. Calculation Results for a Radial Hydropower Plant

Table 2 shows the dependence of torque on turbine diameter for the radial HPP.

Table 2. Dependence of torque on turbine diameter.

Turbine Diameter, mm	Torque, N·m
540 (×0.5)	5.0
810 (×0.75)	16.3
1080	40.4
1350 (×1.25)	91.6

Table 2 shows a nonlinear relationship between the turbine diameter and the torque. Larger turbines yield significantly higher torque, implying greater mechanical advantage or power output. However, the diameter of the turbine is limited by the depth of the river, which is very important when installing in shallow rivers. This relationship is crucial for optimizing turbine design in applications requiring high torque, such as power generation or mechanical drives. Understanding the precise nature of this relationship can help in predicting performance and ensuring efficient and safe design.

Figure 11 presents a contour plot depicting the velocity distribution within a radial turbine during operation. The plot uses a color gradient, with red representing the highest velocities and blue indicating the lowest. The velocity decreases uniformly towards the center of the hydro-turbine, then, in the expanding part, we can see an increase in the velocity of the liquid flow, as evidenced by the blue and light blue areas in Figure 11.

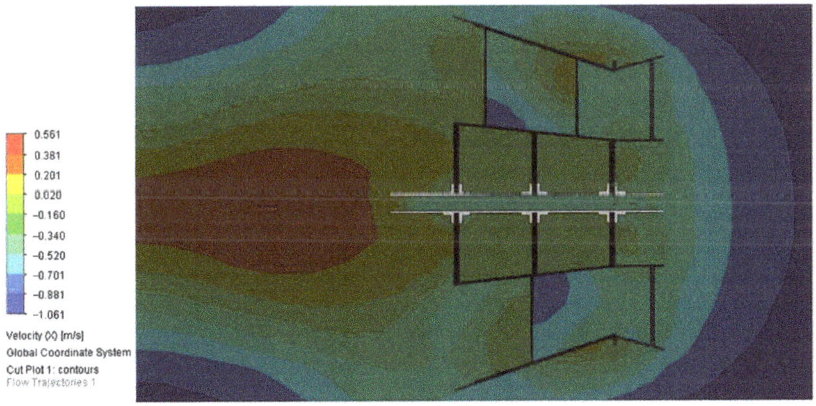

Figure 11. Velocity distribution during operation of a radial turbine.

A distinct region of high velocity, represented by red and orange contours, is observed at the turbine inlet. This indicates that the fluid entering the turbine possesses a significant amount of kinetic energy. As the fluid flows through the turbine, the velocity decreases gradually, as is evident by the progression of colors towards green and then blue. This velocity reduction is a consequence of the turbine blades converting the kinetic energy of the fluid into mechanical energy. The contour lines suggest a consistent flow pattern, with the velocity contours becoming more concentrated towards the center of the turbine. This indicates that the flow is being guided and directed effectively by the turbine blades.

A noticeable velocity gradient exists near the blades, where the contours are closer together. This indicates a rapid change in velocity in this area, which is likely due to the interaction of the fluid with the rotating blades. A region of very low velocity, represented by blue contours, is observed at the exit of the turbine. This indicates that the fluid has lost most of its kinetic energy and is close to a static state. These flow dynamics are crucial for optimizing the design and performance of the radial turbine.

Figure 12 shows the calculated values of the moment of force and the force exerted by the water flow on the blades of the hydraulic turbine. The moment of the force was 40.162 N·m. The total force acting on the hydraulic turbine blades was 30.574 N.

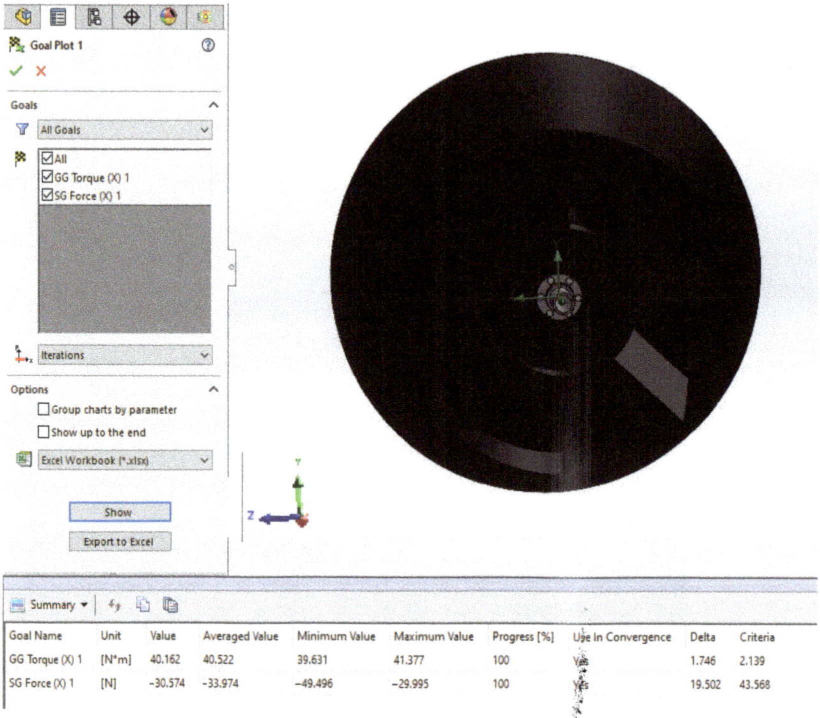

Figure 12. Calculated values of torque and force.

Figure 13 shows the distribution of water flow velocity through the hydraulic turbine channel throughout the entire volume of the structure. The velocity of the water decreases uniformly as it passes through the channel and, due to the decrease, pressure is exerted on the blades of the hydraulic turbine, which leads to rotation of the turbine.

At the inlet (left side of Figure 13), the water flow is relatively uniform and primarily blue, indicating lower velocities. As the fluid approaches the turbine, the velocity starts to vary more significantly. Within the turbine, there are areas with higher velocities (yellow and green) and lower velocities (blue and cyan). The flow becomes more turbulent as it interacts with the turbine components, causing a mix of velocities. At the outlet (right side), the water flow becomes uniform again but with lower velocities compared to the inlet.

There are areas closer to the end of the turbine blades where the flow velocity increases, and the velocity in this area is 0.88 m/s. These velocities are shown in blue in Figure 14. This effect is explained by the fact that when the fluid flow passes from the narrowing channel to the expanding channel, the Venturi effect occurs. If we consider the design of the turbine, we can see the guide channel, which is made in the form of a cone with partitions. Then, water enters the blades of the turbine, the shape of which expands towards the outlet. As a result, we have a typical Venturi pipe design. At the initial moment, the water flow at the inlet to the structure narrows, the pressure drops in this area, and the fluid flow velocity increases according to the Bernoulli equation. But, since the blades of the turbine have a twisted design, the fluid flow, encountering resistance on its way, creates a torque on the shaft of the turbine.

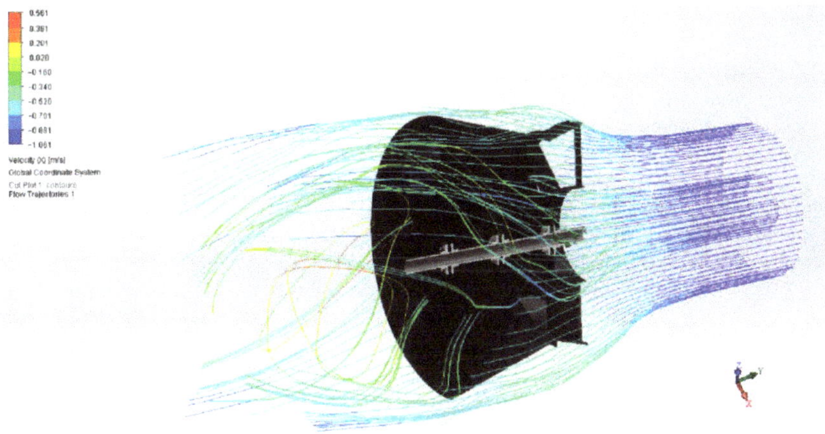

Figure 13. Flow velocity distribution in the hydraulic turbine.

Figure 14. View of the hydraulic turbine from the guide side.

The interaction between the water and the turbine blades creates regions of high and low velocity, indicating the energy transfer between the fluid and the turbine. The variations in flow velocity within the turbine indicate the efficiency of the turbine in converting hydraulic energy into mechanical energy. Uniform flow at the outlet suggests efficient energy extraction, with minimal kinetic energy left in the fluid. The presence of both high and low velocity regions within the turbine indicates effective interaction between the fluid and the turbine blades, suggesting good design and performance of the turbine.

Figure 14 shows a view of the structure from the guide side, which shows the distribution of the force exerted by the water flow on the hydraulic turbine blades.

3.3. Comparison of Pendulum and Radial Hydropower Plants

Both pendulum and radial hydropower plants were compared using the relationship between various parameters and the resulting torque generated by each system.

Considering the pendulum HPP, the torque increases almost linearly with the working angle, indicating a direct proportional relationship. The system exhibits near-linearity in the torque response from -20° to $+40^{\circ}$, suggesting predictable performance. There is a slight asymmetry and non-linearity at negative angles, indicating potential variations in mechanical efficiency or resistance. Doubling the blade area and increasing the lever length significantly increases the torque, suggesting a strong influence of these parameters.

Meanwhile, for the radial HPP, the turbine diameter has a nonlinear impact on the torque, with larger turbines producing significantly higher torque. The velocity field within the turbine shows a clear pattern of high velocity at the inlet, gradual velocity reduction as the fluid flows through the turbine, and a low velocity region at the exit. The velocity gradient near the blades suggests a rapid change in velocity due to the interaction with rotating blades.

Regarding the torque generation parameters, the pendulum system exhibits a near-linear relationship between the torque and the working angle, whereas the radial system shows a nonlinear relationship between the torque and the turbine diameter. The pendulum system's torque is significantly influenced by the blade area and lever length. In contrast, the radial system's torque is primarily dependent on the turbine diameter. The radial system analysis provides insights into the velocity distribution within the turbine, highlighting the energy conversion process and the flow dynamics.

The pendulum system emphasizes the impact of lever geometry, while the radial system focuses on the role of turbine diameter and the fluid dynamics within the turbine. Comparing the two systems provides a broader understanding of the principles governing hydropower plant design and optimization.

4. Conclusions

The characteristics of two types of small hydropower plants were studied based on their design parameters that affect efficiency. However, both types have both advantages and disadvantages. The advantage of radial hydropower plants is the constant rotation and torque. The advantage of the pendulum hydropower plant is high torque, but the disadvantage is cyclic operation. This disadvantage can be mitigated by installing hydropower plants with lever mechanisms, increasing the torque output in proportion to the magnitude of its increase. The efficiency of the pendulum hydropower plant is higher than that of the Lenyov HPP, if we consider specific characteristics [33,34]. At the same time, the pendulum hydropower plant still has high untapped potential, which can be revealed by increasing the lever and blade area.

It has been established that the increase in torque with increasing lever length, without increasing the blade area, is insignificant and amounted to 20 N·m of torque with a blade area of 0.3 m^2. In terms of specific area with a diameter of 1 m, the pendulum hydropower plant has up to 12 times higher torque. It has been established that the pendulum hydropower plant best satisfies the requirements for converting a low river speed into high revolutions of a current generator.

However, the current generator needs higher revolutions; therefore, the system will include a gearbox converting the high torque created on the blade into high revolutions (speed), and transmitting them to the current generator. The current generator uses neodymium magnets, which manage well with the task of generating electric current at low revolutions and can be used in these instances. The optimal construction of such a system, taking into account all the losses on the gearbox, is interesting for further study.

The design of the radial hydropower plant was made in the form of a Venturi tube; there was a narrowing part, a middle part, and an expanding part. According to experiments performed using the SolidWorks CAD program, the Venturi effect can be observed. Initially, the flow rate at the inlet reached 1 m/s, then, the speed decreases to 0.3 m/s and at the outlet there is an increase in the flow rate of the liquid, again to 0.88 m/s. This increase in speed can be explained by the Venturi effect. As a result, an increase in the efficiency of the pendulum hydropower plant is possible: firstly, due to the installation of a larger number of blades, which leads to an increase in the contact area and, consequently, to an increase in torque; and secondly, due to a change in the design of the narrowing and expanding parts of the hydro-turbine. In the future, we will consider all possible basic schemes for turning the blade using the Venturi effect. These issues require separate research and optimization. In the future, we will also consider all possible symmetries and improve the basic design of the blade rotation. The results obtained will be taken into account in the manufacture of prototypes of micro-hydropower plants for further studies.

The development of the pendulum hydropower plant concept should give impetus to the development of design solutions in this area. It is necessary to develop designs that provide a limitation of the lever movement and blade rotation. The concept provides many options for blade combinations and their paired operation. In general, the introduction of damless hydropower plants with high torque can be a promising innovative step in small hydropower.

Author Contributions: Conceptualization, A.Z. and B.A.; methodology, A.Z., I.O., D.D. and B.A.; software, M.A.; validation, A.Z., M.A. and M.N.; formal analysis, M.A., D.D. and M.N.; investigation, A.Z., B.A. and I.O.; resources, A.Z. and B.A.; data curation, I.O., D.D., B.A. and M.N.; writing—original draft preparation, A.Z. and M.A.; writing—review and editing, M.N.; visualization, B.A. and D.D.; supervision, A.Z.; project administration, A.Z.; funding acquisition, A.Z. All authors have read and agreed to the published version of the manuscript.

Funding: This research was funded by the Science Committee of the Ministry of Science and Higher Education of the Republic of Kazakhstan, grant number AP19677819.

Data Availability Statement: The data presented in this study are available on request from the corresponding author.

Conflicts of Interest: The authors declare no conflicts of interest.

References

1. Sternberg, R. Hydropower's future, the environment, and global electricity systems. *Renew. Sustain. Energy Rev.* **2010**, *14*, 713–723. [CrossRef]
2. Bilgili, M.; Bilirgen, H.; Ozbek, A.; Ekinci, F.; Demirdelen, T. The role of hydropower installations for sustainable energy development in Turkey and the world. *Renew. Energy* **2018**, *126*, 755–764. [CrossRef]
3. Xu, J.; Ni, T.; Zheng, B. Hydropower development trends from a technological paradigm perspective. *Energy Convers. Manag.* **2015**, *90*, 195–206. [CrossRef]
4. Jager, H.I.; Efroymson, R.A.; Opperman, J.J.; Kelly, M.R. Spatial design principles for sustainable hydropower development in river basins. *Renew. Sustain. Energy Rev.* **2015**, *45*, 808–816. [CrossRef]
5. Feng, Z.K.; Niu, W.J.; Cheng, C.T.; Liao, S.L. Hydropower system operation optimization by discrete differential dynamic programming based on orthogonal experiment design. *Energy* **2017**, *126*, 720–732. [CrossRef]
6. Barsi, D.; Ubaldi, M.; Zunino, P.; Fink, R. Optimized design of a novel hydraulic propeller turbine for low heads. *Designs* **2021**, *5*, 20. [CrossRef]
7. Masood, Z.; Khan, S.; Qian, L. Machine learning-based surrogate model for accelerating simulation-driven optimisation of hydropower Kaplan turbine. *Renew. Energy* **2021**, *173*, 827–848. [CrossRef]
8. Abeykoon, C. Modelling and optimisation of a Kaplan turbine—A comprehensive theoretical and CFD study. *Clean. Energy Syst.* **2022**, *3*, 100017. [CrossRef]
9. Bhattarai, S.; Vichare, P.; Dahal, K.; Al Makky, A.; Olabi, A.G. Novel trends in modelling techniques of Pelton Turbine bucket for increased renewable energy production. *Renew. Sustain. Energy Rev.* **2019**, *112*, 87–101. [CrossRef]
10. Paish, O. Small hydro power: Technology and current status. *Renew. Sustain. Energy Rev.* **2002**, *6*, 537–556. [CrossRef]

11. Elbatran, A.H.; Yaakob, O.B.; Ahmed, Y.M.; Shabara, H.M. Operation, performance and economic analysis of low head micro-hydropower turbines for rural and remote areas: A review. *Renew. Sustain. Energy Rev.* **2015**, *43*, 40–50. [CrossRef]
12. Hu, H.; Yu, S.S.; Trinh, H. A Review of uncertainties in power systems—Modeling, impact, and mitigation. *Designs* **2024**, *8*, 10. [CrossRef]
13. Chaulagain, R.K.; Poudel, L.; Maharjan, S. A review on non-conventional hydropower turbines and their selection for ultra-low-head applications. *Heliyon* **2023**, *9*, e17753. [CrossRef]
14. Capata, R.; Calabria, A.; Baralis, G.M.; Piras, G. Experimental fitting of efficiency Hill chart for Kaplan hydraulic turbine. *Designs* **2024**, *8*, 80. [CrossRef]
15. Kumar, K.; Saini, R.P. A review on operation and maintenance of hydropower plants. *Sustain. Energy Technol. Assess.* **2022**, *49*, 101704. [CrossRef]
16. Tiwari, G.; Kumar, J.; Prasad, V.; Patel, V.K. Utility of CFD in the design and performance analysis of hydraulic turbines—A review. *Energy Rep.* **2020**, *6*, 2410–2429. [CrossRef]
17. Chitrakar, S.; Solemslie, B.W.; Neopane, H.P.; Dahlhaug, O.G. Review on numerical techniques applied in impulse hydro turbines. *Renew. Energy* **2020**, *159*, 843–859. [CrossRef]
18. Tobo, Y.M.; Ramayya, A.V.; Tibba, G.S. CFD simulation and optimization of very low head axial flow turbine runner. *Int. J. Renew. Energy Dev.* **2015**, *4*, 181–188. [CrossRef]
19. Wilson, D.A.; Pun, K.; Ganesan, P.B.; Hamad, F. Geometrical optimization of a Venturi-type microbubble generator using CFD simulation and experimental measurements. *Designs* **2021**, *5*, 4. [CrossRef]
20. Sedighkia, M.; Abdoli, A. An optimization approach for managing environmental impacts of generating hydropower on fish biodiversity. *Renew. Energy* **2023**, *218*, 119283. [CrossRef]
21. Yu, S.; Zhang, J.; Cheng, C.; Shen, J. Bidding optimization for cascade hydropower plants in multi-regional electricity markets. *J. Clean. Prod.* **2024**, *435*, 140477. [CrossRef]
22. He, M.; Han, S.; Chen, D.; Zhao, Z.; Jurasz, J.; Mahmud, M.A.; Liu, P.; Deng, M. Optimizing cascade Hydropower-VRE hybrid systems: A novel approach addressing whole-process vibration to enhance operational safety. *Energy* **2024**, *304*, 131965. [CrossRef]
23. Zhang, R.; Zhang, S.; Wen, X.; Yue, Z.; Zhou, Y. Optimization of short-term hydropower scheduling with dynamic reservoir capacity based on improved genetic algorithm and parallel computing. *J. Hydrol.* **2024**, *636*, 131238. [CrossRef]
24. Wu, Z.; Fang, G.; Ye, J.; Huang, X.; Yan, M. Optimizing ecological benefits in weekly-regulated hydropower plants: A framework integrating power market information. *Renew. Energy* **2024**, *231*, 120921. [CrossRef]
25. Dao, F.; Zeng, Y.; Zou, Y.; Qian, J. Fault diagnosis method for hydropower unit via the incorporation of chaotic quadratic interpolation optimized deep learning model. *Measurement* **2024**, *237*, 115199. [CrossRef]
26. Kannapiran, U.M.; Bhaskar, A.S. Flood inundation mapping of upstream region in the Adyar River basin: Integrating hydrologic engineering centre's river analysis system (HEC-RAS) approach with groundwater considerations. *Groundw. Sustain. Dev.* **2024**, *24*, 101085. [CrossRef]
27. Almulla, Y.; Zaimi, K.; Fejzić, E.; Sridharan, V.; de Strasser, L.; Gardumi, F. Hydropower and climate change, insights from the integrated water-energy modelling of the Drin Basin. *Energy Strategy Rev.* **2023**, *48*, 101098. [CrossRef]
28. Feng, J.; Duan, T.; Bao, J.; Li, Y. An improved Back Propagation Neural Network framework and its application in the automatic calibration of Storm Water Management Model for an urban river watershed. *Sci. Total Environ.* **2024**, *915*, 169886. [CrossRef]
29. Araoye, T.O.; Ashigwuike, E.C.; Mbunwe, M.J.; Bakinson, O.I.; Ozue, T.I. Techno-economic modeling and optimal sizing of autonomous hybrid microgrid renewable energy system for rural electrification sustainability using HOMER and grasshopper optimization algorithm. *Renew. Energy* **2024**, *229*, 120712. [CrossRef]
30. Vinayakumar, B.; Antony, R.; Binson, V.A.; Youhan, S. Experimental and numerical study on gravitational water vortex power plant for small water bodies. *E-Prime-Adv. Electr. Eng. Electron. Energy* **2024**, *7*, 100460. [CrossRef]
31. Urquiza, G.; Garcia, J.C.; Gonzalez, J.G.; Castro, L.; Rodriguez, J.A.; Basurto-Pensado, M.A.; Mendoza, O.F. Failure analysis of a hydraulic Kaplan turbine shaft. *Eng. Fail. Anal.* **2014**, *41*, 108–117. [CrossRef]
32. Yan, X.; Zhang, F.; Zheng, Y.; Kan, K.; Rossi, M. Numerical investigation of hydraulic instability of pump-turbines in fast pump-to-turbine transition. *J. Energy Storage* **2024**, *96*, 112731. [CrossRef]
33. Zhilkashinova, A.; Abilev, M.; Zhilkashinova, A. Microplasma-sprayed V_2O_5/C double-layer coating for the parts of mini-hydropower systems. *Coatings* **2020**, *10*, 725. [CrossRef]
34. Zhilkashinova, A.; Abilev, M.; Ocheredko, I.; Tuyakbayev, B.; Nurbayev, M.; Azamatov, B. Mini-hydropower plant based on Lenyov hydrobelt and volume-sectional hydraulic engine. *Processes* **2022**, *10*, 368. [CrossRef]

designs

MDPI

Article

Intelligent Inversion Analysis of Surrounding Rock Parameters and Deformation Characteristics of a Water Diversion Surge Shaft

Xing-Wei Zou [1], Tao Zhou [2], Gan Li [3,*], Yu Hu [4], Bo Deng [2] and Tao Yang [2]

[1] Sino-Hydro Bureau 5 Co., Ltd., Chengdu 610200, China; zouxingwei@126.com
[2] Sichuan Qingyuan Engineering Consultants, Chengdu 610015, China; zhoutao@qyec.com.cn (T.Z.); dengbo@qyec.com.cn (B.D.); yanghongkun@qyec.com.cn (T.Y.)
[3] School of Civil & Environmental Engineering and Geography Science, Ningbo University, Ningbo 315000, China
[4] State Key Laboratory of Hydroscience and Engineering, Tsinghua University, Beijing 100084, China; huyu93@tsinghua.org.cn
* Correspondence: ligan303@126.com; Tel.: +86-159-1095-7882

Abstract: The water diversion surge shaft is vital for a hydropower station. However, the complex geological properties of the surrounding rock make it challenging to obtain its mechanical parameters. A method combining particle swarm optimization (PSO) and support vector machine (SVM) algorithms is proposed for estimating these parameters. According to the engineering geological background and support scheme, a three-dimensional model of the water diversion surge shaft is established by FLAC3D. An orthogonal test is designed to verify the accuracy of the numerical model. Then, the surrounding rock mechanical parameter database is established. The PSO-SVM intelligent inversion algorithm is used to invert the optimal values of the mechanical parameters of the surrounding rock. The support for excavating the next layer depends on the mechanical parameters of the current rock layer. An optimized design scheme is then compared and analyzed with the original support scheme by considering deformation and plastic characteristics. The research results demonstrate that the PSO-SVM intelligent inversion algorithm can effectively improve the accuracy and efficiency of the inversion of rock mechanical parameters. Under the influence of excavation, the surrounding rock in the plastic zone mainly fails in shear, with maximum deformation occurring in the middle and lower parts of the excavation area. The maximum deformation of the surrounding rock under support with long anchor cables is 0.6 cm less than that of support without long anchor cables and 4.07 cm less than that of support without an anchor. In the direction of the maximum and minimum principal stress, the maximum depth of the plastic zone under the support with long anchor cables is 1.3 m to 2.6 m less than that of the support without long anchor cables and the support without an anchor. Compared with the support without long anchor cables and support without an anchor, the support with long anchor cables can effectively control the deformation of the surrounding rock and limit the development of the plastic zone.

Keywords: water diversion surge shaft; parameter inversion; numerical simulation; particle swarm optimization; support vector machine

1. Introduction

As an important part of a water diversion hydropower station, the stability of the surrounding rock of the water diversion surge shaft directly affects the safety of its operation [1,2]. Affected by factors such as rock properties and ground stress, the surrounding rock safety of deep underground tunnels will face various safety threats [3–5]. Since surge shafts are usually located in complex geological conditions, the mechanical properties of the surrounding rock usually show nonlinearity, heterogeneity and anisotropy [6–8], resulting in the acquisition of mechanical parameters of surrounding rock, and the prediction of its deformation characteristics is a major problem in engineering design. Although traditional indoor tests and in situ tests can provide some information on surrounding rock mechanical parameters, they are difficult to carry out due to the complex construction environment, large engineering volume and lack of large-scale in situ monitoring [9,10]. Especially in deep underground projects, the variability and uncertainty of surrounding rock parameters make the construction potentially dangerous [11–13]. Therefore, a method that can efficiently and accurately obtain surrounding rock parameters has important application value in practical engineering.

Traditional inversion methods make it difficult to handle complex nonlinear problems and are also inefficient. The intelligent inversion method is data-based, combining traditional numerical simulation and modern machine learning intelligent algorithms. By learning from measured data, the optimal solution can be found in the multidimensional parameter space, and the efficient identification of engineering parameters can thereby be realized. This method not only breaks through the high dependence of traditional inversion methods on initial values and objective functions but also can handle engineering problems under highly nonlinear, multidimensional and complex boundary conditions. However, it has high requirements for computer numerical simulation technology and big data processing capabilities. With the rapid development of artificial intelligence technology and computer science, numerical simulation [14–18] and machine learning algorithms [19–22] have been widely used in tunnel engineering and deep geological engineering. Parameter identification and prediction methods in the field of geotechnical engineering have entered a new stage. Based on the XGBoost model, a macro parameter prediction and micro parameter inversion model is established, which can solve the complicated macro and micro parameter calibration problem in the discrete element method simulation of rock mechanics [23]. To obtain rock creep parameters, a secondary particle swarm optimization (PSO) for rheological parameters inversion is proposed by Yang et al. [24]. According to the wavelet analysis theory, the support vector machine (SVM) method is used to perform inverse calculations on geotechnical mechanics by Ruan et al. [25], and the particle swarm algorithm is applied to optimize the parameter function. An improved adaptive particle swarm optimization with a neural network algorithm is proposed to realize the back analysis of tunnel surrounding rock mechanical parameters based on the displacement information of the engineering site [26]. Among various intelligent algorithms, particle swarm optimization and support vector machine have gradually become the mainstream methods in the inversion of surrounding rock parameters due to their excellent performance in global search and classification capabilities, and many new algorithms have been derived according to engineering needs. Chen et al. [27] use PSO to optimize velocity inversion and improve the imaging accuracy of the surface wave while using a tunneling method for predicting geological structures. A new back-analysis method of rock mechanical parameters combining the LS-SVM model and Littlewood–Paley wavelets is proposed by Wang et al. [28]. Using PSO to estimate the optimal rheological parameter set, Kovacevic et al. [29] establish a prediction model for the medium- and long-term vertical settlement performance of soft rock tunnels based on their self-developed neural network NetRHEO. VGGNet, ResNet and SVM are used by Ma et al. [30] to establish surrounding rock classification models, among which the SVM classifier had the highest accuracy, revealing that the SVM classification model has greater robustness in learning and generalizing small samples and unbalanced samples. Based on the area under the curve method, it is found that the prediction accuracy of the coupled model for landslide susceptibility assessment is significantly better than that of the single information volume model and the SVM model [31]. Zhang et al. [32] developed an intelligent optimization algorithm that combines PSO and Gaussian process (GP) theory. Then, the finite difference method (FDM) is applied to form a PSO-GP-FDM displacement inversion analysis method. Yan et al. [33] propose an improved Gaussian process regression (IGPR) with combinatorial kernel functions by adding two mononuclear functions. Combining PSO with the

IGPR model, the PSO algorithm can be employed directly to search for the optimal rock force parameters.

While many studies focus on inverting rock mechanical parameters using machine learning, there is a lack of research specifically addressing the inversion of these parameters for the stratified excavation of the water diversion surge shaft. To perform rock mechanical parameter inversion more accurately and efficiently, the particle swarm optimization and support vector machine are combined to propose a new rock mechanical parameter inversion method. Using the PSO-SVM rock mechanical parameter inversion method, a three-dimensional numerical model of the water diversion surge shaft excavation process is established by FLAC3D. By inversely analyzing the mechanical parameters of the surrounding rock of the water diversion surge shaft, its deformation and plasticity characteristics will be studied to provide a scientific basis and technical support for the design and safety assessment of deep underground engineering.

2. PSO-SVM Inversion Method of Rock Mechanical Parameters

2.1. PSO Method

Particle Swarm Optimization is a global optimization algorithm based on population intelligence [34]. The principle of PSO is shown in Figure 1. It originated from the study of the foraging behavior of bird flocks and was proposed by Kennedy and Eberhart in 1995 [35]. PSO has several advantages, including its easy implementation, few parameters and strong global search ability. It is suitable for solving complex nonlinear and multi-parameter inverse problems and has been widely used in function optimization, neural network training and combinatorial optimization [36,37].

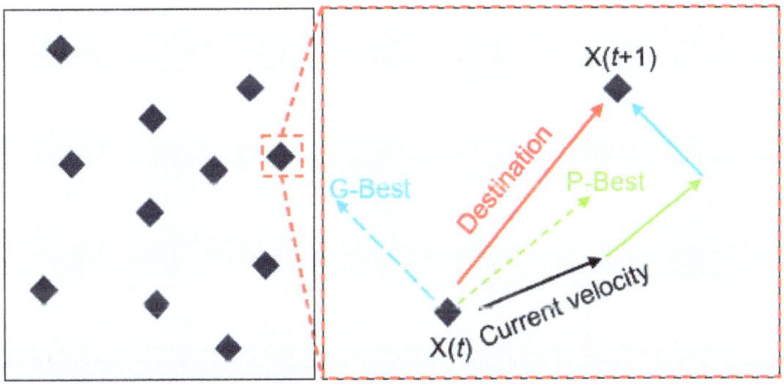

Figure 1. Schematic diagram of particle swarm optimization.

PSO realizes the search for a globally optimal solution through the following four steps.

(1) Generating a set of particles with random initial positions and velocities in the search space, with each particle representing a possible solution.

(2) Calculating the fitness value of each particle and evaluating the advantages and disadvantages of the solution according to the fitness value.

(3) According to the historical and global optimal positions of particles, adjusting the velocity and position of each particle concerning the specific optimization formula.

(4) Judging the fitness of the particles and outputting the optimal solution if it meets the termination condition; otherwise, returning to step (2) to continue the iteration.

2.2. SVM Method

Support Vector Machine is a supervised learning model proposed by Vapnik in 1995 [38]. The principle of SVM is shown in Figure 2. It is mainly used for classification and regression analysis. SVM classifies samples of different classes by constructing an optimal hyperplane. It has good generalization ability and is suitable for high-dimensional and nonlinear datasets, especially in small samples and nonlinear problems. However, it has high computational complexity on large-scale datasets [39–41].

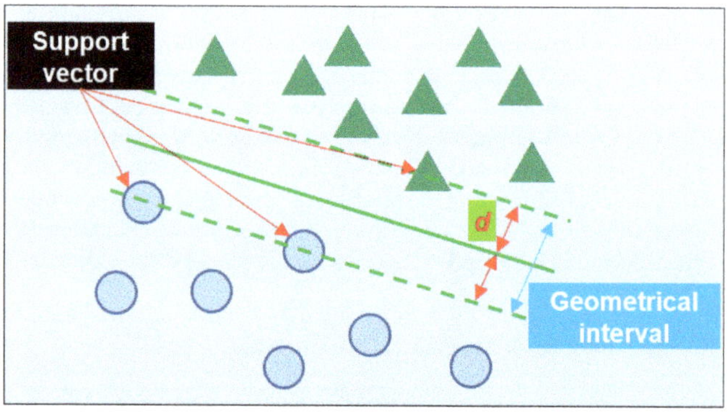

Figure 2. Schematic diagram of support vector machine.

The basic principles of SVM are as follows.

(1) Commonly used kernel functions include linear, polynomial and radial basis functions. The choice of kernel function significantly impacts the classification effect of SVM.

(2) A hyperplane that maximizes the classification interval is found by solving a constrained optimization problem.

(3) The optimal hyperplane is used to predict the classification or regression of new samples.

2.3. Inversion Steps of Rock Mechanical Parameters

Based on the engineering geological background and design scheme, the appropriate surrounding rock constitutive model, support constitutive model and boundary conditions are selected. The three-dimensional numerical model of the water diversion surge shaft is established by FLAC3D. According to the long-term monitoring data of the dome on site, the influence of each parameter on the maximum deformation is analyzed by SVM, the parameter selection is optimized and the parameter orthogonal analysis table is formulated. The in situ stress inversion is carried out for the monitoring and measurement of the displacement. The calculated displacement of the numerical model is compared to the monitored displacement to verify accuracy. Then, the deformation parameter database of the surrounding rock is established. After assigning initial values to rock mass parameters according to preliminary geological characteristics, the three-dimensional numerical model is excavated and supported in layers, and deformation characteristic data are recorded. Rock mechanical parameter inversion is performed through a particle swarm optimization–support vector machine algorithm. Through repeated updates and iterations, the minimum error value is continuously reduced until the accuracy requirements are met and the optimal value of the required parameter is obtained. Finally, assuming that the variability of rock mass parameters is relatively consistent, the excavation of the next layer is based on the rock mechanical parameters of that layer. The different support schemes are compared and analyzed, and the above steps are repeated until the excavation is completed. The inversion steps of rock mechanical parameters are shown in Figure 3.

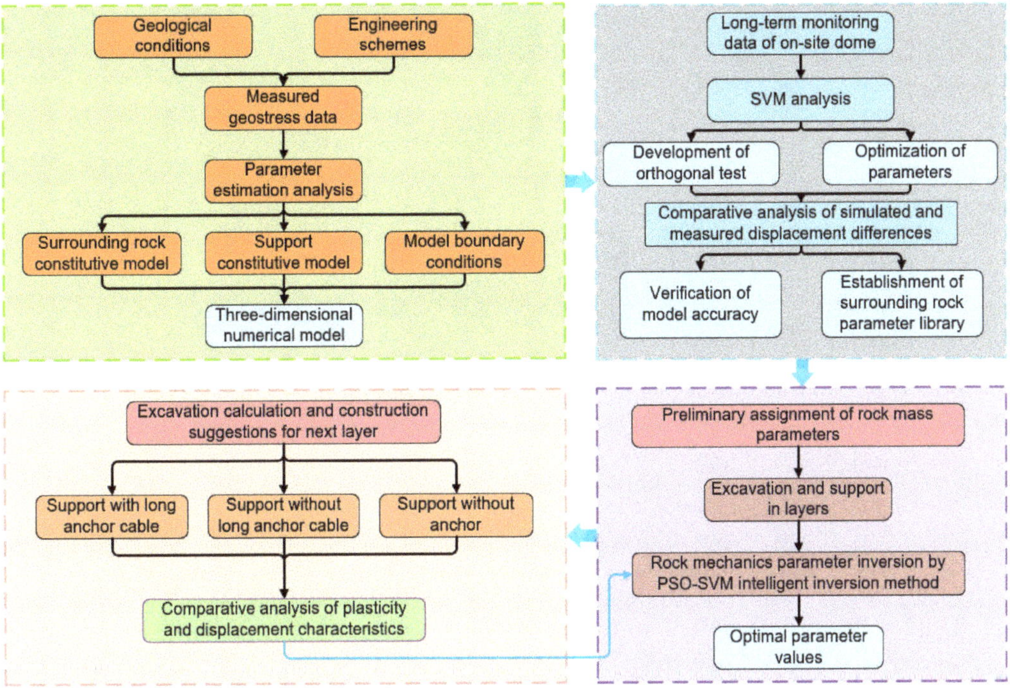

Figure 3. Inversion steps of rock mechanical parameters.

3. Inversion of Surrounding Rock Parameters of Water Diversion Surge Shaft

3.1. Engineering Situations

The shape of the water diversion surge shaft based on a water conservancy project is a circular impedance surge shaft. The dome height is 14 m, the maximum excavation diameter is 35 m, the excavation diameter of the shaft section is 33.40 m and the height is 79.50 m. The thickness of the overlying rock mass on the top arch of the surge shaft, which is constructed in the fine-grained biotite monzogranite of the Yanshanian period, is 135 m. Geological surveys and low-elevation exploration tunnels reveal that there are a small number of randomly developed NW-trending steeply dipping small faults or compressive fracture zones in the surrounding rock. The geological structure is mainly characterized by joint fissures. The weathering degree of rock mass is generally slightly weathered with four groups of dominant joints (N40°~60°W/SW(NE)∠60°~80°, N55°~80°E/SE∠65°~80°, SN/E∠60°~80°, N30°~45°W/SW(NE)∠10°~20°) whose spacing is large. The rock mass has good integrity and is characterized by numerous block and sub-block structures.

3.2. Construction of the Numerical Model

According to the layout of the surge chamber, the numerical model of the cavern is established by using the finite difference software FLAC3D V7.0 (Figure 4) to realize the dynamic simulation of excavation. According to the Saint Venant principle [42], to ensure that the initial stress state is always maintained around the model, the size of the model should be about three to five times the maximum size of the excavation object. Therefore, the transverse (X-axis direction) width of the model is 340 m, the longitudinal (Y-axis direction) width is 200 m and the height from the bottom to the arch bottom (Z-axis direction) is 190 m. The Shell element is employed to simulate the supporting structure, and the Cable element is used to simulate the steel bar, prestressed anchor cable, anchor bolt, etc. The total number of model elements is 3,078,024.

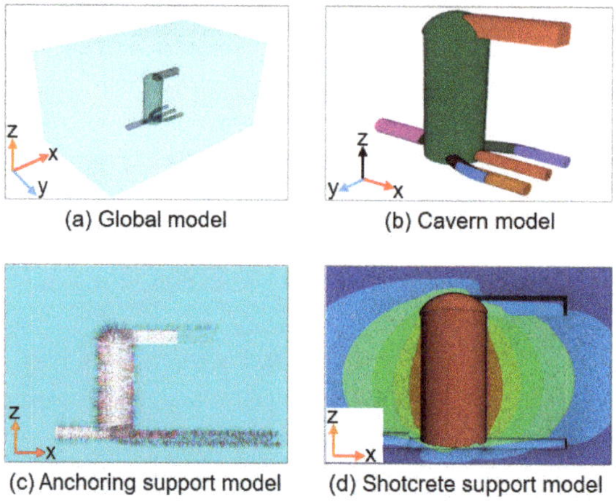

Figure 4. Numerical model.

3.3. Parameter Inversion Results

Based on the dome monitoring data, the parameter orthogonal test plan (Table 1) is developed, and the influence of each parameter on the maximum deformation is analyzed by SVM, optimizing the mechanical parameters. The analysis results shown in Figure 5 demonstrate that with the increase in cohesion, the upper limit of the maximum deformation of the surrounding rock remains unchanged, but the lower limit decreases. With the increase in the elastic modulus of the surrounding rock, the upper and lower limits of the maximum deformation of the surrounding rock decrease. The range of the maximum deformation of the surrounding rock increases with the increase in the surrounding rock parameters.

Table 1. Orthogonal analysis table of mechanical parameters.

Experiment Number	Elastic Modulus (GPa)	Internal Friction Angle (°)	Cohesive Force (MPa)	Maximum Deflection (mm)
0	20	45	3	4.73
1	15	35	2.5	5.41
2	20	40	3	4.5
3	25	45	3.5	3.11
4	15	40	3.5	5.69
5	20	45	2.5	5.01
6	25	35	3	3.22
7	15	45	3	5.22
8	20	35	3.5	4.17
9	25	40	2.5	3.8

By using the numerical simulation analysis method, the deformation characteristics of surrounding rock under different working conditions are obtained (Figure 6). Furthermore, the rock mechanical parameter inversion method based on the PSO-SVM algorithm (Figure 7) is employed to obtain the characteristics of rock mechanical parameters at an elevation of 2940 m, which demonstrates that the elastic modulus is 22 GPa, the internal friction angle is 43°, and the cohesion is 0.3 MPa. According to the assumption that the variability of rock parameters is minimal, the rock parameters obtained this time are used to guide the excavation of the cavern at the next elevation, and the rock mechanical parameters at the elevations of 2922 m and 2898 m are obtained by inversion. The inversion of surrounding rock at an elevation of 2922 m demonstrates that the elastic modulus is 23 GPa, the internal friction angle is 41° and the cohesion is 0.3 MPa. The inversion of surrounding rock at an elevation of 2898 m demonstrates that the elastic modulus is 24 GPa, the internal friction angle is 43° and the cohesion is 0.4 MPa.

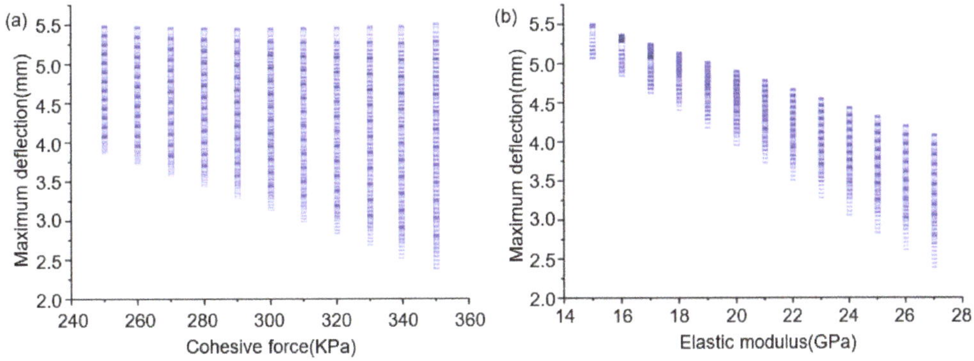

Figure 5. Influence of different mechanical parameters of surrounding rock on deformation: (**a**) The influence of cohesion of surrounding rock on deformation; (**b**) The influence of elastic of surrounding rock on deformation.

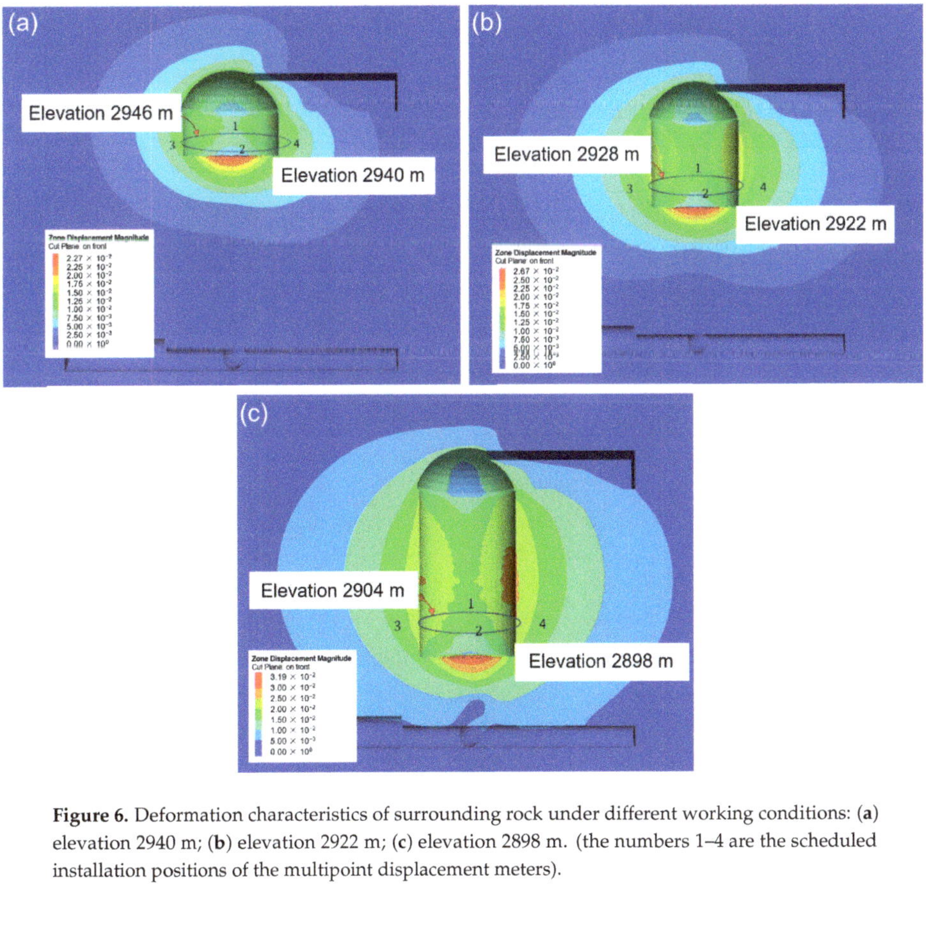

Figure 6. Deformation characteristics of surrounding rock under different working conditions: (**a**) elevation 2940 m; (**b**) elevation 2922 m; (**c**) elevation 2898 m. (the numbers 1–4 are the scheduled installation positions of the multipoint displacement meters).

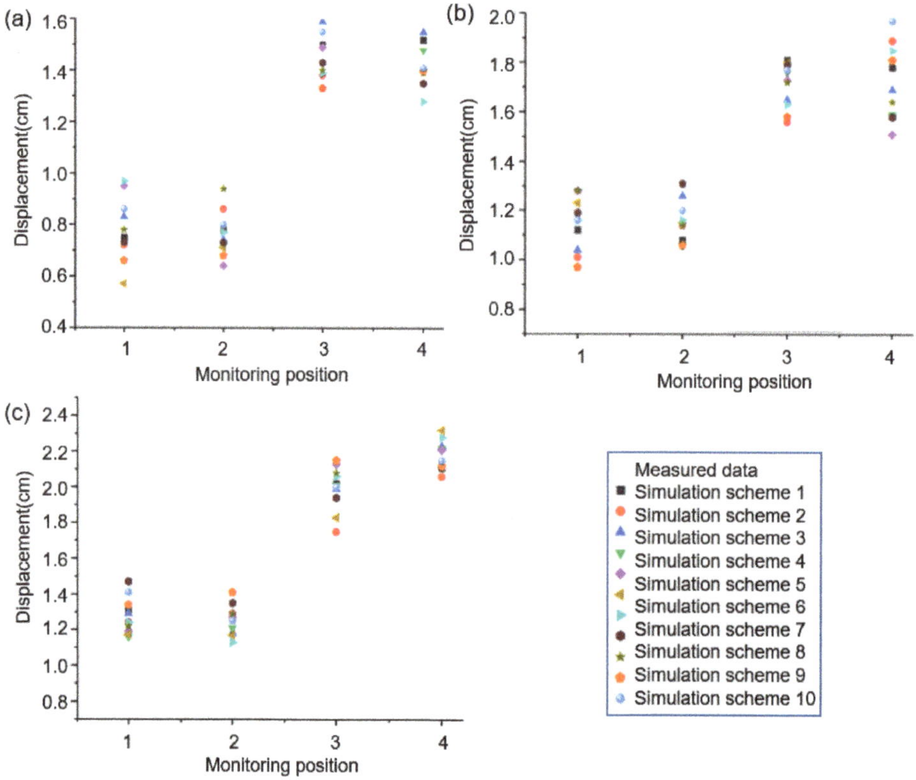

Figure 7. PSO-SVM parameter inversion: (**a**) elevation 2940 m; (**b**) elevation 2922 m; (**c**) elevation 2898 m.

3.4. Comparison of Field Monitoring Data and Numerical Simulation Results

Multi-point displacement meters are installed on the profile near an elevation of 2926 m. Taking the deepest point of the anchor head as the relatively fixed point, the absolute displacement values of the surface, depth 10 m and depth 5 m, are calculated. Its location, number and installation process are shown in Figure 8.

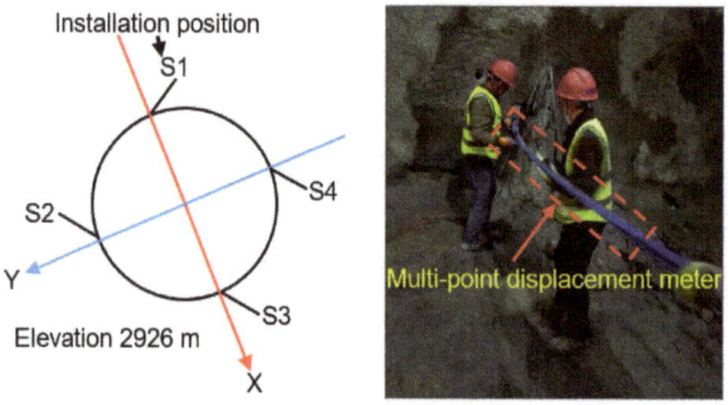

Figure 8. Installing a multi-point displacement meter.

To verify the accuracy of the numerical model, the absolute displacement–time diagram is drawn based on the displacement data (Figure 9). As shown in Figure 9, during the monitoring period, the overall temperature change is relatively gentle and without sudden changes. The average temperature change in the first two weeks is 0.39 °C per day, so the impact on the displacement monitoring of each anchor head is negligible.

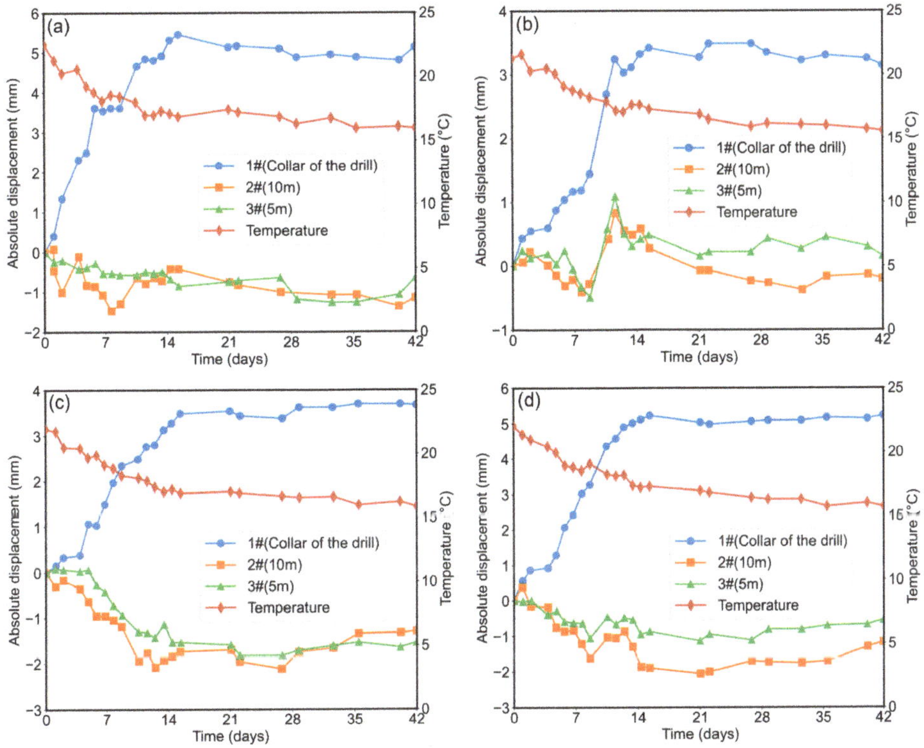

Figure 9. Absolute displacement–time diagram: (**a**) S1; (**b**) S2; (**c**) S3; (**d**) S4.

The absolute displacement values of the 1# anchor heads of each multi-point displacement meter showed a trend of increasing first and then stabilizing, among which S1-1# reached a maximum value of 5.4 mm on the sixteenth day, S2-1# reached a maximum value of 3.47 mm on the twenty-third day, S3-1# reached a maximum value of 3.71 mm on the thirty-seventh day and S4-1# reached a maximum value of 5.32 mm on the forty-second day. Except for the S2-2# and S2-3# anchor heads, the absolute displacement values of the 2# and 3# anchor heads of the other multi-point displacement meters also showed a trend of first decreasing and then tending to be stable. During the second week of monitoring, the sudden increase in the absolute displacement values of each S2 anchor head may be related to the heterogeneity of the surrounding rock. Since the elastic modulus of the surrounding rock in the buried direction of the S2 multi-point displacement meter is relatively high, the surrounding rock only has surface displacement in the early stage of excavation. As the distance between the surface and the depth of the surrounding rock increases, the tensile stress on the surrounding rock also increases. Eventually, the rock is destroyed, and the absolute displacement of each anchor head increases accordingly.

Based on the measured displacement data recorded by the multi-point displacement meter, the maximum relative displacements between the surface of the cavern monitored by S1, S2, S3 and S4 and the depth of 10 m are calculated, which are 3.6 mm, 5.0 mm, 6.0 mm and 6.7 mm, respectively, with an average maximum relative displacement of 5.32 mm. As shown in Figure 10, the displacement of the surrounding rock on the surface of the cave is about 22.0 mm, and the displacement of the deep surrounding rock at 10 m is 17.5 mm, resulting in a relative displacement of 4.5 mm. The numerical simulation results of the relative displacement of the surrounding rock are compared with the average value of the measured results; the difference between them is 1.82 mm, indicating that the numerical simulation results are consistent with the measured results and proving that the numerical model has high accuracy.

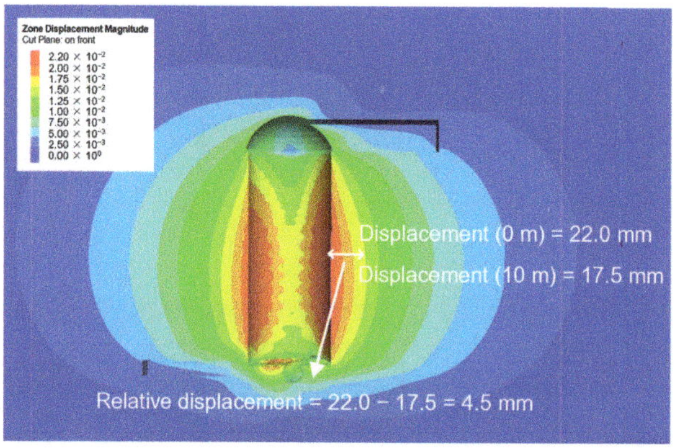

Figure 10. Displacement of water diversion surge shaft in numerical simulation.

4. Analysis of Deformation Characteristics of Water Diversion Surge Shaft Under Different Supports During Layered Excavation

According to the construction documents, the water diversion surge shaft is excavated in layers. The profile of the support scheme along the water flow direction is shown in Figure 11. (The red lines in the figure are long anchor cables that form an angle of 15° with the horizontal plane. They are arranged in five rows at an elevation of 2916 to 2891, with 17 cables in each row and a row spacing of 6 m. The short anchor bolts are shown in the subsequent support schemes.) The support schemes in different elevations are shown in Figure 12. A numerical simulation support model is established based on the support schemes (Figure 13).

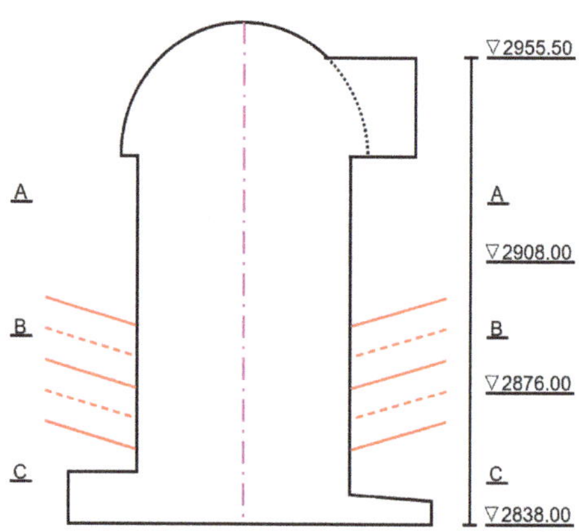

Figure 11. Profile of the support scheme along the water flow direction. (A, B, C are marks for dividing the general position profile of the water diversion surge shaft).

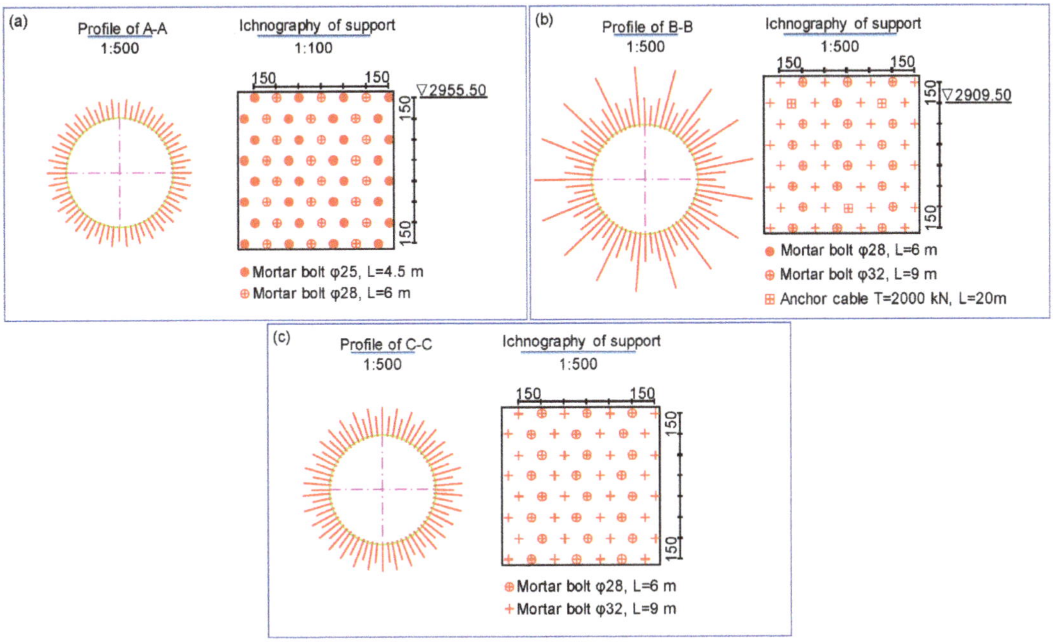

Figure 12. Support schemes in different elevations: (**a**) 2916~2955 m; (**b**) 2890~2916 m; (**c**) 2876~2890 m.

Figure 13. Profile of the support scheme of the numerical simulation.

To study the differences between different support schemes, the deformation characteristics of the water diversion surge shaft under the conditions of support with a long anchor cable, support without a long anchor cable and support without an anchor are analyzed based on FLAC3D. The research results are shown in Figures 14–16. By comparing Figures 14–16, it is found that when the excavation depth is small, the maximum displacement of the surge shaft mainly occurs in the middle part directly below the excavation area. As the excavation depth increases, the maximum displacement of the side walls on both sides of the water diversion surge shaft and the depth of the plastic zone increases significantly, and the failure form of the plastic zone is mainly shear failure.

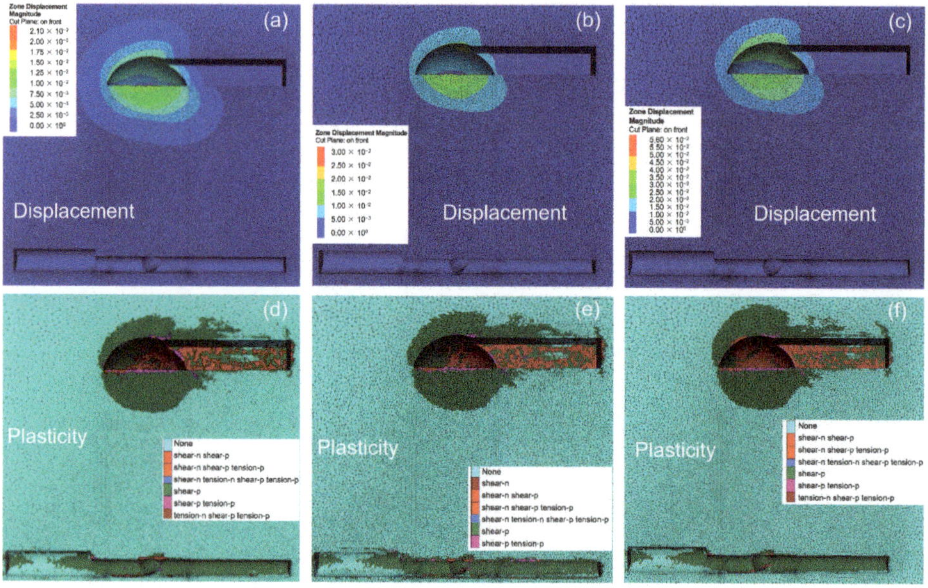

Figure 14. Displacement and plastic characteristics of water diversion surge shaft in the dome: (**a**) support with long anchor cable; (**b**) support without long anchor cable; (**c**) support without anchor; (**d**) support with long anchor cable; (**e**) support without long anchor cable; (**f**) support without anchor.

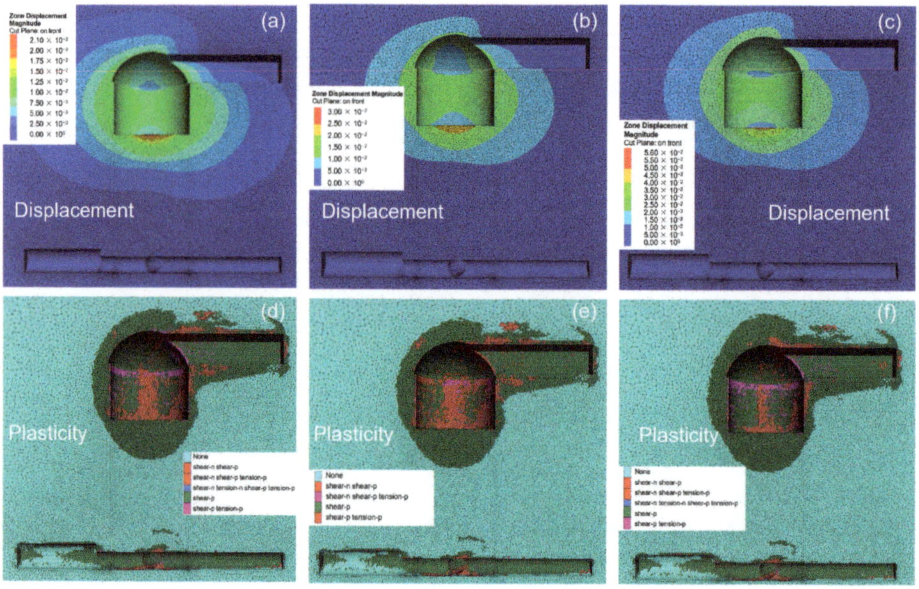

Figure 15. Displacement and plastic characteristics of water diversion surge shaft in elevation 2931: (**a**) support with long anchor cable; (**b**) support without long anchor cable; (**c**) support without anchor; (**d**) support with long anchor cable; (**e**) support without long anchor cable; (**f**) support without anchor.

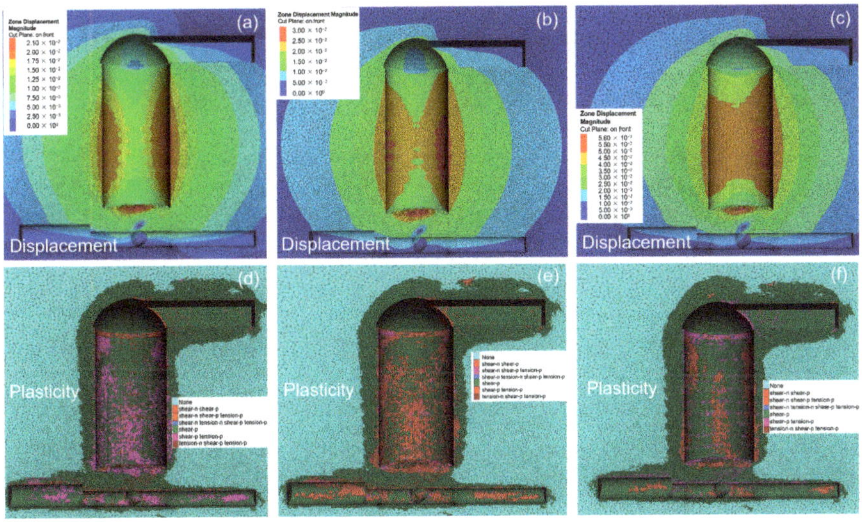

Figure 16. Displacement and plastic characteristics of water diversion surge shaft in elevation 2886: (**a**) support with long anchor cable; (**b**) support without long anchor cable; (**c**) support without anchor; (**d**) support with long anchor cable; (**e**) support without long anchor cable; (**f**) support without anchor.

According to the displacement characteristics, before excavation at the elevation of 2931 m, the interaction between the lower tunnel and the water diversion surge shaft is weak. When excavated to the elevation of 2886 m, the interaction between the lower tunnel and the water diversion surge shaft is greatly intensified, and attention should be paid to observing the stability of the lower cavern. The influence on the dome is relatively weak during the increase in excavation depth. After excavation at an elevation of 2886 m, under the action of the extended anchor cable, the displacement of the side walls on both sides of the water diversion surge shaft is relatively uniform, and the displacement on the right side is slightly larger than that on the left side. Without the long anchor cable support, the unevenness of the displacement of the side walls on both sides becomes more obvious, and the displacement field is connected in the middle of the water diversion surge shaft. Without anchor support, the displacement of the right side wall is significantly greater than that of the left side. The support effect of the left support device is weak. The right side wall displacement should be paid special attention to, and corresponding support measures should be set up. Regarding the plastic characteristics, before excavation at the elevation of 2886, there is no significant difference in the plastic zone under the effect of the support with long anchor cables and the support without long anchor cables. When excavating the dome, the shear failure plastic zone is concentrated on the right side of the dome. When excavating to the 2931 elevation, the plastic zone is concentrated in the middle of the surge shaft.

When comparing the displacement and plastic characteristics of the water diversion surge shaft excavation process under three support schemes, it is evident that with anchor support, the surrounding rock is influenced by the combined action of anchor bolts/cables and shotcrete. This increases overall stability and effectively controls displacement and deformation, resulting in a relatively uniform stress distribution area between the rock mass and the support structure. Compared with the surrounding rock supported without long anchors, the surrounding rock supported with long anchors has a smaller overall displacement and significantly reduced plastic zone area. Compared with the surrounding rock without anchor support, the surrounding rock with anchor support has improved overall bearing capacity and a significantly reduced plastic zone area under the tension of anchor bolts/cables and the constraint of shotcrete, thus reducing plastic deformation.

5. Response Characteristics of Surrounding Rock of Water Diversion Surge Shaft at the Excavation Completion Stage Under Different Support Schemes

After the completion of excavation in the water diversion surge shaft, the comprehensive response characteristics of the surrounding rock under different support schemes need to be further studied. FLAC3D is used to establish three-dimensional numerical models of the schemes of support with long anchor cables, support without long anchor cables, and support without anchor. The displacement and plastic zone characteristics of the surrounding rock at the excavation completion stage under different support schemes are compared and analyzed under the maximum principal stress and the minimum principal stress.

5.1. Comparative Analysis of Surrounding Rock Displacement of Water Diversion Surge Shaft Under Different Support Schemes

As shown in Figure 17, in the direction of maximum principal stress, under the influence of ground stress and excavation disturbance, the maximum deformation of the surrounding rock under the long anchor cable support is 2.15 cm, and the maximum deformation of the surrounding rock after the long anchor cable is removed is 2.75 cm. The maximum deformation of the surrounding rock under no anchor support is 6.22 cm. Compared with the scheme of support without long anchor cables, the maximum deformation of the surrounding rock under no anchor support increases significantly, with an increase of 3.47 cm. In the direction of minimum principal stress, the maximum deformation of the surrounding rock under each support scheme is smaller than the maximum deformation of the surrounding rock in the direction of maximum principal stress. The maximum deformation of the surrounding rock under support with long anchor cables is 2.0 cm, and the maximum deformation of the surrounding rock under support without long anchor cables is 2.25 cm. The maximum deformation of the surrounding rock without anchor support is 6.1 cm, which increases by 3.85 cm compared with the support without long anchor cables. The surrounding rock deformation degree of each support scheme shows a trend of increasing first and then decreasing with increasing depth, and the maximum surrounding rock deformation occurs in the middle and lower sections of the surge shaft. After comparison, the support with long anchor cables has the best control effect on the surrounding rock displacement, followed by the support without long anchor cables, and the support without an anchor cannot effectively control the surrounding rock displacement. Comparing Figure 17a,b, it can be seen that the displacement of the surrounding rock in the direction of the maximum principal stress is greater than the displacement of the surrounding rock in the direction of the minimum principal stress. To prevent the instability of the surge shaft, the displacement of the surrounding rock in the direction of the maximum principal stress should be focused on according to the displacement characteristics.

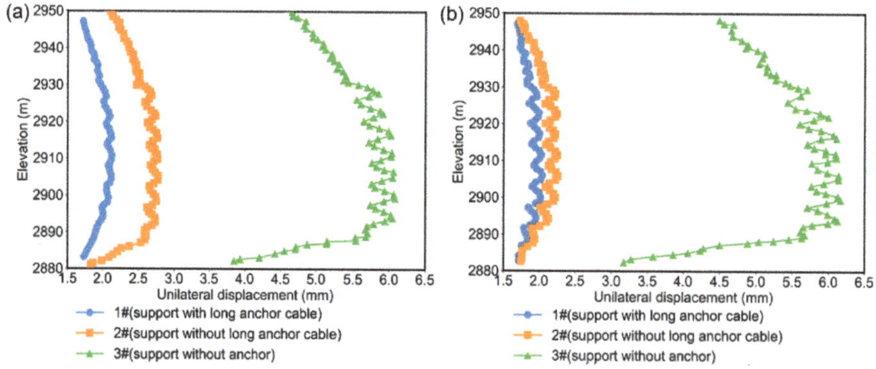

Figure 17. Displacement characteristics under different elevations: (**a**) maximum principal stress direction; (**b**) minimum principal stress direction.

5.2. Comparative Analysis of Plastic Zone of Surrounding Rock of Water Diversion Surge Shaft Under Different Support Schemes

As shown in Figure 18, in the direction of maximum principal stress, the maximum depth of the plastic zone under the support with long anchor cables is 5.3 m. Under the support without long anchor cables, the maximum depth of the plastic zone is 6.6 m. The maximum depth of the plastic zone without anchor support is 7.9 m, which is 1.3 m higher than that without long anchor support. As shown in Figure 19, in the direction of minimum principal stress, the maximum depth of the plastic zone is 16.2 m under the effect of the support with long anchor cables, and the maximum depth of the plastic zone is 18.6 m under the support without long anchor cables. The maximum depth of the plastic zone without anchor support is 21.2 m, and the depth of the plastic zone increases by 2.6 m compared with the support without long anchor cables. The plastic zones of the three surrounding rocks are mainly shear failure. It can be seen that the surge shaft has been in a high-stress environment, and the excavation leads to various plastic states, resulting in the instability of the surrounding rock damage. The plastic zone in the direction of the minimum principal stress is much larger than that in the direction of the maximum principal stress. Especially in the absence of long anchor cable support or anchorless support, the plastic zone spreads to the entire surge shaft. To ensure the stability of the surge shaft, the support with long anchor cables is essential and the plastic deformation in the direction of the minimum principal stress should be focused on.

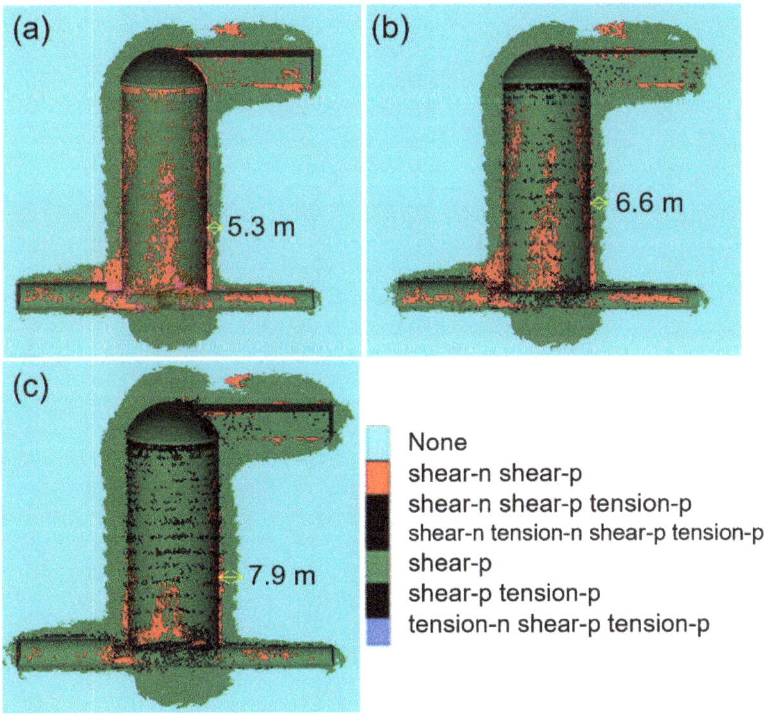

Figure 18. Plastic characteristics of maximum principal stress direction: (**a**) support with long anchor cable; (**b**) support without long anchor cable; (**c**) support without anchor cable.

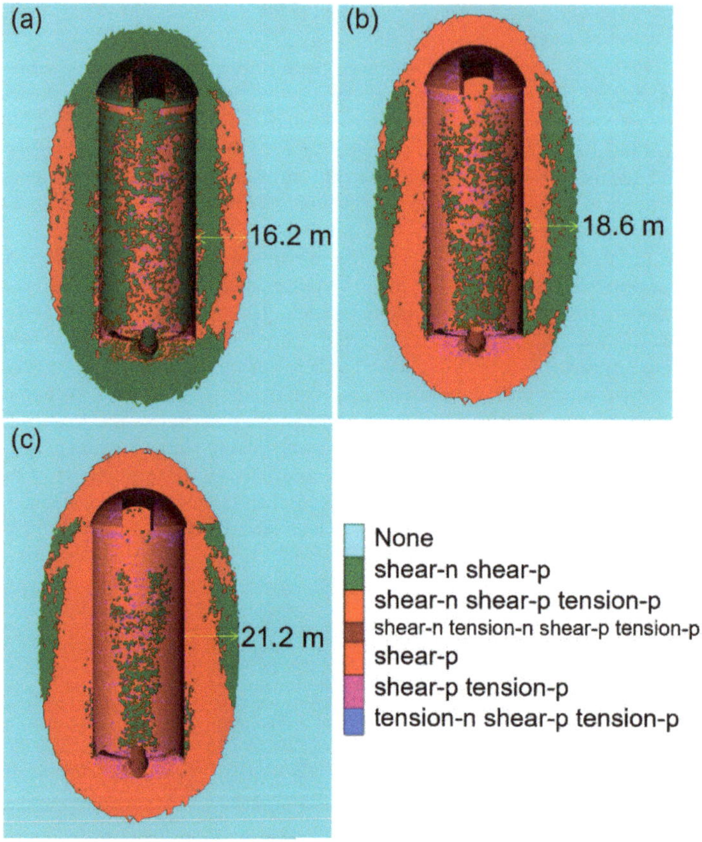

Figure 19. Plastic characteristics of minimum principal stress direction: (**a**) support with long anchor cable; (**b**) support without long anchor cable; (**c**) support without anchor cable.

6. Conclusions

In this paper, a PSO-SVM surrounding rock mechanical parameter inversion algorithm is established according to the geological conditions and monitoring data of a water diversion surge shaft excavation project. Based on the SVM-PSO intelligent inversion algorithm, the surrounding rock mechanical parameters are obtained. Using three-dimensional numerical simulation software, the displacement and plasticity characteristics of the surrounding rock during the excavation of the water diversion surge shaft are analyzed. The following conclusions are drawn:

(1) A rock mechanical parameter inversion method is proposed, combining particle swarm optimization and support vector machine. An orthogonal analysis is conducted on monitoring data to establish a database of surrounding rock mechanical parameters. The particle swarm optimization is employed to globally search for and optimize these parameters. Finally, the support vector machine is applied to perform local refined fitting, which effectively enhances the accuracy and efficiency of the parameter inversion process.

(2) By analyzing the plastic zone and displacement field of the surrounding rock under the layered excavation, it is found that the main form of destruction of the plastic zone of the surrounding rock is shear failure. With the increase in excavation depth, the support from the upper surrounding rock decreases, increasing the stress of the lower surrounding rock. The surrounding rock after stress redistribution produces the maximum deformation in the middle and lower parts. Therefore, special attention should be paid to

the shear failure area of the surrounding rock and the deformation of the middle and lower surrounding rock during the excavation, taking corresponding support measures.

(3) After excavation, in the maximum principal stress direction, compared with the long anchor cable support, the deformation increased by 0.6 cm under the support without long anchor cables, and the deformation increased by 4.07 cm without anchor support. In the minimum principal stress direction, compared with the long anchor cable support, the deformation increased by 0.25 cm under the support without long anchor cables, and the deformation increased by 3.99 cm without anchor support. In the maximum principal stress direction, compared with the support with long anchor cables, the plastic zone depth increased by 1.3 m under the support without long anchor cables. In the minimum principal stress direction, compared with the long anchor cable support, the plastic zone depth increased by 2.4 m under the support without long anchor cables. In terms of controlling the deformation of the surrounding rock and limiting the development of the plastic zone, the support with long anchor cables is significantly better than the support without long anchor cables.

(4) Complex geological conditions will inevitably be encountered during the construction process. Taking into account the variability of surrounding rock parameters, how to invert the mechanical parameters of expansive soft rock, a rock mass that is strongly affected by hydraulic properties, will become the focus of our subsequent research.

Author Contributions: Conceptualization, X.-W.Z. and G.L.; methodology, X.-W.Z.; software, T.Z.; validation, X.-W.Z., T.Z., G.L., Y.H., B.D. and T.Y.; formal analysis, X.-W.Z. and Y.H.; investigation, X.-W.Z. and B.D.; resources, G.L.; data curation, X.-W.Z.; writing—original draft preparation, X.-W.Z., T.Z., G.L., Y.H. and T.Y.; writing—review and editing, G.L. and Y.H.; visualization, X.-W.Z. and Y.H.; supervision, G.L. and B.D.; project administration, G.L.; funding acquisition, G.L. All authors have read and agreed to the published version of the manuscript.

Funding: The research was founded by the Zhejiang Provincial Natural Science Foundation (LQ24D020001), the open Fund of State Key Laboratory of High-speed Railway Track Technology (2022YJ127-1) and the Open Project of Key Laboratory of Geotechnical and Underground Engineering of the Ministry of Education (Tongji University) (KLE-TJGE-B2306).

Data Availability Statement: Data are contained within the article.

Acknowledgments: Thanks to the help of three anonymous reviewers and journal editors.

Conflicts of Interest: Xing-Wei Zou was employed by the Sino-Hydro Bureau 5 Co., Ltd. The remaining authors declare that the research was conducted in the absence of any commercial or financial relationships that could be construed as a potential conflict of interest.

References

1. Li, X.; Wang, H.; Jia, J.; Yang, C.; Hu, G.; Xue, Z. Ultimate displacement discrimination of stability and reliability analysis of surrounding rocks of tunnel and underground engineering. *Rock. Soil. Mech.* **2005**, *26*, 850–854.
2. Xie, G.; Li, C.; Wang, L. Mechanical characteristics and practical application on stress shell of roadway surrounding rock. *J. China Coal Soc.* **2016**, *41*, 2986–2992.
3. Fu, Q.; Yang, J.; Gao, Y.B.; Li, C.J.; Song, H.X.; Liu, Y.X.; Wu, X. Combined blasting for protection of gob-side roadway with thick and hard roof. *J. Rock. Mech. Geotech.* **2024**, *16*, 3165–3180. [CrossRef]
4. Ren, F.Q.; Zhu, C.; Karakus, M.; He, M.C. Rockburst mitigation mechanisms of pressure relief borehole and rock bolt support: Insights from granite true triaxial unloading rockburst tests. *Eng. Geol.* **2024**, *336*, 107571. [CrossRef]
5. Zhu, C.; Xing, X.S.; He, M.C.; Tang, Z.C.; Xiong, F.; Ye, Z.Y.; Xu, C.S. Failure behavior and strength model of blocky rock mass with and without rockbolts. *Int. J. Min. Sci. Technol.* **2024**, *34*, 747–762. [CrossRef]
6. Xu, C.; Ren, Q.; Li, R. Entropy catastrophe criterion of surrounding rock stability. *Chin. J. Rock. Mech. Eng.* **2004**, *23*, 1992–1995.
7. Zhao, S.J.; Zhang, Q.; Miao, Y.S.; Zhang, W.Z.; Zhao, X.B.; Xu, W. Sub-Homogeneous peridynamic model for fracture and failure analysis of roadway surrounding rock. *Cmes-Comp. Model. Eng.* **2024**, *139*, 3167–3187. [CrossRef]
8. Zhou, P.; Shen, Y.; Zhao, J.; Zhang, X.; Gao, B.; Zhu, S. Research on disaster-induced mechanism of tunnels with steeply dipping phyllite strata based on an improved ubiquitous-joint constitutive model. *Chin. J. Rock. Mech. Eng.* **2019**, *38*, 1870–1883.
9. Cheng, Q.; Guo, X. Study on comprehensive selection of mechanical parameters of rock mass in deep-buried tunnel anchor. *Chin. J. Undergr. Space Eng.* **2021**, *17*, 479–487.

10. Gao, M.; Ye, S.; Yang, B.; Liu, Y.; Li, J.; Liu, J.; Xie, H. Progress in research on deep in situ rock mechanics. *Bull. Natl. Nat. Sci. Found. China* **2021**, *35*, 895–903.
11. Xu, J.C.; Yang, C.B. Probabilistic back analysis based on adam, bayesian and multi-output gaussian process for deep soft-rock tunnel. *Rock Mech. Rock Eng.* **2023**, *56*, 6843–6853. [CrossRef]
12. Sun, J.L.; Wu, S.C.; Wang, H.; Wang, T.; Geng, X.J.; Zhang, Y.J. Inversion of surrounding rock mechanical parameters in a soft rock tunnel based on a hybrid model eo-lightGBM. *Rock. Mech. Rock. Eng.* **2023**, *56*, 6691–6707. [CrossRef]
13. Peng, Z.; Wang, Z.; Hong, C.; Li, K.; Li, A. Influence of parameter uncertainty on stability of underground water-sealed cavern. *J. Shandong Univ. Eng. Sci.* **2024**, *54*, 126–135.
14. Li, S.B.; Zhang, Y.G.; Cao, M.Y.; Wang, Z.N. Study on excavation sequence of pilot tunnels for a rectangular tunnel using numerical simulation and field monitoring method. *Rock. Mech. Rock. Eng.* **2022**, *55*, 3507–3523. [CrossRef]
15. Luo, Y.B.; Chen, J.X.; Chen, Y.; Diao, P.S.; Qiao, X. Longitudinal deformation profile of a tunnel in weak rock mass by using the back analysis method. *Tunn. Undergr. Sp. Tech.* **2018**, *71*, 478–493. [CrossRef]
16. Qin, S.L.; Zhao, X.D.; Song, J.Y.; Wang, C. Mechanical mechanism and theoretical analysis of anchor net support based on model test and numerical simulation. *Geomat. Nat. Haz Risk* **2024**, *15*, 2350484. [CrossRef]
17. Li, H.B.; Pan, W.P.; Hua, X.Z.; Luan, B.; Huang, Z.J. Instability characteristics of surrounding rock and surrounding rock control technology of deep coal roadway crossing the fault: A case study of Zhuxianzhuang coal mine. *Geomat. Nat. Haz Risk* **2024**, *15*, 2366376. [CrossRef]
18. Zhu, C.; He, M.C.; Karakus, M.; Zhang, X.H.; Tao, Z.G. Numerical simulations of the failure process of anaclinal slope physical model and control mechanism of negative Poisson's ratio cable. *Bull. Eng. Geol. Environ.* **2021**, *80*, 3365–3380. [CrossRef]
19. Zhang, J.L.; Mei, M.; Wang, J.; Shang, G.P.; Hu, X.F.; Yan, J.; Fang, Q.; Plebankiewicz, E. The construction and application of a deep learning-based primary support deformation prediction model for large cross-section tunnels. *Appl. Sci.* **2024**, *14*, 912. [CrossRef]
20. Liu, L.; Li, T.B.; Ma, C.C. Research on 3D geological modeling method based on deep neural networks for drilling data. *Appl. Sci-Basel.* **2024**, *14*, 423. [CrossRef]
21. Mei, H.; Wang, Q.Y.; Yu, L.; Zeng, Q. A deep learning-based algorithm for intelligent prediction of adverse geologic bodies in tunnels. *Meas. Sci. Technol.* **2024**, *35*, 096119. [CrossRef]
22. Zhang, J.W.; Han, S.; Li, M.C.; Li, H.; Zhao, W.C.; Wang, J.; Liang, H. CasMDN: A deep learning-based multivariate distribution modelling approach and its application in geotechnical engineering. *Comput. Geotech.* **2024**, *168*, 106164. [CrossRef]
23. Zhou, Z.Q.; Bai, S.S.; Chu, K.W.; Li, J.L.; Sun, J.W.; Wang, M.X.; Sun, Y.; Li, M.H.; Liu, Y.H. Calibration of DEM macro and micro parameters via XGBoost method. *Granul. Matter.* **2022**, *24*, 106. [CrossRef]
24. Yang, W.; Zhang, Q.; Li, S.; Wang, G.; Li, Y. Application of particle swarm optimization in time-dependent parameters inversion. *J. Cent. South Univ. Sci. Technol.* **2013**, *44*, 282–288.
25. Ruan, Y.; Gao, C.; Liu, K.; Jia, R.; Ding, H. Inversion of rock and soil mechanics parameters based on particle swarm optimization wavelet support vector machine. *Rock Soil. Mech.* **2019**, *40*, 3662–3669.
26. Ling, T.; Qin, J.; Song, Q.; Hua, F. Intelligent displacement back-analysis based on improved particle swarm optimization and neural network and its application. *J. Railw. Sci. Eng.* **2020**, *17*, 2181–2190.
27. Chen, L.; Meng, J.D.; Li, Z.Z.; Xu, X.J.; Hao, L.; Ren, Y.X.; Zhao, Y. An optimized observation system and inversion method for fault detection based on surface-wave while tunneling. *J. Appl. Geophys.* **2024**, *228*, 105472. [CrossRef]
28. Wang, F.; Li, X.; Miao, L.; Xu, P. Mechanical parameters identification of surrounding rock based on wavelet SVM. *J. Hydroelectr. Eng.* **2010**, *29*, 184–190.
29. Kovacevic, M.S.; Bacic, M.; Gavin, K.; Stipanovic, I. Assessment of long-term deformation of a tunnel in soft rock by utilizing particle swarm optimized neural network. *Tunn. Undergr. Sp. Tech.* **2021**, *110*, 103838. [CrossRef]
30. Ma, J.J.; Li, T.B.; Yang, G.; Dai, K.K.; Ma, C.C.; Tang, H.; Wang, G.W.; Wang, J.F.; Xiao, B.; Meng, L.B. A real-time intelligent classification model using machine learning for tunnel surrounding rock and its application. *Georisk* **2023**, *17*, 148–168. [CrossRef]
31. Wei, W.; Jia, Y.; Sheng, Y.; Xu, G.; Yang, Y.; Zhang, D. Research on landslide susceptibility evaluation model based on I, SVM and I-SVM. *Saf. Environ. Eng.* **2023**, *30*, 136–144.
32. Zhang, Y.; Su, G.S.; Li, Y.; Wei, M.D.; Liu, B.C. Displacement back-analysis of rock mass parameters for underground caverns using a novel intelligent optimization method. *Int. J. Geomech.* **2020**, *20*, 04020035. [CrossRef]
33. Yan, H.H.; Liu, K.Y.; Xu, C.; Zheng, W.B. A novel method for identifying geomechanical parameters of rock masses based on a PSO and improved GPR hybrid algorithm. *Sci. Rep.* **2022**, *12*, 5670. [CrossRef]
34. Shen, Y.; Guo, B.; Gu, T. Particle swarm optimization algorithm and comparison with genetic algorithm. *J. Univ. Electron. Sci. Technol. China* **2005**, *34*, 696–699.
35. Kennedy, J.; Eberhart, R.C. Particle swarm optimization. In Proceedings of the IEEE International Conference on Neural Networks (ICNN 95), Perth, Australia, 27 November–1 December 1995.
36. Tang, J.; Liu, G.; Pan, Q.T. A review on representative swarm intelligence algorithms for solving optimization problems: Applications and trends. *IEEE-CAA. J. Automatic.* **2021**, *8*, 1627–1643. [CrossRef]
37. Jiang, A.N.; Wang, S.Y.; Tang, S.L. Feedback analysis of tunnel construction using a hybrid arithmetic based on Support Vector Machine and Particle Swarm Optimisation. *Autom. Constr.* **2011**, *20*, 482–489. [CrossRef]
38. Cortes, C.; Vapnik, V. Support-vector networks. *Mach. Learn.* **1995**, *20*, 273–297. [CrossRef]

39. Zhou, S.T.; Zhang, Z.X.; Luo, X.D.; Huang, Y.F.; Yu, Z.; Yang, X.W. Predicting dynamic compressive strength of frozen-thawed rocks by characteristic impedance and data-driven methods. *J. Rock Mech. Geotech.* **2024**, *16*, 2591–2606. [CrossRef]
40. Huang, Y.F.; Lei, Y.; Luo, X.D.; Fu, C. Prediction of compressive strength of rice husk ash concrete: A comparison of different metaheuristic algorithms for optimizing support vector regression. *Case Stud. Constr. Mat.* **2023**, *18*, e02201. [CrossRef]
41. Huang, Y.F.; Zhou, Z.K.; Li, M.Y.; Luo, X.D. Prediction of ground vibration induced by rock blasting based on optimized support vector regression models. *Cmes-Comp. Model. Eng.* **2024**, *139*, 3147–3165. [CrossRef]
42. Ran, Z.; Wu, J.X. Common energy decay indices of Saint-Venant's principle. *J. Eng. Mech.* **2008**, *25*, 14.

designs

MDPI

Article

Innovative Road Maintenance: Leveraging Smart Technologies for Local Infrastructure

Laura Fabiana Jáuregui Gallegos, Rubén Gamarra Tuco and Alain Jorge Espinoza Vigil *

School of Civil Engineering, Universidad Católica de Santa Maria, San Jose Urbanization, Yanahuara District, Arequipa 04013, Peru; 76651677@ucsm.edu.pe (L.F.J.G.); rgamarrat@ucsm.edu.pe (R.G.T.)
* Correspondence: aespinozav@ucsm.edu.pe

Abstract: Roads are essential for economic development, facilitating the circulation of services and resources. This research seeks to provide local governments with a comprehensive framework to enhance road maintenance, focusing on the surface and functional evaluation of pavements. It compares the conventional methods International Roughness Index (IRI) and the Pavement Condition Index (PCI) with novel methodologies that employ smart technologies. The efficiency of such technologies in the maintenance of local roads in Peru is analyzed, taking as a case study a 2 km section of the AR-780 highway in the city of Arequipa. The International Roughness Index (IRI) obtained through the Merlin Roughness Meter and the Roadroid application were compared, finding a minimum variation of 4.0% in the left lane and 8.7% in the right lane. Roadroid turned out to be 60 times faster than the conventional method, with a cost difference of 220.11 soles/km (USD $57.92/km). Both methods classified the Present Serviceability Index (PSI) as good, validating the accuracy of Roadroid. In addition, the Pavement Condition Index (PCI) was evaluated with conventional methods and a DJI Mavic 2 Pro drone, finding a variation of 6.9%. The cost difference between the methodologies was 1047.73 soles/km (USD $275.72/km), and the use of the drone proved to be 10 times faster than visual inspection. This study contributes to closing the knowledge gap regarding the use of smart technologies for better pavement management on local roads, so the actors in charge of such infrastructure make decisions based on science, contributing to the well-being of the population.

Keywords: road management; maintenance road management; PCI; IRI; local roads; smart road maintenance; smart technologies

1. Introduction

Road infrastructure is a fundamental pillar for the development and efficient functioning of global economies, facilitating the flow of goods, services, and resources between different regions. In this context, the proper maintenance of transportation routes, especially pavements, is crucial [1]. This process not only optimizes the comfort and safety of users but also preserves the functionality of the pavement, ensuring essential features such as skid resistance and a suitable appearance, while also reducing associated operational costs [2]. Consequently, it is imperative to design, build, and maintain road infrastructures appropriately, guaranteeing their long-term durability and safety [3].

In the specific case of Peru, despite global technological advancements, the country lacks an efficient monitoring system that allows for the accurate assessment of its infrastructure [4], including roadways. Some methodologies have been proposed to evaluate the condition of bridges [5–7], mainly in response to hydrological phenomena such as El Niño, which significantly affects infrastructure [8]. However, there is little literature on pavement conservation plans to properly prioritize maintenance and potential pavement redesign, which should occur periodically to monitor its condition and attempt to predict its future state [9]. This deficiency has significant repercussions, particularly in the economic realm, as it affects the durability of roads and increases associated operational costs [10].

Pavement evaluation provides essential information for both its design and management. Through this evaluation, it is possible to determine the necessary interventions to ensure proper functionality, whether through maintenance, rehabilitation, or reconstruction [11]. This process also allows for a more accurate estimation of the costs associated with the required interventions, thus optimizing planning and resource allocation [12].

In this research, two types of evaluations were conducted: functional and surface evaluations. The functional evaluation focuses on analyzing the International Roughness Index (IRI), which measures the smoothness of the pavement surface [2]. Knowing this index is crucial for defining conservation or rehabilitation actions, as surface irregularities not only affect vehicle performance but also increase pavement deformation, leading to higher rehabilitation and maintenance costs [13]. On the other hand, the surface evaluation is based on analyzing the Pavement Condition Index (PCI). Quantifying this index allows for efficient management of the necessary interventions for pavement treatment and maintenance, thereby ensuring its optimal conservation [14]. Both indices are essential for road evaluation, and their effectiveness is enhanced by using modern and low-cost techniques [15]. These tools enable efficient monitoring of pavement conditions, which optimizes municipal governments' use of resources in infrastructure management. This approach is crucial for ensuring sustainable and effective road network maintenance [1].

The use of innovative maintenance methods for important infrastructure, such as roads, faces several challenges. Particularly in developing countries, breaking engineering paradigms to adopt non-conventional methods is important if resilient infrastructure is to be implemented. Therefore, this research aims to address this challenge and bridge this gap through a Peruvian case study that evaluates the impact of conventional and smart methods, analyzing the savings in terms of cost and time, and comparing the precision of the results. As such, a framework for the implementation of smart technologies in the management of local road maintenance is proposed.

It is worth noting that the responsibility for these roads falls on local governments or municipalities [16]. From this context, the research seeks to answer the question: What does the use of smart technologies for local road maintenance management involve? To address this question, five specific objectives have been defined:

- Determine the roughness value using the traditional Merlín Roughness Meter method (method commonly used in Peru because of its low cost, but it takes more time to collect the data) and compare it with Roadroid.
- Determine the pavement quality using the Present Serviceability Index (PSI) with the Merlín Roughness Meter and Roadroid.
- Determine the PCI through a conventional failure inspection and compare it with an inspection using drones.
- Establish the intervention range for local roads based on the PCI obtained through visual failure inspections and drone flights.
- Evaluate the efficiency of using smart technologies in the management of local road maintenance.

The research is based on an exhaustive literature review, which is presented in Section 2, detailing the state of the art. Section 3 presents the main research methods used, including the case study, and also contains the flow charts corresponding to the four methodologies employed to meet the specific objectives. Section 4 presents the results obtained during the research, while Section 5 is dedicated to discussing those results. Section 6 shows the conclusions drawn from the study. Finally, Section 7 presents the limitations and future lines of research.

2. Background Information

To propose a framework for the application of smart technologies in local road maintenance management, it is necessary to identify the performance of modern technologies applied to road infrastructures, which requires a literature review to find the most efficient methods. Table 1 outlines key related studies in chronological order, the methods employed, and the relevance of each study to the case study.

Table 1. Research background.

Author	Year	Research Methods	Relevance	Case Study
[9]	2017	Visual Inspection	Calculate PCI and provide maintenance proposals.	Ramón Castilla Avenue, Chulucanas, Piura.
[12]	2018	Roadroid	IRI value obtained, and SDI value determined.	Jenderal Sudirman Kalianget Road–Indonesia
[17]	2018	Roadroid and Visual Inspection for PCI	Correlation of pavement PCI with IRI.	Magetan District Road–Indonesia.
[18]	2019	Roadroid and Visual Inspection for PCI	Determine PCI and serviceability level.	Lucre-Huacarpay Avenue–Peru.
[1]	2019	Drone and Visual Inspection	Compare PCI between both methodologies.	Separadora Industrial Avenue, blocks 8–10, Ate, Lima–Peru.
[19]	2021	Abakal and Roadroid	Calculate IRI and PSI.	Antunez de Mayolo Avenue–Peru.
[14]	2021	Roadroid and ROOGA equipment	Compare IRI using the application and ROOGA equipment with leveling.	Covadonga Road, Ayacucho.
[15]	2021	DJI Phantom 4 Pro V2.0 Drone	Determine PCI.	Los Conquistadores Avenue, Lima.
[11]	2021	Laser profilometer and Roadroid	Compare IRI and PSI.	Concession No. 3: Mocupe, Lambayeque.
[13]	2021	Systematic Literature Review	Evaluate the efficiency of PCI and IRI for assessing serviceability through various studies.	Literature review of different indexed articles.
[20]	2022	Roadroid and Merlin Roughness Meter Method	Compare IRI and PSI with their respective correlation coefficients.	Prolongation Iquitos Avenue, Javier Prado Avenue, and Petit Thouars Avenue, Lima.
[21]	2023	Systematic Literature Review	Evaluate current trends in the implementation of smart technologies for road construction and maintenance.	Literature review of different indexed articles.

Regarding the table above, it is noteworthy that several authors have used the Roadroid application in their research, making comparisons with various methodologies, such as the PCI method through visual inspection, the use of the Abakal application for android cellphones, the ROOGA equipment, and the laser profilometer. Each of these approaches aims to calculate the surface regularity of roads. The main differences between these studies lie in the type of roads analyzed (highways, roads, local roads, and urban streets), while a notable similarity is the high effectiveness reported for the Roadroid application. This finding represents a positive precedent for the present research, which seeks to identify smart technologies for the efficient management of road infrastructures.

Additionally, the table presents the evaluation of the Pavement Condition Index (PCI) using various methodologies, such as visual inspection and the use of advanced technology, specifically the DJI Phantom 4 drone. According to [15], the most notable difference between both techniques is that smart technology, represented by the drone, significantly reduces evaluation time while showing accuracy comparable to that of the conventional method. This finding is highly relevant to the present research, as it establishes a positive precedent in the search for efficient alternatives for the management of local roads.

3. Methodology

The research methodology includes methodological schemes, detailed below through four figures, which address the first four specific objectives, focusing on the case study of the AR-780 local road in Arequipa. The general data for the AR-780 road in Polobaya, Arequipa, are as follows: pavement type: flexible, length: 2 km, road width: 7.6 m. The price change from soles to dollars was considered with a value of 3.80 soles per dollar.

Figure 1, corresponding to the first flowchart, summarizes the methodology for determining the IRI (International Roughness Index) using the Merlin Roughness Meter. The procedure begins with calibrating the equipment by placing it on a horizontal surface and adjusting its support points until the pointer indicates 25. If it doesn't, the lower bolts are adjusted until this is achieved. Then, two readings are taken: an initial one (Li) and another with a calibration block under the movable foot (Lf) to determine the correction factor. Three people are required: an operator, an assistant to record the readings and a spotter. The operator moves the equipment until the wheel completes one rotation (1.98 m), places it on its three support points, and records the pointer's position in a field form, repeating the process until 200 consecutive readings are completed. The recording form has 20 rows and 10 columns, filled from top to bottom and left to right.

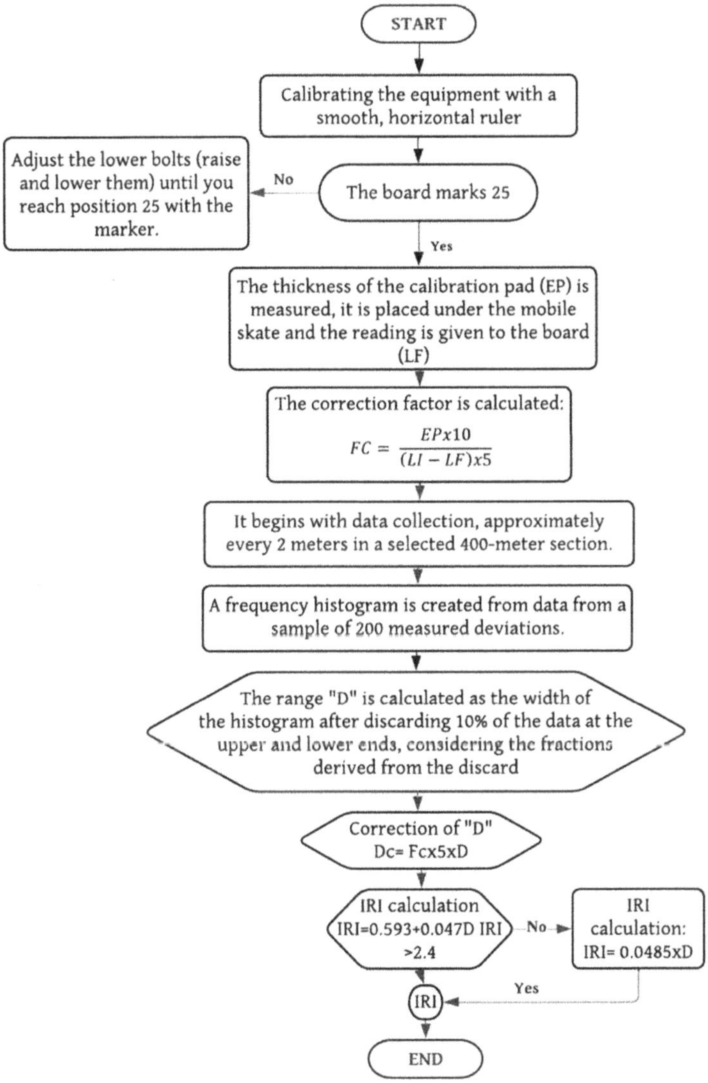

Figure 1. Flowchart of the Rugosimeter Merlin Equipment method for determining the IRI.

Figure 2, corresponding to the second flowchart, summarizes the methodology for determining the IRI (International Roughness Index) using the Roadroid application. The procedure begins by creating an account on the Roadroid website, where personal data and the phone's IMEI are registered. The license can be free for educational purposes or purchased on the site. After registration, the phone is calibrated to ensure no data is lost. A mount for the phone must be installed on the vehicle's windshield, ensuring a horizontal position without obstructions. To perform the IRI (International Roughness Index) diagnosis, the GPS and Wi-Fi or mobile data are activated. The app settings, such as vehicle type, are adjusted, and the phone is calibrated. Once calibrated, the diagnostic process begins by pressing the camera button, and the vehicle should move at 50 km/h. The data is uploaded to the Roadroid system using Wi-Fi. These can be viewed on a virtual map after 15–20 min, and reports can be generated and downloaded in formats like KML or TXT for analysis in programs like ArcGIS or Excel [22].

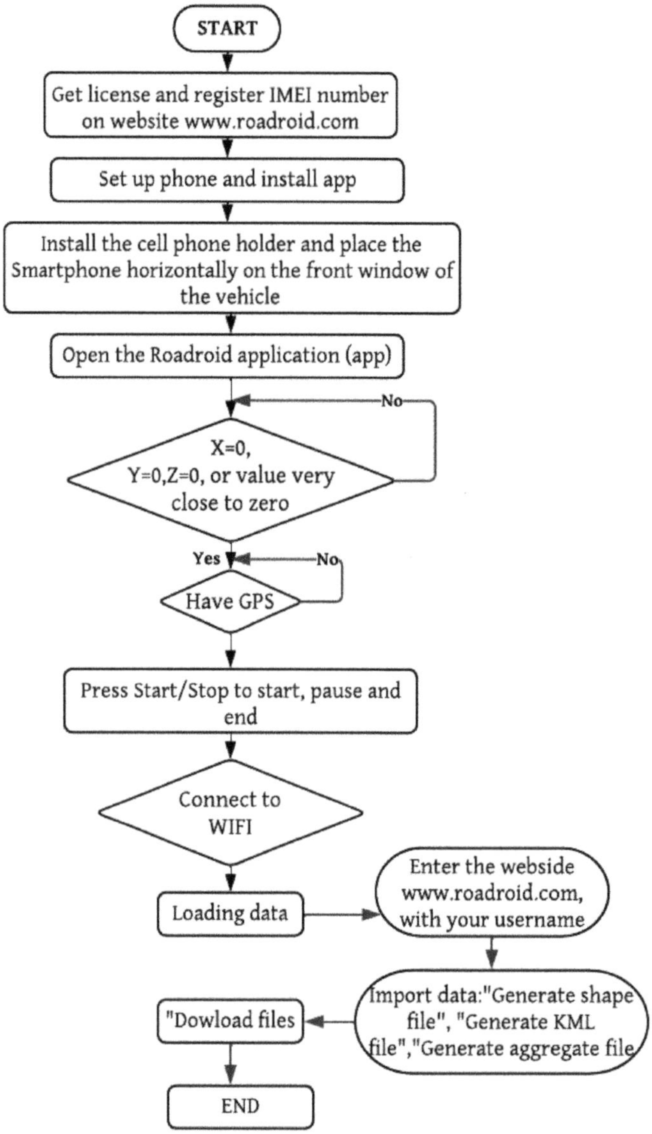

Figure 2. Flowchart of the Roadroid method for determining the IRI.

Figure 3, corresponding to the third flow chart, summarizes the procedure for measuring the PCI (Pavement Condition Index) through visual inspection. To calculate it, the failure data obtained during the on-site inspection according to the 19 specified failures is required. These values are then subjected to the following steps: the Deduct Value (DV) is calculated by recording the sum of each type of failure according to its severity and quantity. The density of the failures is then used to determine the deducted value for each type of damage. Next, the maximum allowable deducted values (m) are calculated using an equation that considers the maximum deducted value for each sampling unit. The total deducted value is adjusted using a correction curve, thus obtaining the maximum Corrected Deduct Value (CDV). Finally, the PCI is calculated by subtracting the maximum CDV from 100, which indicates the pavement's condition. The PCI is classified according to various failures such as deformations, cracks, and erosion in the pavement [23].

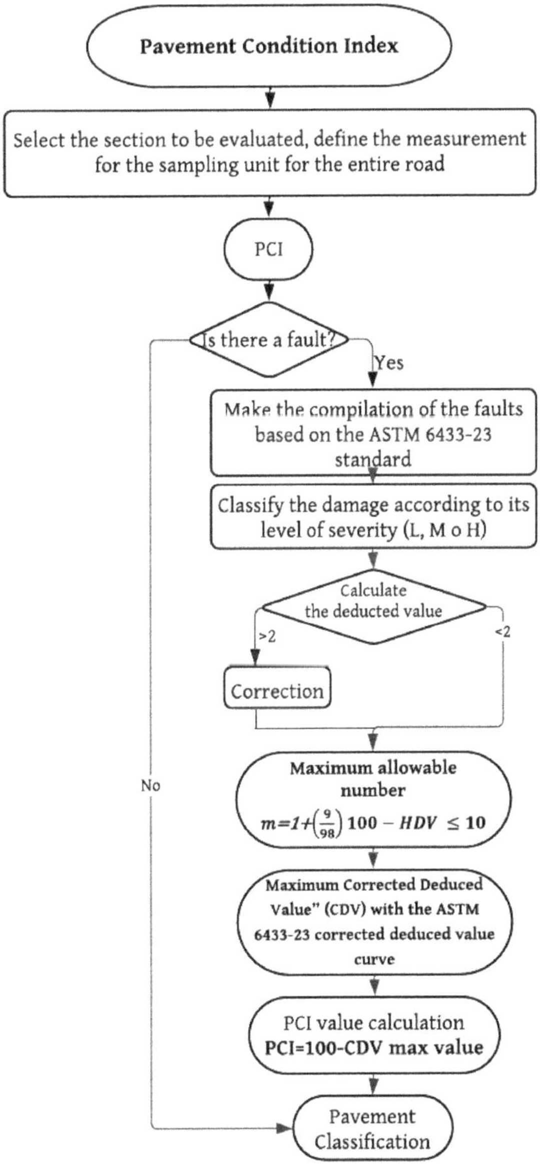

Figure 3. Procedure for measuring PCI (Pavement Condition Index) by visual inspection [23].

Figure 4, corresponding to the fourth flow chart, summarizes the procedure for measuring the PCI (Pavement Condition Index) using the DJI Mavic 2 Pro drone, programmed via a mobile phone with the DJI PILOT app. The flight was conducted at 11 a.m. at an altitude of 50 m to capture high-resolution images, taking photos every 2 s with a 70% overlap. A linear flight plan was generated, and the flight parameters, such as takeoff speed (15 m/s) and flight speed (5.5 m/s), were configured. The drone followed the programmed route, returning to the takeoff point for battery replacement three times. All information was stored in the drone's memory and then processed using Agisoft Metashape 2.1.3 software, where the images were imported, oriented in high quality, and a dense point cloud was created for the final analysis.

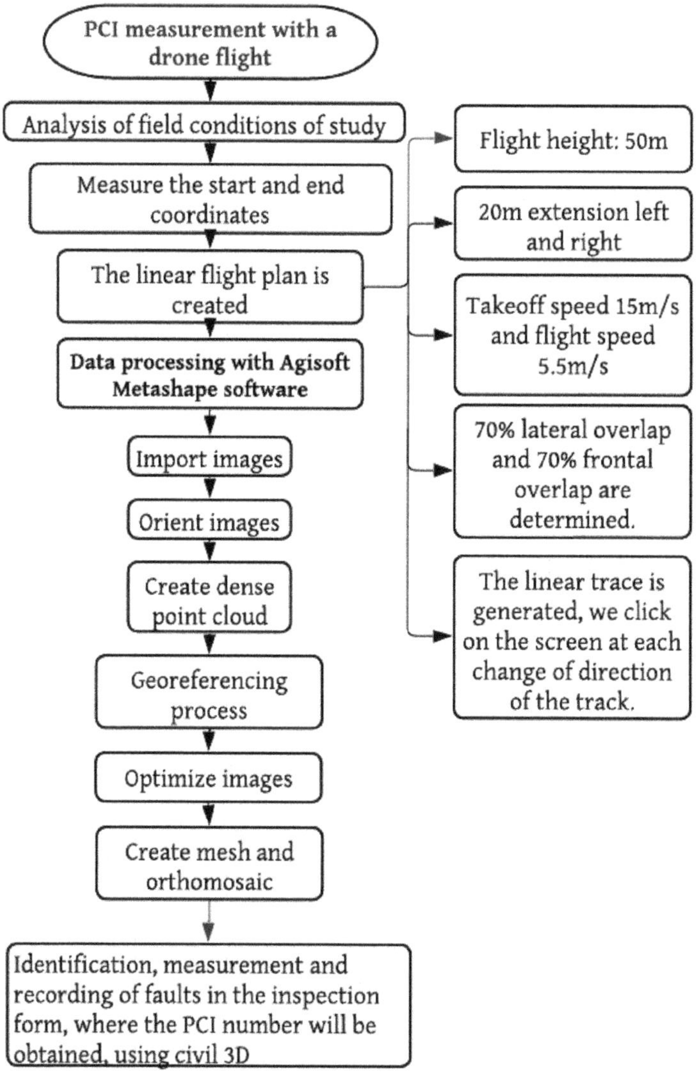

Figure 4. PCI (Pavement Condition Index) evaluation by flying the DJI Mavic 2 pro Drone.

4. Results

4.1. International Roughness Index Evaluation

The Merlin Roughness Meter test was conducted on both lanes of a 2 km road section. Measurements began at kilometer marker 0+000 km and extended to 2+000 km, with records taken every 400 m. On the right lane, readings were taken first on the right track and then on the left track, with a total of 200 readings per section. The same procedure was replicated on the left lane, maintaining the same interval and measurement methodology.

As for the evaluation using the Roadroid application, the analysis sections were also defined every 400 m, with 200 continuous readings per lane. The data obtained was electronically processed, following the established methodology to segment the pavement into 5 evaluation sections for each lane and traffic direction. Since using the Roadroid Pro 3 app (version 3.0.8) requires a license, an investigative request was made, resulting in the acquisition of a temporary free license.

4.1.1. IRI with Merlin Roughness Meter

Figure 5 shows the results for the IRI for the right lane evaluated using the Merlin Roughness Meter Series 696, Model TM171, Manufacturer Tamiequipos, Bogota, Colombia. The highest IRI value, 1.85, occurs in the section from 0+800 to 1+200.

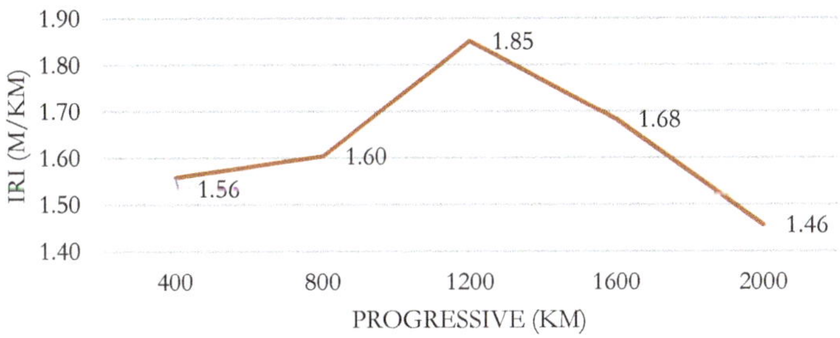

Figure 5. IRI right lane with Merlın Roughness tester.

Figure 6 shows the results for the IRI for the left lane evaluated using the Merlin Roughness Meter. The highest IRI value, 1.88, occurs in the section from 0+000 to 0+400.

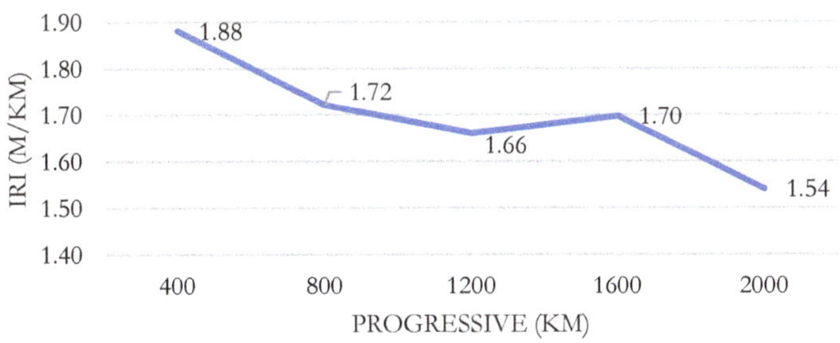

Figure 6. IRI left lane with Merlin roughness tester.

4.1.2. IRI with the Roadroid Aplication

Figure 7 shows the results for the IRI for the right lane evaluated using the Roadroid app. The highest IRI value, 1.72, occurs in the section from 0+800 to 1+200.

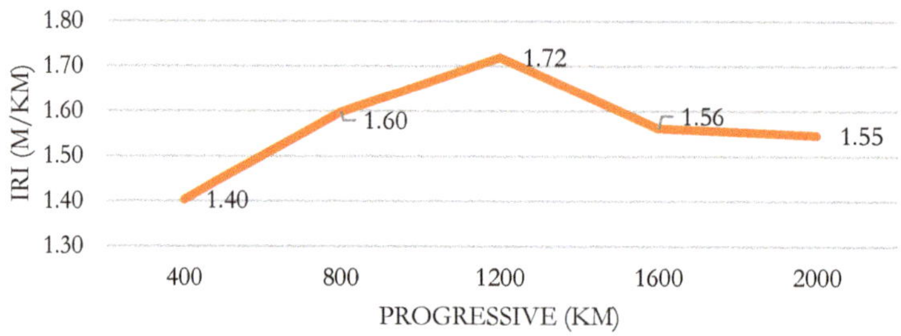

Figure 7. IRI right lane using Roadroid.

Figure 8 shows the results for the IRI for the left lane evaluated using the Roadroid app. The highest IRI value, 1.60, occurs in the section from 0+800 to 1+200.

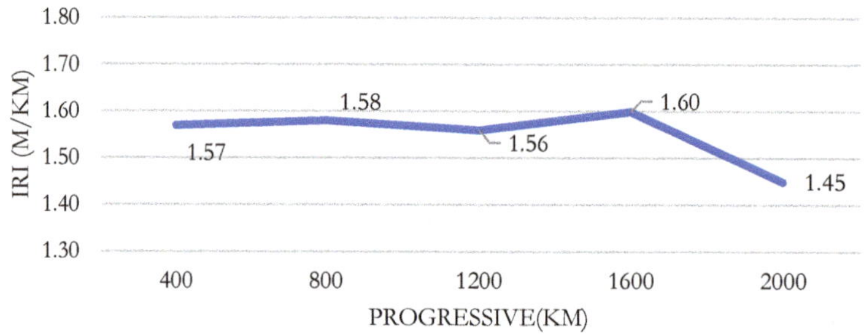

Figure 8. IRI left lane using Roadroid.

4.1.3. Comparison of IRI Using Merlin Roughness Meter and Roadroid

Table 2 shows that the average roughness value measured using the Merlin Roughness Meter for the right lane was 1.63 m/km, while for Roadroid it was 1.57 m/km, with an average variation of 4.007%. It can be concluded that there is a high correlation between the IRI with Merlin and the IRI with Roadroid found in the right lane. Table 3 shows that the average roughness value measured using the Merlin Roughness Meter for the left lane was 1.7 m/km, while for Roadroid it was 1.56 m/km, with an average variation of 8.7%. It can be concluded that there is a high correlation between the IRI with Merlin and the IRI with Roadroid found in the left lane.

Table 2. Comparative table between the IRI results with both teams—Right Lane.

Progression		MERLIN	ROADROID	
Start	**End**	**IRI**	**IRI**	**Variation**
(km)	**(km)**	**(m/km)**	**(m/km)**	**(%)**
0+000	0+400	1.558	1.403	9.949
0+400	0+800	1.604	1.597	0.422
0+800	1+200	1.851	1.716	7.282
1+200	1+600	1.682	1.562	7.174
1+600	2+000	1.456	1.547	6.222
Average		1.630	1.565	4.007

It is concluded that using the Roadroid application yields results like those obtained with the Merlin method, making it a reliable option for monitoring road conditions.

Additionally, Figure 9 shows the scatter plot of IRI vs. eIRI for the right lane data. The best-fit line to the point cloud with the origin point (0,0) is displayed, with the IRI with

Merlin values on the "x" axis and eIRI with Roadroid values on the "y" axis. A coefficient of determination (R^2) of 0.9967 is obtained, indicating a high correlation (Figure 3).

Table 3. Comparative table between the IRI results with both teams—Left Lane.

Progression		MERLIN	ROADROID	
Start	End	IRI	IRI	Variation
(km)	(km)	(m/km)	(m/km)	(%)
0+000	0+400	1.88	1.57	16.53
0+400	0+800	1.72	1.58	8.22
0+800	1+200	1.66	1.56	6.05
1+200	1+600	1.70	1.60	5.72
1+600	2+000	1.54	1.45	5.89
Average		1.700	1.552	8.71

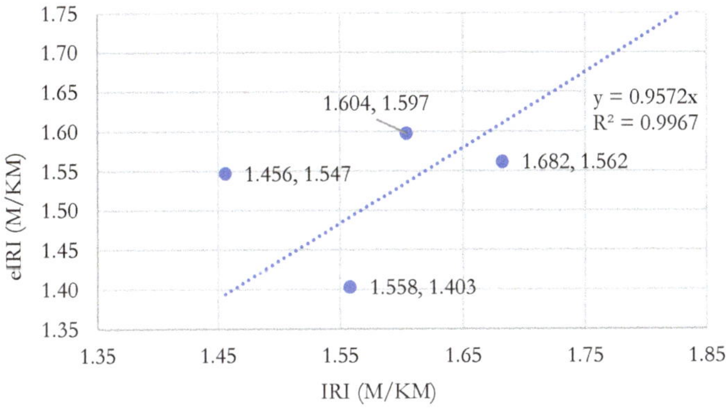

Figure 9. IRI vs. eIRI dispersion table—Right Lane.

Figure 10 shows a linear regression for the left lane data. The best-fit line to the point cloud with the origin point (0,0) is displayed, with the IRI values on the "x" axis and eIRI values on the "y" axis. A coefficient of determination (R^2) of 0.998 is obtained, also indicating a high correlation.

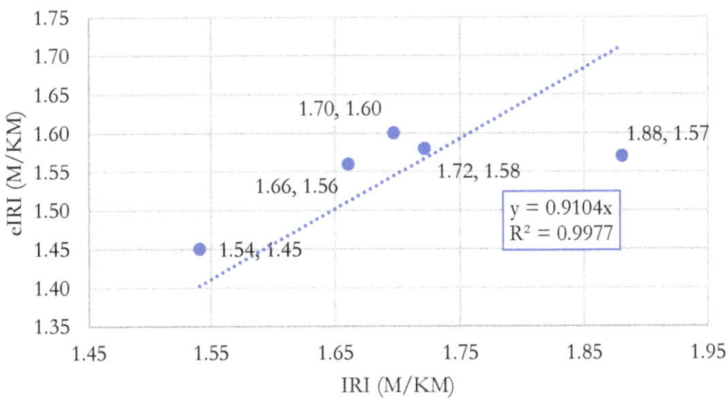

Figure 10. IRI vs. eIRI dispersion table—Left Lane.

4.1.4. Comparison of PSI (Present Serviceability Index) Using the Merlin Roughness Meter and Roadroid

The present serviceability index (PSI) is the capacity of the pavement to handle traffic. The serviceability index ranges from 1 (Terrible) to 5 (Excellent). With the IRI values already determined, the Present Serviceability Index of the pavement (PSI) is calculated, which represents the degree of comfort or safety perceived by the user when traveling on the road, also understood as a safety parameter. The Paterson equation is used to calculate the PSI: $PSI5 * e^{-\frac{IRI}{5.5}}$ [24].

The PSI results using both devices for the right lane are shown in Table 4. For the right lane, employing the Merlin Roughness Meter, the quantitative PSI value was 3.72, categorized as GOOD on the qualitative scale. Using the Roadroid application, the quantitative PSI value was 3.76, also categorized as GOOD. The proximity of these values and their identical qualitative scale (GOOD) demonstrate a strong correlation between the PSI measurements for the right lane, with a maximum difference of 1.1% in their quantitative values.

Table 4. Comparative table between the PSI results with both teams—Right Lane.

Progression		**MERLIN**	**ROADROID**	**Serviceability**			
Start	**End**	**IRI**	**IRI**	**Value**		**Scale**	**Scale**
(km)	**(km)**	**(m/km)**	**(m/km)**	**PSI-MERLIN**	**PSI-ROADROID**		
0+000	0+400	1.56	1.40	3.77	3.87	GOOD	GOOD
0+400	0+800	1.60	1.60	3.74	3.74	GOOD	GOOD
0+800	1+200	1.85	1.72	3.57	3.66	GOOD	GOOD
1+200	1+600	1.68	1.56	3.68	3.76	GOOD	GOOD
1+600	2+000	1.46	1.55	3.84	3.77	GOOD	GOOD
			Average	3.72	3.76	GOOD	GOOD

The PSI results for both devices on the left lane are also shown (Table 5). For the left lane, the Merlin Roughness Meter yielded a quantitative value of 3.67, within the GOOD qualitative scale, while the Roadroid application gave a quantitative value of 3.77, also within the GOOD qualitative scale. These similar values and the identical qualitative scale (GOOD) demonstrate a good correlation between the PSI measurements for the left lane, with a maximum difference of 3% in their quantitative values.

Table 5. Comparative table between the PSI results with both teams—Left Lane.

Progression		**MERLIN**	**ROADROID**	**Serviceability**			
Start	**End**	**IRI**	**IRI**	**Value**		**Scale**	**Scale**
(km)	**(km)**	**(m/km)**	**(m/km)**	**PSI-MERLIN**	**PSI-ROADROID**		
0+000	0+400	1.88	1.57	3.55	3.76	GOOD	GOOD
0+400	0+800	1.72	1.58	3.66	3.75	GOOD	GOOD
0+800	1+200	1.66	1.56	3.70	3.77	GOOD	GOOD
1+200	1+600	1.70	1.60	3.67	3.74	GOOD	GOOD
1+600	2+000	1.54	1.45	3.78	3.84	GOOD	GOOD
			Average	3.67	3.77	GOOD	GOOD

4.1.5. Analysis of Unit Prices for Obtaining IRI with Roadroid and the Merlin Roughness Meter

The unit prices required to obtain the IRI with Roadroid and with Merlin Roughness Meter for the 2 km of study are shown. There are cost differences between both methods, for example, to use the Roadroid method requires a civil engineer (specialist), vehicle, gasoline, cell phone and Roadroid license (Table 6), however, for the Merlin Roughness Meter method it requires a laborer (watchman), a technician, a civil engineer and also the Merlin roughness meter equipment (Table 7).

Table 6. Unit price to obtain IRI with Roadroid.

Item	Specification	Unit	Quantity	Unit Price	Subtotal
Labor	Civil Engineer (specialist)	Hh	2.00	500.00	1000.00
Vehicle	Medium car rental	Day	1.00	356.00	356.00
Gasoline	84 octanes	Glb	0.07	13.86	0.91
Equipment	Android cell phone	Day	1.00	2.47	2.47
	Annual license for 1 user per day of use	Day	1.00	3.19	3.19
				TOTAL	1362.57
				UP (Soles/km)	S/681.29
				UP (Dollar/km)	USD $179.29

Table 7. Unit price to obtain IRI with the Merlin Roughness Tester.

Item	Specification	Unit	Quantity	Unite Price	Subtotal
Field Work	Laborer (watchman)	Hh	8	39.1	312.96
	Technician	Hh	8	54.98	439.84
	Civil Engineer	Hh	2	150	300
Equipment	Merlin Roughness Tester	Day	1	750	750
				TOTAL	1802.80
				UP (Soles/km)	S/901.40
				UP (Dollar/km)	USD $237.21

For evaluating 1 km of the AR-780 road, there is a difference in unit costs. The cost for evaluating roughness with the Roadroid application is 681.29 Peruvian soles, compared to 901.40 Peruvian soles for the Merlin Roughness Meter, resulting in a positive profitability margin of 24.42%.

The Roadroid application's performance is 1.08 km/min, as the vehicle traveled at an approximate speed of 75 km/h. In contrast, the traditional Merlin Roughness Meter has a performance of 1 km/h. This indicates that the data collection process using the Roadroid application is 60 times faster. Therefore, it is concluded that the use of the Roadroid application is not only more economical but also more cost-effective.

Figure 11 shows the Cost-Benefit analysis between these two methods, indicating that the Roadroid application is more efficient in terms of time and cost for measuring IRI over 1 km of road, with a time of 1 min and a cost of 681.29 peruvian soles per km (USD $179.29), compared to 60 min and a cost of 901.4 soles per km (USD $237.21) using the Merlin Roughness Meter.

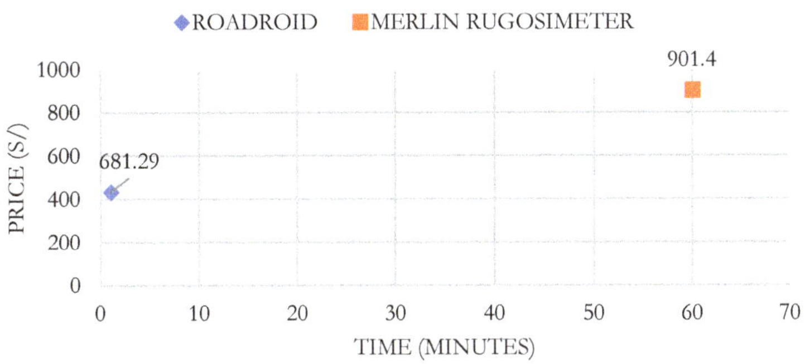

Figure 11. Cost Benefit in time between the Merlin test and Roadroid.

4.2. Pavement Condition Index (PCI) Evaluation

In the study conducted along 2 km of the AR-780 Polobaya road to evaluate the Pavement Condition Index (PCI) through visual inspection, 67 sample units were used. Each unit measured 30 m in length and 7.6 m in width, covering an area of 228 m^2, except for unit UM-67, which covered 152 m^2.

For the PCI evaluation using a drone, a DJI Mavic 2 Pro was employed, programmed with a mobile phone. The flight took place at 11 a.m., following the guidelines of the NTC 001–2015 standard [25]. A height of 50 m was set, allowing high-resolution images to be captured without interference on the road. Photographs were taken every 2 s, with a 70% crosswise and lengthwise overlap, ensuring adequate image overlap to obtain enough homologous points. The drone's software automatically calculated the number of points, the number of photos, and the flight time.

4.2.1. Comparison Between Pavement Condition Index (PCI) Using Visual Inspection and Drone Flight

Figure 12 shows the PCI data for each 30 m sampling unit using both methodologies.

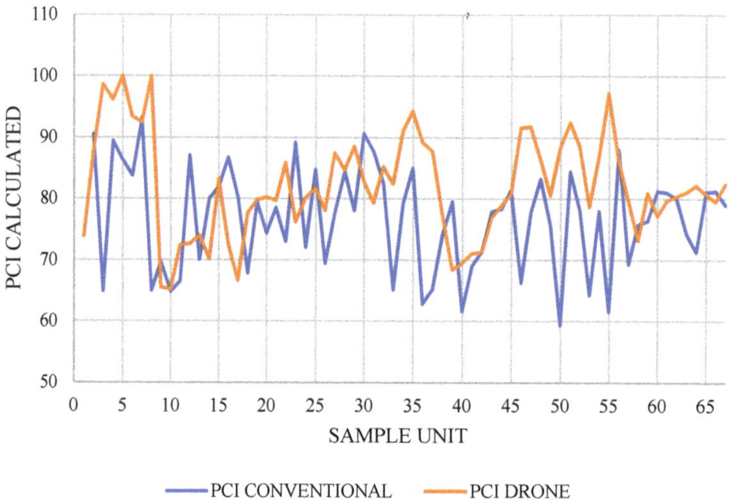

Figure 12. PCI values calculated for both methodologies.

PCI methodology relates to a numerical scale with values from zero (0), for a failed pavement condition, to one hundred (100), for an excellent pavement condition according to ASTM 6433 [23]. Figure 13 shows a bar graph of the average PCI value for both methodologies obtaining a classification of "VERY GOOD" in both cases.

In response to the values presented in the comparison of the PCI between both methodologies—through the conventional method of visual inspection of failures and via drone flight (DJI Mavic 2 Pro)—for the AR-780 road, the average PCI value through the conventional visual inspection method was 76.75, and for the DJI Mavic 2 Pro drone, it was 82.04, with an average variation of 6.89%.

In response to the values presented in the comparison of the intervention range between both methodologies—through the conventional method of visual inspection of failures and via drone flight (DJI Mavic 2 Pro)—for the AR-780 road, the PCI of the AR-780 local road was determined through conventional failure inspection, obtaining 50 sample units that require maintenance and 17 units that require rehabilitation (2, 7, 9, 10, 12, 17, 25, 32, 35, 36, 39, 40, 45, 49, 52, 54, and 56). Meanwhile, using the drone method for PCI, 62 sample units were found to require maintenance and 5 units to require rehabilitation

(Figure 14). It is notable that both methods coincide with 5 sample units requiring rehabilitation intervention: UM-09, 10, 17, 39, and 40; and 50 sample units suggested maintenance intervention, with 12 sample units differing in their intervention range.

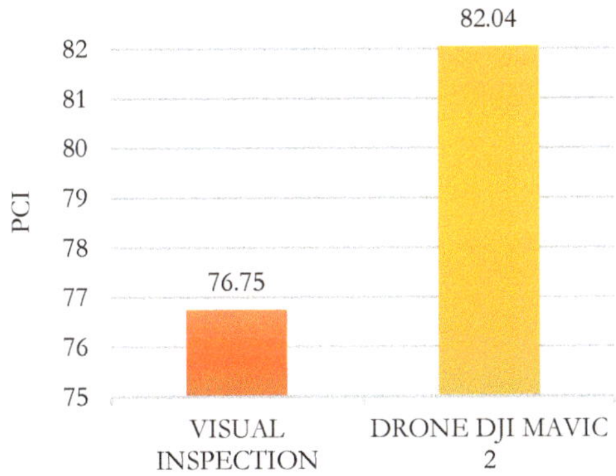

Figure 13. PCI values calculated for both methodologies.

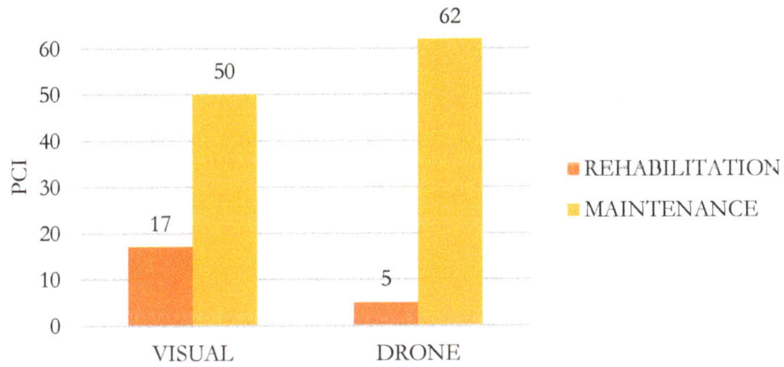

Figure 14. PCI for both types of evaluation and their respective classification.

It can be concluded that PCI measurement using the traditional method provides greater accuracy and detail since it is conducted in the field and yields better results for pavement failures compared to the method using the DJI Mavic 2 Pro drone.

4.2.2. Unit Price Analysis to Obtain PCI Using Drone Flight and Visual Inspection

Table 8 shows the unit cost analysis for the PCI test through conventional visual inspection for 2 km, with a cost of 1422.73 soles (USD $374.40 dollars) per kilometer of road. Table 9 shows the price of the PCI test using a drone for 2 km with a cost per kilometer of 375 soles (USD $98.68 dollars). Comparing both costs, a reduction of 73.6% is achieved using the drone evaluation method. Both tables show the difference in labor, materials and equipment used in each method.

In Figure 15, a cost-benefit analysis between the traditional system and the method using a drone is observed, obtaining lower costs and time using a drone, which is 375 soles ($98.68 dollars) and 3 h, compared to the traditional method, which is 1422.73 soles ($374.40 dollars) and 30 h.

Table 8. PCI measurement using conventional visual inspection for 2 km.

PCI Measurement Service by Conventional Visual Inspection for 2 km					
Item	**Specification**	**Unit**	**Quantity**	**U.P.**	**Subtotal**
Labor	Civil Engineer (specialist)	Hh	30	80.00	2400
	Assistant	Hh	30	10.00	300
Materials	Spray	Und	3	15.5	46.5
	Laminated board 0.60 m × 0.80 m	Und	1	53	53
	Whiteboard markers	Und	3	3.6	10.8
Equipment	28″ Safety cones	Und	1	29.5	29.5
	50 m measuring tape	Day	1	2.47	2.47
	8 m measuring tape	Day	1	3.19	3.19
				TOTAL	2845.46
				UP (Soles/km)	S/1422.73
				UP (Dollar/km)	USD $374.40

Table 9. PCI measurement using the application for 2 km.

Item	**Specification**	**Unit**	**Quantity**	**U.P.**	**Total**
Labor	Civil Engineer (specialist)	Hh	3	116.67	350
	Assistant	Hh	3	16.67	50
Equipment	Mavic 2 Pro Drone and Cell Phone	Glb	1	350	350
				TOTAL	750.00
				UP (Soles/km)	S/375.00
				UP (Dollar/km)	USD $98.68

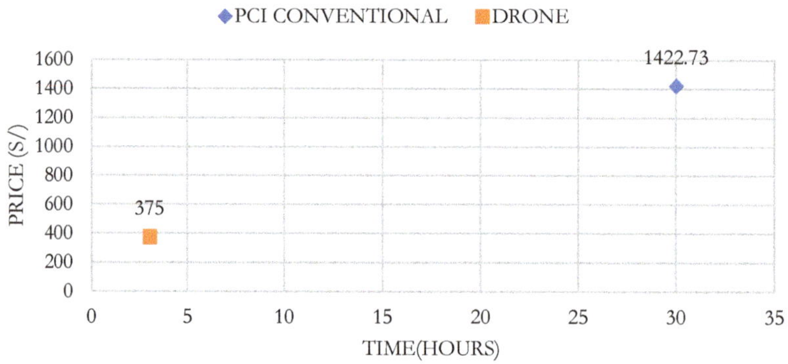

Figure 15. Cost-benefit analysis between the traditional system and the method using drones.

4.3. Proposal for Maintenance Management on Local Roads

In proposing a maintenance management plan for local roads, it is worth mentioning that local governments are responsible for local roads. By applying Smart technologies, they can measure the IRI (International Roughness Index) and PCI (Pavement Condition Index) to get an overview of the roads needing maintenance and thus have information about the current functional condition of their roads and determine what type of maintenance is required.

In the department of Arequipa, there are 149.64 km of paved local roads managed by their respective local governments, according to the update of [26], which involve execution costs and operational expenses for their corresponding routine maintenance.

Therefore, this technical and economic proposal aims to intervene in local roads to facilitate and shorten the evaluation time and reduce costs.

By operating for the 149.64 km of local roads in the city of Arequipa at the unit cost for evaluating IRI obtained by the Roadroid method, the result is 101,947.68 soles. For the Merlin Roughness Meter method, the result is 134,885.50 soles, achieving savings of 32,937.82 soles with the Roadroid method, the equivalent in U.S. dollars is 8667.85.

Also, by operating for the 149.64 km of all local roads in Arequipa at the unit cost to evaluate PCI, the total cost obtained is 202,897.32 soles using the conventional method and 56,115.00 soles using the drone method, generating savings of 146,782.32 soles (38,626.93 U.S. dollars) for the respective entity if this method is used.

5. Discussion

This research has considered it essential to divide the discussion into five key terms that reflect the relevance and scope of this study.

5.1. Roughness Indicator (IRI)

The IRI results obtained in this research present a moderate relationship compared to the findings of Castillo and Morales [27]. In this study, the IRI variation between the Merlin roughness meter and the Roadroid application ranges from 0% to 8.71%, while Castillo and Morales [27] report a difference of up to 27.5%. This discrepancy can be attributed to the conditions under which each study was conducted. Castillo's research was carried out on a busy avenue, subject to factors such as variable speed, the presence of speed bumps, and fluctuations in traffic flow. In contrast, this research was conducted on a local road, free from such influences, which may have favored greater consistency in the IRI results.

5.2. Serviceability Indicator (PSI)

The PSI results in this research show a low correlation compared to those obtained by Mamani A. and Lozano R. [19]. In my study, the average PSI was 3.7 with both the Roadroid application and the Merlin roughness meter, while Mamani reported a PSI of 1.99. This suggests that the flexible pavement of the AR-780 local road is in better condition than that of Antúnez de Mayolo Avenue in Tacna. Additionally, the importance of comparing PSI data with a conventional method to ensure greater accuracy in the results is highlighted. In this regard, the Merlin roughness meter is a recommended method for such comparison.

5.3. Pavement Condition Indicator (PCI)

Compared to the research by Cubas [15], the Pavement Condition Index (PCI) results in this study show a high correlation. In this research, the conventional and drone-based PCI evaluations yielded results of 76.75 and 82.04, respectively, while Cubas [15] reported values of 59 for the conventional evaluation and 61 for the drone evaluation. These results reflect an agreement between both studies in that conventional visual inspection proves to be more accurate in evaluating the pavement condition PCI. This difference is because drones, in the context of this study, which require millimetric precision in the photogrammetry, do not provide the necessary accuracy on the z-axis.

5.4. Intervention Range Indicator

Compared to the research by Cubas [15], the Pavement Condition Index (PCI) results in this study show a high correlation. In this research, the PCI values obtained through conventional and drone methods indicate the need for intervention in the maintenance range. In contrast, the results of Cubas [15] suggest that both the conventional and drone-based evaluations correspond to an intervention range for rehabilitation and structural reinforcement. This suggests that the local road analyzed in this study requires corrective maintenance due to its incipient failures, while the avenue studied by Cubas [15] needs medium-term rehabilitation.

5.5. Maintenance Management Indicator

In this research, the evaluation using drones takes 3 h, while the traditional method requires 30 h to cover the entire study area. This demonstrates that the use of drones is approximately 10 times faster in terms of time. Regarding cost, the evaluation with drones incurs a cost of 375.00 soles per kilometer, compared to 1422.73 soles per kilometer with the traditional method, representing a 73.64% reduction in costs. Therefore, drone inspection is significantly more economical than conventional visual inspection.

The research conducted by Colque [28] reports that the total drone usage time to evaluate two avenues was 4 h, with a total cost of 4650.25 soles.

In conclusion, both this research and that of Colque [28] show a significant reduction in costs and execution time when using drones for pavement evaluation.

6. Conclusions

This research determined the roughness value using the traditional Merlin Roughness Meter method and the Roadroid application. The average roughness value using the Merlin Roughness Meter test for the right lane was 1.63 m/km, and for Roadroid, it was 1.57 m/km, with an average variation of 4.007% and a correlation coefficient R^2 of 0.997. It can be concluded that there is a high relationship between the IRI and eIRI found in the right lane. The average roughness value using the Merlin Roughness Meter test for the left lane was 1.7 m/km, and for Roadroid, it was 1.56 m/km, with an average variation of 8.71% and a correlation coefficient R^2 of 0.998. It can be concluded that there is a high relationship between the IRI and eIRI found in the left lane. Therefore, it is concluded that using the Roadroid application provides results like those obtained with the Merlin method, making it a reliable option for monitoring road conditions.

The Present Serviceability Index (PSI) for the AR-780 road using the Merlin Roughness Meter test for the right lane had an average quantitative value of 3.72, and using the Roadroid application, it was 3.76, both falling within the qualitative scale of GOOD. The PSI value for the left lane using the Merlin Roughness Meter test had an average quantitative value of 3.67, and using the Roadroid application, it was 3.77, both falling within the qualitative scale of GOOD. It is concluded that the Roadroid application provides reliable data because the PSI values obtained are close to those obtained by the Merlin method, with a maximum difference of 3% in their quantitative values for the road under study.

The PCI value was determined using the conventional method of visual inspection of failures and through the flight of the DJI Mavic 2 Pro drone. The average PCI value using the conventional visual inspection method was 76.75, and for the DJI Mavic 2 Pro drone, it was 82.04, with an average variation of 6.89%. The PCI of the AR-780 local road was determined through a conventional failure inspection, obtaining 50 sample units that require maintenance and 17 units that require rehabilitation (2, 7, 9, 10, 12, 17, 25, 32, 35, 36, 39, 40, 45, 49, 52, 54, and 56). Using the drone method for PCI, 62 sample units were found to require maintenance and 5 units to require rehabilitation. It can be observed that in both methods, there is a coincidence in 5 sample units that require rehabilitation intervention: UM-09, 10, 17, 39, and 40; and 50 sample units suggest maintenance intervention, with 12 sample units not matching in their intervention range. Therefore, it can be concluded that the PCI measurement using the traditional method provides greater accuracy and detail, as it is conducted in the field and yields better results for pavement failures compared to the method used by the DJI Mavic 2 Pro drone. The results obtained by the drone can be optimized by complementing them with topographic leveling. The drone did not offer higher precision because it is not the appropriate equipment for z-axis measurement, which requires millimetric precision for this study.

The efficiency of the smart technologies used in this research for local road maintenance management was evaluated. In terms of time and cost when implementing the Roadroid application compared to the Merlin method, it was found that with the application, 1 km/min can be covered, while with the Merlin method, it is 1 km/h, indicating that the application is 60 times faster in data collection than the Merlin method. Regarding cost, with Roadroid, a Unit Price Analysis per kilometer of 681.29 soles (USD $179.29) was obtained, and with the Merlin method, it was 901.40 soles (USD $237.21), making the use of the Roadroid application more economical with a profitability percentage of 24.42%. Through the Cost-Benefit analysis, it is concluded that the Roadroid application is more efficient in terms of data collection time and costs, achieving similar precision in the results. As for PCI measurement on the road, using the drone, it takes 3 h for data collection and analysis time, and with the traditional method, it takes 30 h for the entire study area, indicating that the use of drones is 10 times faster in terms of data collection and analysis time. Regarding cost, with the drone, a Unit Price Analysis per kilometer of 375.00 soles (USD $98.68) was obtained, and with the traditional method, it was 1422.73 soles (USD $374.40), making it more economical with a profitability percentage of 73.64%. Through the Cost-Benefit analysis, it is concluded that the use of drones is more efficient in terms of data collection and analysis time and in costs. However, the use of the traditional method to obtain the PCI of the road is more accurate than that obtained by the drone.

7. Limitations and Future Line of Research

7.1. Reasearch Limitations

Among the limitations of the Roadroid application, it must be considered that it depends on external factors such as: speed of travel, traffic, number of singularities (potholes, breakwaters), distance between singularities, geometry of the road under study and elements outside the road surface such as branches, small stones and others that may be on the roadway. Also, the use of the application requires a constant minimum speed of displacement, which in sharp curves can cause risk of accidents.

As for the limitations in the use of the drone, it does not allow millimeter measurements in the z-axis, however, it shows results close to those of the field inspection. Regarding its use, it is recommended to make a previous evaluation of the area and the schedule in which the equipment is going to be flown, since the presence of flora such as large trees produces shadows on the evaluated road, which does not allow a correct evaluation.

7.2. Future Line of Research

In this research focused on local roads, the road under study is a remote road with little traffic, however, if you want to apply the methodology in another road with different conditions, it should be taken into account that its effectiveness and efficiency may vary if they are busy roads or roads with many potholes and curves and if it has excessive vegetation this may prevent the observation of the road and prevent the use of this type of technology. This research gives rise to future investigations that wish to explore this phenomenon such as: Interviewing the road maintenance area to evaluate the implementation of drones in the road maintenance procedure, use of these technologies in important and high traffic roads in a city and comparison of different applications and equipment such as drones to evaluate their effectiveness and costs in the results.

Author Contributions: Conceptualization, L.F.J.G., R.G.T. and A.J.E.V.; methodology, L.F.J.G., R.G.T. and A.J.E.V.; software, L.F.J.G.; validation, L.F.J.G.; formal analysis, L.F.J.G.; investigation, L.F.J.G.; resources, L.F.J.G.; data curation, L.F.J.G.; writing—original draft preparation, L.F.J.G.; writing—review and editing, L.F.J.G., R.G.T. and A.J.E.V.; visualization, L.F.J.G.; supervision, L.F.J.G., R.G.T. and A.J.E.V.; project administration, L.F.J.G. All authors have read and agreed to the published version of the manuscript.

Funding: This research received no external funding.

Data Availability Statement: The findings of this research are supported by the data available from the corresponding author, L.F.J.G., upon reasonable request.

Acknowledgments: The authors thank Manuel Luque Delgado for his unconditional support and constant motivation, Diego Gómez Quispe for his support during the formatting of the document, Alan Huarca Pulcha for his constant revision and suggestions, and Universidad Católica de Santa

María for providing useful insight and assistance with the research reported, particularly the research group DROP (Disruptive Research Opportunity).

Conflicts of Interest: The authors declare no conflicts of interest.

References

1. Cruz Toribio, J.; Gutiérrez Lázares, J. *Cálculo del Índice de Condición del Pavimento con Imágenes del Vehículo Aéreo no Tripulado*; Universidad de Ingeniería: Lima, Peru, 2018.
2. Scholotjes, M.R.; Visser, A.; Bennet, C. *Evaluation of a Smartphone Roughness Meter*; University of Petroria: Pretoria, South Africa, 2014. Available online: http://hdl.handle.net/2263/45571 (accessed on 14 December 2024).
3. Llapasi. La Importancia del Matenimiento de las Carreteras. Available online: https://llapasi.com/sin-categoria/la-importancia-del-mantenimiento-de-las-carreteras/ (accessed on 13 September 2024).
4. Espinoza Vigil, A.J.; Carhart, N.J. Local infrastructure governance in Peru: A systems thinking appraisal. *Infrastruct. Asset Manag.* **2024**, *11*, 145–160. [CrossRef]
5. Ccanccapa Puma, J.; Hidalgo Valdivia, A.V.; Espinoza Vigil, A.J.; Booker, J. Preserving Heritage Riverine Bridges: A Hydrological Approach to the Case Study of the Grau Bridge in Peru. *Heritage* **2024**, *7*, 3350–3371. [CrossRef]
6. Espinoza Vigil, A.J.; Booker, J. Hydrological Vulnerability Assessment of Riverine Bridges: The Bajo Grau Bridge Case Study. *Water* **2023**, *15*, 846. [CrossRef]
7. Huarca Pulcha, A.; Espinoza Vigil, A.J.; Booker, J. Prioritizing Riverine Bridge Interventions: A Hydrological and Multidimensional Approach. *Designs* **2023**, *7*, 117. [CrossRef]
8. Espinoza Vigil, A.J.; Booker, J.D. Building national disaster resilience: Assessment of ENSO-driven disasters in Peru. *Int. J. Disaster Resil. Built Environ.* **2023**, *14*, 423–433. [CrossRef]
9. Sánchez, J.C. *Evaluación del Estado del Pavimento de la Av. Ramón Castilla, Chulucanas, Mediante el Método PCI*; Universidad de Piura: Piura, Peru, 2017.
10. ECOASFALT. Consecuencias del Asfalto en Mal Estado de Conservación. Available online: https://www.ecoasfalt.es/consecuencias-del-asfalto-en-mal-estado-de-conservacion/ (accessed on 13 September 2024).
11. Velásquez, J.C. *Caracterización y Análisis del Método ROADROID en la Obtención del Índice de Rugosidad Internacional (IRI) Para la Gestión de Pavimentos*; Universidad Nacional de IngenieríaL: Lima, Peru, 2021.
12. Arianto, T.; Suprapto, M.; Syafi'i. Pavement Condition Assessment Using IRI from Roadroid and Surface Distress Index Method on National Road in Sumenep Regency. *IOP Conf. Ser. Mater. Sci. Eng.* **2018**, *303*, 012091. [CrossRef]
13. Oblitas Gastelo, B.E.; Medina Cardozo, I.I.; Paredes Asalde, C.R. Índice de regularidad internacional e índice de condición de pavimento para definir niveles de serviciabilidad de pavimentos. *ITECKNE Innovación Investig. Ing.* **2021**, *18*, 170–175. [CrossRef]
14. Sarabia, W. *Evaluación de Fallas Superficiales del Pavimento Flexible Por el Método PCI y Empleo del Dron, Carretera Covadonga—Mollepata, Ayacucho*; Universidad César Vallejo: Ayacucho, Peru, 2021.
15. Cubas, J. *Uso de un Dron Para Optimizar la Evaluación Superficial del Pavimento Flexible Por el Método PCI en la Av. Los Conquistadores, Distrito de San Isidro, Lima*; Universidad Privada del norte: Lima, Peru, 2021.
16. Ministerio de Transportes y Comunicaciones. Clasificador de Rutas del Sistema Nacional de Carreteras—SINAC (D.S. N° 044-2008-MTC); Gobierno del Perú 2008. Available online: https://www.gob.pe/institucion/mtc/normas-legales/1922194-044-2008-mtc (accessed on 17 September 2024).
17. Putra, D.A.; Suprapto, M. Assessment of the road based on PCI and IRI roadroid measurement. *MATEC Web Conf.* **2018**, *195*, 04006. [CrossRef]
18. Mozo Barrantes, W.V.; Quispe Mendoza, A.J. *Evaluación Superficial de la Condición y Serviciabilidad del Pavimento Flexible Por el Método Pavement Condition Index (PCI) y ROADROID en el Circuito Humedal Lucre-Huacarpay*; Universidad Andina del Cusco: Cusco, Peru, 2019.
19. Mamani Arratia, D.E.; Lozano Requelme, J.L. *Evaluación de la Rugosidad en el Pavimento Utilizando Teléfonos Inteligentes Para Mejorar la Transitabilidad Vehicular en la Av. Antúnez de Mayolo, Distrito Gregorio Albarracín Lanchipa, Tacna*; Universidad Privada de Tacna: Tacna, Peru, 2021.
20. Almenara, C. *Aplicación de Telefonos Inteligentes Para Determinar la Rugosidad de Pavimentos Urbanos en Lima*; Pontificia Universidad Católica del Perú: Lima, Peru, 2015.
21. Gabbar, H.A.; Chahid, A.; Isham, M.U.; Grover, S.; Singh, K.P.; Elgazzar, K.; Mousa, A.; Ouda, H. HAIS: Highways Automated-Inspection System. *Technologies* **2023**, *11*, 51. [CrossRef]
22. Roadriod. Available online: https://www.roadroid.com/map (accessed on 13 September 2024).
23. *ASTM D6433-23*; Standard Practice for Roads and Parking Lots Pavement Condition Index Surveys. ASTM International: West Conshohocken, PA, USA, 2023. [CrossRef]
24. Carey, W.; Irick, P. The Pavement Serviceability-Performance concept. National Academic of Science-National Research Council (U.S.). *Highw. Res. Board* **1960**, *250*, 40–58. Available online: https://onlinepubs.trb.org/Onlinepubs/hrbbulletin/250/250-003.pdf (accessed on 13 December 2024).

25. Ministerio de Transportes y Comunicaciones. *Norma Técnica Complementaria: Requisitos para las Operaciones de Sistemas de Aeronaves Pilotadas a Distancia NTC 001-2015*; Gobierno del Perú: Lima, Perú, 2015. Available online: https://www.gob.pe/institucion/mtc/informes-publicaciones/321488-ntc-001-2015-requisitos-para-las-operaciones-de-sistemas-de-aeronaves-pilotadas-a-distancia (accessed on 20 September 2024).
26. Ministerio de Transportes y Comunicaciones. *Convenio 207-2023-MTC/21: Convenio de Gestion Para la Ejecucion del Mantenimiento Rutinario de Vías Vecinales Entre Provias Descentralizado y la Municipalidad Provincial de Arequipa*; Gobierno del Perú: Lima, Perú, 2023. Available online: https://www.gob.pe/institucion/pvd/normas-legales/5853899-00207-2023-mtc-21 (accessed on 17 September 2024).
27. Castillo, R.; Morales, Y. *Optimización de la Evaluación Funcional del Pavimento Flexible Mediante el Software Roadroid y el Método Merlón de la Av. Prolongación Iquitos-Lince-Lima-2022*; Universidad Privada del Norte: Lima, Peru, 2023.
28. Colque, J. *Aplicación del Método PCI en Pavimentos Fléxibles Empleando Drones*; Universidad Peruana los Andes: Huancayo, Peru, 2023.

designs

MDPI

Article

Prediction of Shear Strength in Anisotropic Structural Planes Considering Size Effects

Wei-Bin Ma [1,2], Wen-Hao Zou [1,2], Jin-Long Zhang [1,2] and Gan Li [3,*]

1 Railway Engineering Research Institute, China Academy of Railway Sciences Corporation Limited, Beijing 100081, China; dwangfei@163.com (W.-B.M.); zouwenhao0308@163.com (W.-H.Z.); zjl525@outlook.com (J.-L.Z.)
2 State Key Laboratory of High-Speed Railway Track System, Beijing 100081, China
3 Institute of Rock Mechanics, School of Civil and Environmental Engineering and Geography Science, Ningbo University, Ningbo 315000, China
* Correspondence: ligan303@126.com; Tel.: +86-159-1095-7882

Abstract: It is essential to elucidate the shear mechanical behavior of structural planes to assess the risk to rock masses and protect them from shear failure. Current research on shear mechanical behavior is focused on isotropic structural planes with the same lithology on both sides. However, anisotropic structural planes, commonly found in nature, may exhibit unique mechanical behavior that differs from isotropic structural planes. Therefore, it is necessary to study the factors affecting the shear strength of the anisotropic structural planes. In this paper, the direct shear numerical tests on anisotropic structural planes were carried out using the three-dimensional distinct element code (3DEC) based on the laboratory test. The numerical test results illustrate that the error between the peak shear strength of the numerical test and the laboratory test is basically within 10%. The shear stress-displacement curves of the numerical and laboratory tests are similar, which verifies the accuracy of the numerical test. According to the Barton standard sections, anisotropic structural plane models with different roughness and size were established, and the direct shear numerical tests with different normal stresses were carried out. To predict the peak shear strength of the anisotropic structural planes, one hundred and eighty-one sets of direct shear numerical test data were selected. Normal stress, roughness, compressive strength of soft and hard rock masses, basic friction angle of soft and hard rock masses, and structural plane size were used as input parameters to establish a back propagation (BP) neural network model. The research results show that, under identical conditions, the shear strength of the anisotropic structural planes decreases as the structural plane size increases. On the contrary, the shear strength increases with the increasing structural plane roughness and normal stress. For the BP neural network prediction model, the root mean square error (RMSE) and coefficient of determination (R^2) of the training set are 0.441 and 0.957. For the test set, the RMSE is 0.489, and R^2 is 0.947, which indicates that the predicted values are in good agreement with the actual values.

Keywords: anisotropic structural plane; shear strength; discrete element method; size effect; neural network

1. Introduction

Structural planes are specialized surfaces formed within rock masses during geological evolution. These features typically exhibit low shear strength, making them the weakest zones within the rock mass. As a result, the structural plane plays a critical role

https://doi.org/10.3390/designs9010017

in influencing the stability of rock slopes and tunnels [1–4]. Scientists have indicated that the thickness, joint roughness coefficient (JRC), and basic friction angle of the structural plane have a great influence on the shear mechanical behaviors contributing to the instability of rock slope [5–8]. There were numerous documented cases of engineering failures attributed to insufficient structural plane strength. In 2019, a significant landslide occurred in Shuicheng County, China. Investigations revealed that weak structural planes were one of the main causes of rainfall-induced landslides [9]. Additionally, in 2021, a large landslide took place in Tiejiangwan, where the failure was also linked to the presence of weak structural planes [10]. In 2022, the Luding earthquake in China triggered landslides in accumulated deposits. The primary cause was the presence of numerous discontinuities in the hard rock regions, which were prone to sliding along discontinuities and fracturing under seismic loading, leading to rock mass instability [11].

Numerous scholars have conducted in-depth studies on the factors influencing the shear strength of rock mass structural planes, with structural plane roughness being recognized as a critical determinant. Natural structural planes are typically structurally complex and have irregular surface undulations. Their roughness exhibits significant size effect as well as heterogeneity and anisotropy, posing substantial challenges for the accurate description and quantitative analysis of their mechanical behavior [12,13]. Over the past 50 years, scholars have proposed various methods to quantify the roughness of rock mass structural planes, including statistical parameters, fractal theory, and 3D scanning techniques. Barton [14], through experimental studies, derived an empirical formula for the peak shear strength of rock mass structural planes, and the joint roughness coefficient (JRC) was introduced. Due to the simplicity of JRC, it has been widely adopted for evaluating the mechanical properties of structural planes [15]. Barton et al. [16] proposed a nonlinear criterion for the shear strength of rock mass structural planes and subsequently refined its shear strength component. This improvement incorporated new parameters, such as JCS (Joint Compressive Strength), providing significant enhancements to account for the scale effect within the criterion. In addition, the visual comparison method was proposed for estimating the JRC values of structural planes. By characterizing 136 natural structural planes from seven different rock types, they ultimately selected 10 representative structural planes as typical examples. In 1978, the visual comparison method was adopted as the standard approach for evaluating the roughness of structural planes by the International Society for Rock Mechanics. However, this method heavily relies on the subjective judgment of the evaluator, leading to significant errors. To reduce these errors and improve the accuracy of JRC assessments, scholars have proposed various modifications to the visual comparison method. Considering that undulation amplitude and frequency are two critical parameters influencing roughness, Ünlüsoy et al. [17] proposed evaluating JRC values by measuring the similarity between the surface profile of samples and the Barton standard sections in the spatial frequency domain, using a third-order polynomial function and power spectral density. To address the challenge of estimating the roughness of longer structural planes using Barton standard sections, Barton and Bandis [18] introduced the straightedge method for determining JRC values. Du et al. [19] rigorously reformulated the Barton straightedge method, proposing a simplified formula more suitable for field measurements. Considering the fractal dimension and magnitude distribution of the asperities, Stigsson et al. [20] developed a multilinear model developed to estimate JRC objectively.

In recent years, scholars have carried out numerous laboratory experiments on the shear strength of rock mass structural planes. Failure characteristics of structural planes were studied under various loading conditions and structural plane parameters. Wang et al. [21] studied the relationship between the shear mechanical properties of structural planes and acoustic emission parameters under different roughness and anchorage conditions, revealing four stages of the ideal cumulative influence curve during the shear of the structural plane. It is found that the models with 0–60 degree inclination direction had poor stability. Feng et al. [22] conducted shear tests with varying roughness and normal stress, revealing how mechanical properties change due to these two factors. The relationship equations with normal stress and roughness were established for various mechanical parameters. Meng et al. [23] conducted shear tests on completely occlusive granite structural planes formed by tensile splitting under various normal stresses and monitored the acoustic emission during the shear process. The stress drop caused by the uneven surface of the granite rigid structure plane was reproduced, and a method to predict stress drop and fault sliding rock burst was established based on the acoustic emission b value. Du et al. [24] developed a shear strength size effect direct shear testing system for the structural planes, analyzing the influence of size effect on the ratio of the tangential load peak value to the normal load. According to the Coulomb-Navier criterion, Mashhadiali et al. [25] proposed a theoretical model for predicting the shear strength of anisotropic rocks, in which failure depends on the inclination of inherent structural weak planes. The shear sliding tests under unloading normal stress regarding various initial normal stresses and numbers of shearing cycles were conducted by Yin et al. [26], illustrating how JRC, normal stress, and numbers of shearing cycles affected the shear behavior of structural planes.

However, as the primary method for obtaining the shear strength of rock mass structural planes, laboratory direct shear tests still have limitations. These tests struggle to fully replicate the original stress conditions of rock samples and cannot accurately reproduce the complex geometries of natural rock mass structural planes. Furthermore, large-scale laboratory direct shear tests are time-consuming and labor-intensive. While full-scale in situ tests can avoid sampling challenges and maintain the original state of specimens, they involve complex and hazardous testing environments, difficult equipment operations, and significantly higher costs. In recent years, the rapid advancement of information and computer technologies has increased reliance on simulation software for mechanical numerical tests [27–30]. Additionally, machine learning algorithms are being increasingly applied to analyze experimental data [31–35]. These methods, which have gained widespread recognition in the academic community, help to overcome the time and cost constraints associated with traditional laboratory experiments. Based on the physical and numerical modeling, Menhendra et. al found a new geological phenomenon that there is an important relationship between the direction of the joint and the lateral strain ratio [36]. The failure modes of shallow-buried long-span tunnels were analyzed by Chen et al. [37] using the universal distinct element code. Their study revealed that weak structural planes are the primary cause of shear failure in the overlying strata, which ultimately leads to the occurrence of staggered traction slip collapses. Using particle flow code, Yuan et al. [38] established the correlation between the failure mode of the structural plane and the intensity of energy release, finding that the AE b-value can be used to predict the shear failure of the rock structural plane. Numerical tests of anisotropic deformation around a tunnel were conducted by Wang and Huang [39]. The results show that when a rock mass contains two sets of structural planes with unfavorable orientations, the regions experiencing structural plane sliding failure can exacerbate each other, leading to significant anisotropic deformation. To avoid stress concentration when using PFC2D to simulate the shear behavior of the rock mass structural plane, Guo et al. [40] proposed the method of particle group contact discrimination on both sides of the structural plane.

In natural rock masses, anisotropic structural planes with significantly different lithology on both sides are widely distributed, in addition to isotropic structural planes. However, the current research on the shear strength of rock mass structural planes mainly focuses on isotropic structural planes, while the research on anisotropic structural planes is relatively small. Especially when considering the size effect, the systematic research on the shear strength of anisotropic structural planes is still insufficient, which limits the understanding and effective prediction of the shear behavior of structural planes in complex rock mass engineering. Hence, based on 3DEC and the backpropagation (BP) neural network model, the size effect of the shear strength of rock structural planes was studied. Taking the roughness, size, and normal stress as the variables, direct shear numerical tests of anisotropic structural planes were carried out. The shear characteristics of anisotropic structural planes were studied. The peak shear strength was predicted and analyzed. The research results can provide a scientific reference for predicting the mechanical behavior of structural planes in complex rock mass engineering.

2. The 3DEC Models for Shear Testing of Anisotropic Structural Planes

2.1. Overview of 3DEC

The 3DEC is based on the discrete element method to describe the mechanical behavior of discrete media. It inherits the basic core idea of UDEC and is essentially the result of extending the description of the two-dimensional discrete media mechanics to three-dimensional space. The 3DEC is widely recognized for its ability to model the mechanical interactions of discrete blocks. Considering not only the deformability of individual blocks but also the complex mechanical behavior of joints, the structural response under various loading conditions can be studied in detail by the 3DEC. Therefore, 3DEC is used to analyze the role of joints, faults, and fractures in the overall response of rock mass.

2.2. Model Establishment and Grid Division

A flat structural plane and two Barton standard sections, whose JRC were 0, 2.8, and 10.8, were selected as the anisotropic structural plane models. According to JRC, from small to large, these structural planes were numbered R1, R2, and R3. Their morphologies are shown in Figure 1.

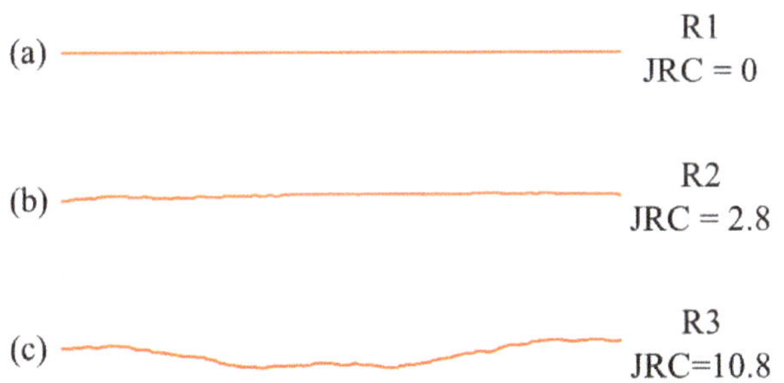

Figure 1. Morphologies of structural plane: (**a**) R1; (**b**) R2; (**c**) R3.

The structural plane section line was drawn in the CAD drawing software; the model was exported in the STL format and imported into the Rhino software. A 10 cm × 10 cm × 10 cm block of the same size as the direct shear laboratory test was created in Rhino. The boolean operation separation command divided the cubic block into two identical rectangular blocks using the structural plane section line. The structural plane models were meshed with an element size of 5 mm, and the unstructured tetrahedral elements were created in the Griddle. The established Rhino models were imported into the 3DEC software in the 3dgrid format for modeling. The established models are shown in Figure 2.

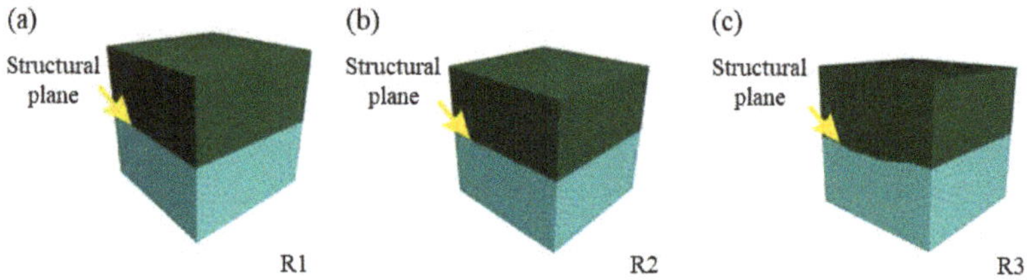

Figure 2. Numerical model of anisotropic structural planes with different roughnesses: (**a**) R1; (**b**) R2; (**c**) R3.

2.3. Material Assignment and Boundary Condition Setting

Sandstone, granite, and cement mortar were used as direct shear laboratory test materials. The cement mortar samples were prepared by mixing cement, sand, and water. The materials mentioned above were assumed to be uniform, with no apparent discontinuities or defects. The basic mechanical parameters of the three materials were obtained through laboratory tests, as shown in Table 1.

Table 1. Basic mechanical parameters of sandstone, granite, and cemented mortar rock samples.

Materials	Density (g/cm^3)	Elasticity Modulus (GPa)	Bulk Modulus (GPa)	Shear Modulus (GPa)	Cohesion (MPa)	Basic Friction Angle (°)	Tensile Strength (MPa)	Compressive Strength (MPa)
Sandstone	2.42	18.85	10.47	7.86	3.07	29.60	2.13	95.55
Granite	2.81	57.92	38.61	23.17	10.33	35.00	6.47	261.55
Cement mortar	2.11	22.33	12.41	9.30	5.03	31.50	1.81	32.61

In the numerical test, the Mohr-Coulomb model was used for the block, and the Coulomb slip joint model was used for the structural plane. To verify the accuracy of the numerical test, the normal stress and shear rate were consistent with those in the laboratory test. During the numerical test, four normal stress levels were applied, namely, 0.5 MPa, 1 MPa, 2 MPa, and 3 MPa. The normal stress was maintained constant throughout the single shearing process, while a tangential load was applied to control the shear rate at 0.5 mm/min. The boundary conditions of the models are shown in Figure 3.

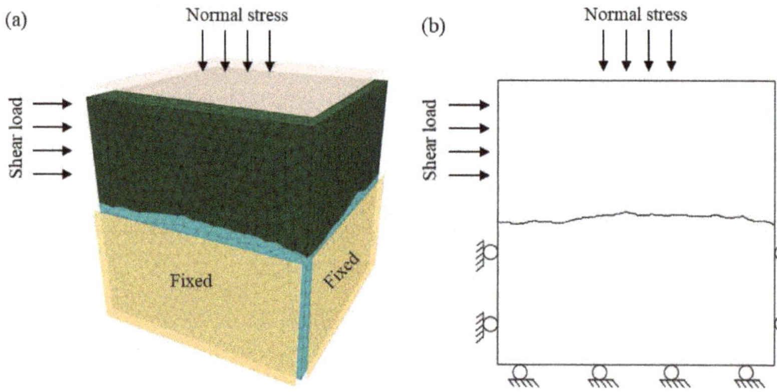

Figure 3. Boundary conditions of the established model: (**a**) Three-dimensional boundary conditions; (**b**) Two-dimensional boundary conditions.

2.4. Analysis and Accuracy Verification of Numerical Results of the Direct Shear Test on the Anisotropic Structural Plane

The shear strength data of the anisotropic structural planes used for accuracy verification are taken from the direct shear laboratory test of rock–mortar anisotropic structural planes conducted by Zhou [41]. The results of the direct shear numerical tests on the structural plane are further derived and compared with the shear strength obtained from the laboratory test. The shear displacement–stress curves of the structural planes with different roughness conditions are given in Figures 4 and 5.

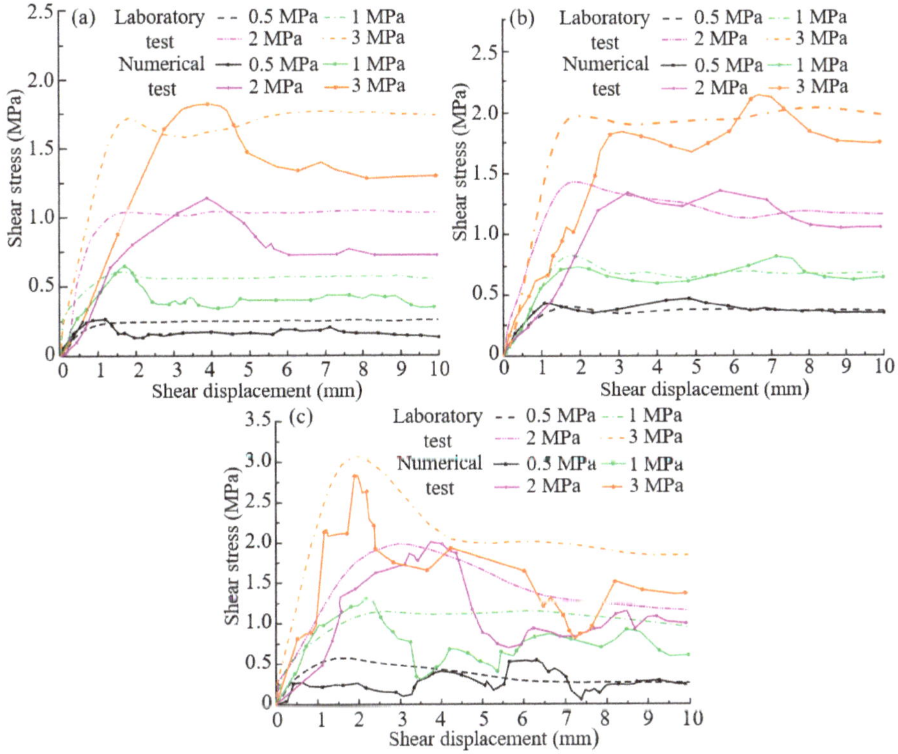

Figure 4. The relationship between shear stress and shear displacement of sandstone-mortar structural planes: (**a**) R1; (**b**) R2; (**c**) R3.

By assessing the shear displacement–stress curve of the anisotropic structural plane, it is found that the shear displacement–stress curve of the anisotropic structural plane can be roughly divided into an ascending stage, a descending stage, and a stable stage. The numerical results show roughly the same trend as the shear displacement–stress curve obtained from the laboratory test. Under the same normal stress conditions, the fluctuation amplitude of the structural plane morphology has a significant effect on the shear behavior. When the fluctuation amplitude of the structural plane morphology is low, the roughness is relatively small and does not contribute significantly to shear resistance. As a result, the fluctuation of the shear displacement–stress curve of the structural plane is small, entering a relatively stable softening deformation stage. The shear strength of the structural plane is close to its residual strength. On the contrary, when the fluctuation amplitude of the structural plane is high, the pronounced roughness creates significant interlocking effects during shearing. These roughness features resist displacement, causing abrupt stress redistributions and localized failures as they deform or break. This mechanism results in multiple stress peaks on the shear displacement–stress curve, indicating intermittent asperity failure and re-engagement, making it difficult to identify the residual strength value of the structural plane. Table 2 shows the comparison results of the peak shear strength of the anisotropic structural plane with different roughnesses and normal stress conditions. The error, which is basically stable within 10%, between the numerical results and the test results is small.

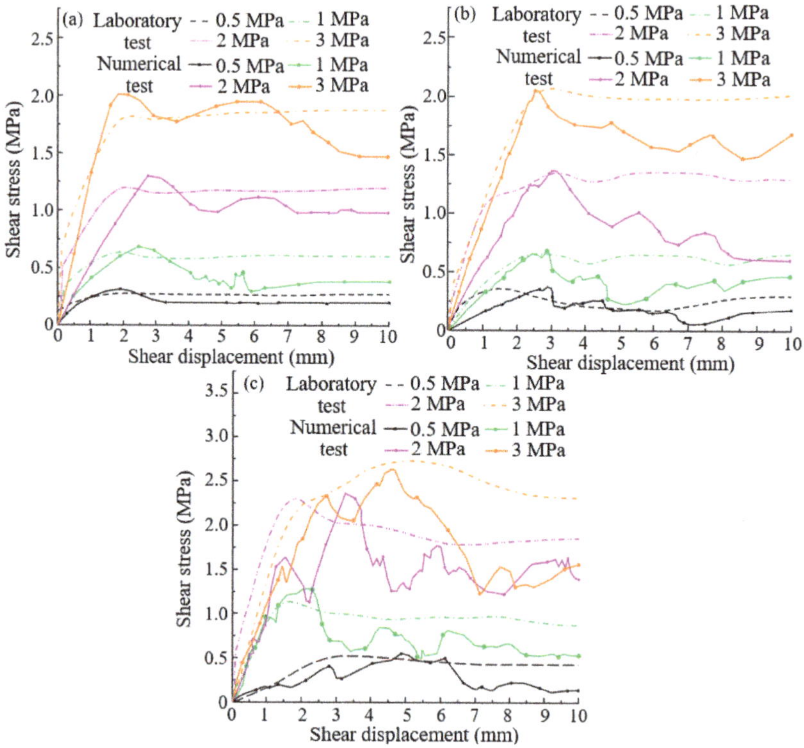

Figure 5. The relationship between shear stress and shear displacement of granite-mortar structural planes: (**a**) R1; (**b**) R2; (**c**) R3.

Table 2. Comparison between laboratory test and numerical test for peak shear strength of anisotropic structural planes.

Anisotropic Structural Plane Types	JRC	Normal Stress (MPa)	Peak Shear Strength (MPa)		Error (%)
			Laboratory Test	Numerical Test	
Sandstone-Cement	R1(JRC = 0)	0.5	0.26	0.26	2.48
		1.0	0.61	0.65	6.01
		2.0	1.05	1.14	8.40
		3.0	1.77	1.89	6.23
	R2(JRC = 2.8)	0.5	0.41	0.45	10.30
		1.0	0.82	0.82	0.19
		2.0	1.43	1.38	3.60
		3.0	2.04	2.20	7.60
	R3(JRC = 10.8)	0.5	0.58	0.55	5.50
		1.0	1.15	1.30	12.80
		2.0	1.99	2.01	1.30
		3.0	3.06	2.84	7.30

Table 2. *Cont.*

Anisotropic Structural Plane Types	JRC	Normal Stress (MPa)	Peak Shear Strength (MPa)		Error (%)
			Laboratory Test	Numerical Test	
Granite-Cement	R1(JRC = 0)	0.5	0.29	0.31	9.10
		1.0	0.64	0.68	6.50
		2.0	1.20	1.30	8.10
		3.0	1.88	2.07	10.50
	R2(JRC = 2.8)	0.5	0.36	0.37	3.70
		1.0	0.66	0.68	2.90
		2.0	1.35	1.35	0.17
		3.0	2.07	2.06	0.54
	R3(JRC = 10.8)	0.5	0.52	0.55	5.90
		1.0	1.13	1.27	12.30
		2.0	2.29	2.37	3.40
		3.0	2.73	2.62	4.10

3. Numerical Shear Testing of Anisotropic Structural Planes with Size Effect

3.1. Design of Direct Shear Numerical Test

The 3rd, 5th, 8th, and 10th sections of Barton standard sections, which were numbered B1 to B4 from small to large according to JRC, were selected. The Rhino modeling software was used to establish the numerical models of the structural planes based on B1 to B4. The rock masses in the upper and lower plates were given strength parameters of different sizes to establish the anisotropic structural plane. The anisotropic structural plane models are shown in Figure 6.

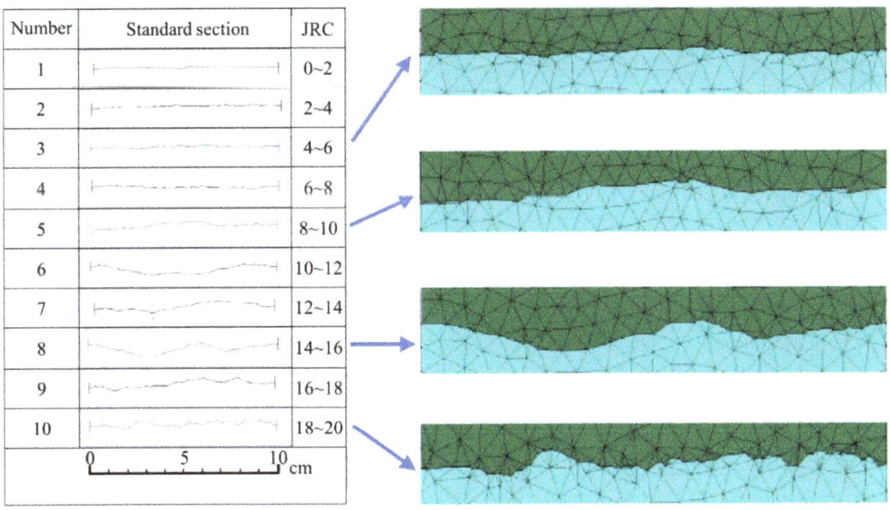

Figure 6. B1–B4 structural plane topography.

To investigate the effects of normal stress, roughness, and size on the peak shear strength of anisotropic structural planes, numerical models of various structural plane roughness were developed with the following dimensions: 10 cm × 10 cm; 30 cm × 30 cm; 50 cm × 50 cm; 70 cm × 70 cm; and 100 cm × 100 cm. Normal stresses of 0.5 MPa, 1 MPa, 2 MPa, 3 MPa, and 5 MPa were applied to the structural planes of each model incrementally. The typical shear displacement–stress curve of the anisotropic structural plane obtained from the experiment is shown in Figure 7. The shear process of the structural plane could be divided into three stages: elastic; failure; and residual. In the elastic stage, the shear stress is proportional to the shear displacement. In the failure stage, brittle failure occurs inside the rock mass, whose shear strength reaches a peak value, with the shear stress value fluctuating greatly. During the residual stage, the shear stress drops sharply to the residual strength, tending to stabilize and remain constant.

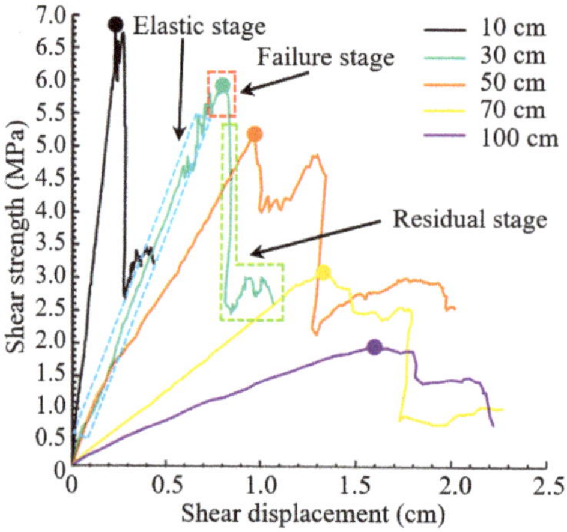

Figure 7. The relationship between shear strength and shear displacement of B1 structural plane with normal stress of 5 MPa.

3.2. Size Effect on Peak Shear Strength of Anisotropic Structural Planes

The size effect of the structural plane has an important influence on the strength and deformation capacity of the rock mass. Figure 8 demonstrates the relationship between the peak shear strength and the size of the anisotropic structural plane. The peak shear strength of all four structural planes (B1 to B4) decreases with increasing size, demonstrating a clear size effect on peak shear strength. Comparing the peak shear strength of the rock mass structural planes from B1 to B4, it can be seen that, for the same normal stress, the peak shear strength is negatively correlated with the size of the structural plane. Under the normal stress of 0.5 MPa, the peak shear strength of the B1 structural plane decreases by 0.022 MPa per 1 cm increase in size. When the normal stress is increased to 5 MPa, the peak shear strength decreases by 0.045 MPa for each 1 cm size increment. Under the normal stress of 0.5 MPa, the peak shear strength of the B2 structural plane decreases by 0.025 MPa per 1 cm increase in size. When the normal stress is increased to 5 MPa, the peak shear strength decreases by 0.053 MPa for each 1 cm size increment. Under the normal stress of 0.5 MPa, the peak shear strength of the B3 structural plane decreases by 0.027 MPa per 1 cm increase in size. When the normal stress is increased to 5 MPa, the peak shear strength decreases by 0.053 MPa for each 1 cm size increment. Under the normal stress of 0.5 MPa, the peak shear strength of the B4 structural plane decreases by 0.031 MPa per 1 cm increase in size. When the normal stress is increased to 5 MPa, the peak shear strength decreases by 0.066 MPa for each 1 cm size increment. The peak shear strength of structural planes decreases slowly with increasing size under low normal stress, while the downward trend increases under high normal stress.

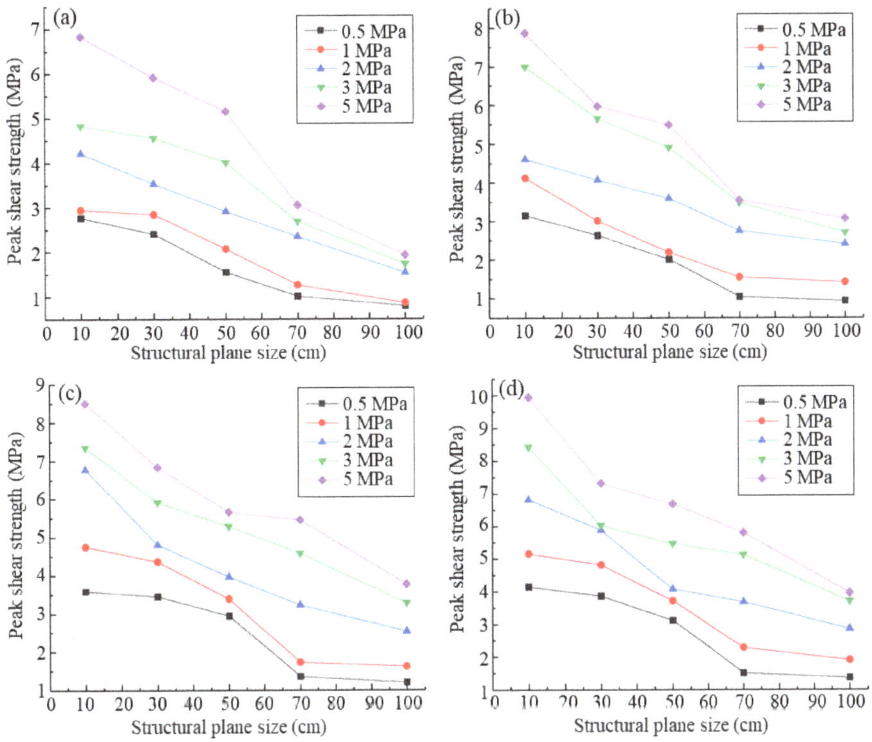

Figure 8. The relationship between peak shear strength and size of the anisotropic structural planes with different roughnesses: (**a**) B1; (**b**) B2; (**c**) B3; (**d**) B4.

3.3. Effect of Normal Stress on Peak Shear Strength of Anisotropic Structural Planes

In the direct shear test, the normal stress also has an important influence on the shear strength of the rock mass structural plane. Figure 9 shows the relationship between the peak shear strength and normal stress of the anisotropic structural plane. The peak shear strength of an anisotropic structural plane increases with higher normal stress for the same size. For the B1 structural plane, when its size is 10 cm, the peak shear strength increases by 0.904 MPa for every 1 MPa increase in normal stress. When the structural plane size reaches 100 cm, the peak shear strength increases by 0.253 MPa per 1 MPa rise in normal stress. For the B2 structural plane, when its size is 10 cm, the peak shear strength increases by 1.049 MPa for every 1 MPa increase in normal stress. When the structural plane size reaches 100 cm, the peak shear strength increases by 0.473 MPa per 1 MPa rise in normal stress. For the B3 structural plane, when its size is 10 cm, the peak shear strength increases by 1.093 MPa for every 1 MPa increase in normal stress. When the structural plane size reaches 100 cm, the peak shear strength increases by 0.571 MPa per 1 MPa rise in normal stress. For the B4 structural plane, when its size is 10 cm, the peak shear strength increases by 1.287 MPa for every 1 MPa increase in normal stress. When the structural plane size reaches 100 cm, the peak shear strength increases by 0.580 MPa per 1 MPa rise in normal stress. At low normal stress, the peak shear strength of structural planes of different sizes tends to be similar. However, as normal stress rises, the differences in peak shear strength become more pronounced, indicating that normal stress significantly influences the peak shear strength of the structural plane.

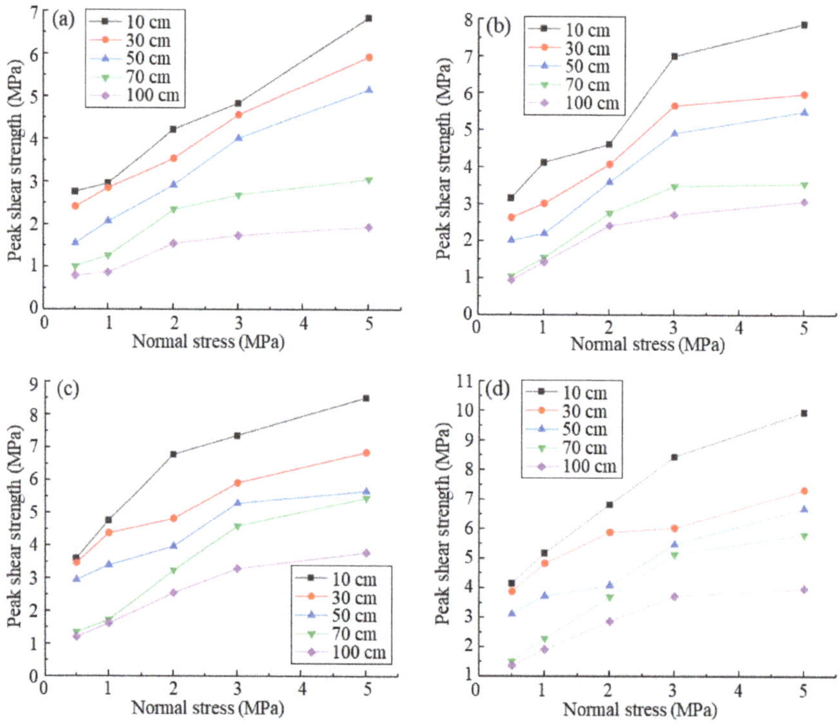

Figure 9. The relationship between peak shear strength and normal stress of the anisotropic structural planes with different roughnesses: (**a**) B1; (**b**) B2; (**c**) B3; (**d**) B4.

3.4. Effect of Roughness on Peak Shear Strength of Anisotropic Structural Planes

The roughness of the structural plane of the rock mass is an important component for its shear strength, especially in the presence of filled joints, where the roughness of the structural plane can lead to an increase in expansibility, local stress, and permeability [42,43]. Table 3 shows the peak shear strength calculated by numerical tests of anisotropic structural planes with different sizes. Under the same normal stress, the peak shear strength of the structural plane increases with the increasing roughness. The tooth-cutting effect is enhanced because the increasing roughness reduces the climbing and expansion effect of the structural plane. In addition, the peak shear strength of the structural plane decreases with increasing size because smaller blocks have greater degrees of freedom and can perceive larger fluctuations in the structural plane. Hence, the roughness decreases when the size increases, which reduces the peak shear strength.

Table 3. Peak shear strength values of anisotropic structural planes with different roughnesses.

Size (cm)	Roughness	Normal Stress (MPa)				
		0.5	1	2	3	5
10	B1	2.76	2.95	4.21	4.83	6.83
	B2	3.15	4.12	4.61	6.99	7.87
	B3	3.59	4.76	6.77	7.35	8.51
	B4	4.15	5.16	6.81	8.43	9.94
30	B1	2.41	2.85	3.53	4.55	5.92
	B2	2.63	3.01	4.07	5.65	5.97
	B3	3.46	4.37	4.81	5.90	6.84
	B4	3.87	4.82	5.86	6.02	7.31

Table 3. *Cont.*

Size (cm)	Roughness	Normal Stress (MPa)				
		0.5	1	2	3	5
50	B1	1.55	2.07	2.92	4.01	5.15
	B2	2.01	2.19	3.59	4.90	5.49
	B3	2.95	3.40	3.97	5.28	5.66
	B4	3.12	3.72	4.08	5.47	6.67
70	B1	1.01	1.26	2.34	2.68	3.05
	B2	1.04	1.55	2.75	3.48	3.54
	B3	1.35	1.73	3.23	4.58	5.45
	B4	1.50	2.28	3.69	5.13	5.79
100	B1	0.79	0.86	1.54	1.73	1.93
	B2	0.94	1.42	2.41	2.71	3.07
	B3	1.20	1.62	2.54	3.28	3.77
	B4	1.36	1.90	2.87	3.72	3.97

4. BP Neural Network Model for Predicting Peak Shear Strength of Anisotropic Structural Planes

4.1. Overview of BP Neural Network

A backpropagation neural network is a multi-layer feedforward neural network, including an input layer, a hidden layer, and an output layer. Figure 10 shows a typical three-layer BP neural network structure diagram. The training samples are input into the hidden layer, and after the training function is calculated, the actual output is sent to the output layer. After one propagation, the error between the actual output and the target output is calculated to determine whether it is less than the target error. If the actual error is greater than the target error, the neural network starts to propagate information backward. The actual output is propagated toward the input layer, and neurons in each layer adjust and correct weights and thresholds based on the information received. The neural network will continue to iterate until the calculation result meets the target error. Each connection weight and threshold in the initial training is assigned a random value in the interval (0, 1).

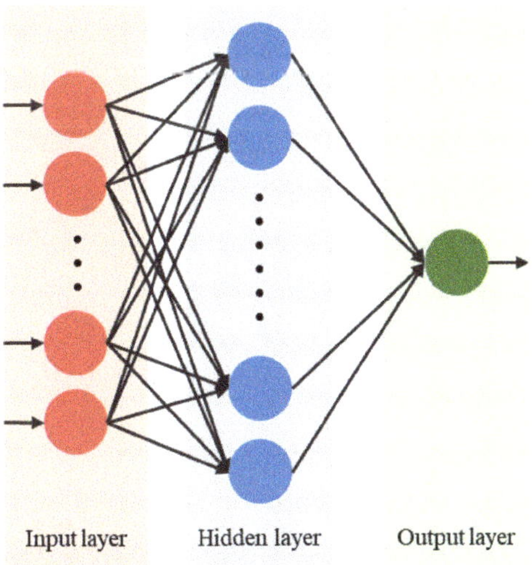

Figure 10. BP neural network structure diagram.

4.2. Design of BP Neural Network

To quantify the JRC value of the Barton standard sections, the morphological analysis method is adopted. The modified straight edge method formula Equations (1) and (2) and the basic roughness scale shown in Figures 11 and 12 are employed to obtain the JRC values of structural planes of different sizes. Table 4 lists the JRC, which is calculated by Equations (1) and (2), of anisotropic structural planes with different roughness.

$$JRC = 0.8589e^{0.6444/L}\arctan 8R_A \quad (1)$$

$$R_A = \frac{A}{L} \quad (2)$$

where A is the height of the larger sawtooth or the fluctuation amplitude of the structural plane, and L is the sampling length of the structural plane.

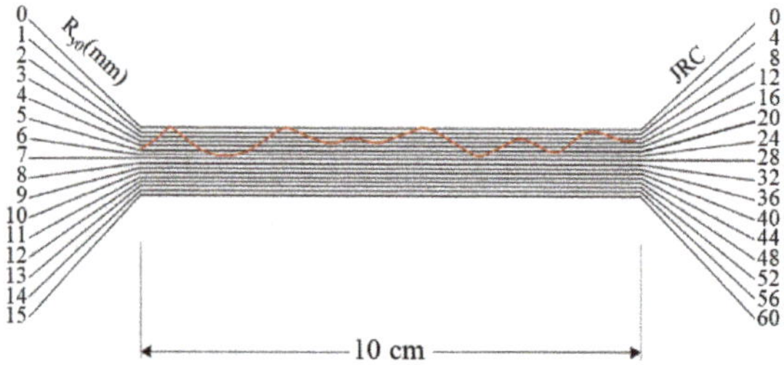

Figure 11. Basic roughness ruler.

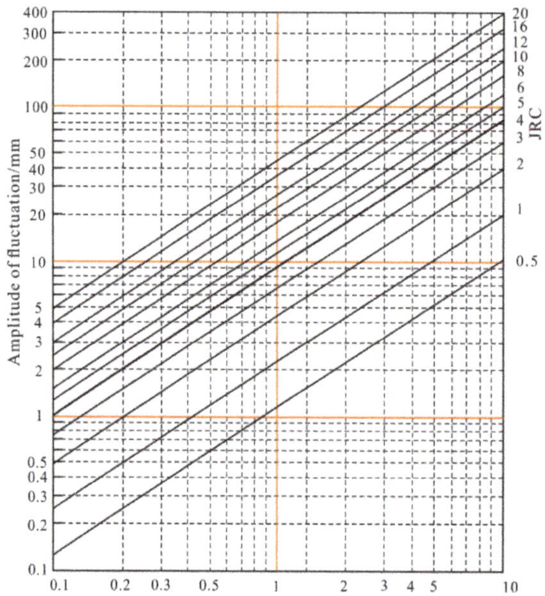

Figure 12. Schematic diagram of the straight edge method for calculating the roughness of structural planes [44].

Table 4. JRC values calculated using the modified straight-edge method.

Number of Structural Planes	Amplitude of Fluctuation	JRC
B1	0.17	0.67
B2	0.65	2.69
B3	1.26	5.24
B4	1.60	6.65
B5	2.27	9.39
B6	2.73	11.25
B7	3.37	13.79
B8	3.88	15.76
B9	4.26	17.21
B10	4.76	19.07

The Barton standard sections proposed by Barton are selected to establish structural planes with different roughness. Eighty-one groups of orthogonal tests are designed using SPSS software. Then, the peak shear strength values of 81 groups of anisotropic structural planes are calculated using 3DEC. The 81 groups of orthogonal test data and 100 groups of test data of B1–B4 structural planes under different sizes and normal stresses are divided into training data and test data according to the proportion imported into the neural network for training. The input parameters of the neural network are normal stress, roughness (JRC), compressive strength of soft and hard rock mass, basic friction angle of soft and hard rock masses, and structural plane size. The output parameter is the peak shear strength of the structural plane. The hidden layer is set to one layer, whose number of units is 11. The neural network training process is shown in Figure 13.

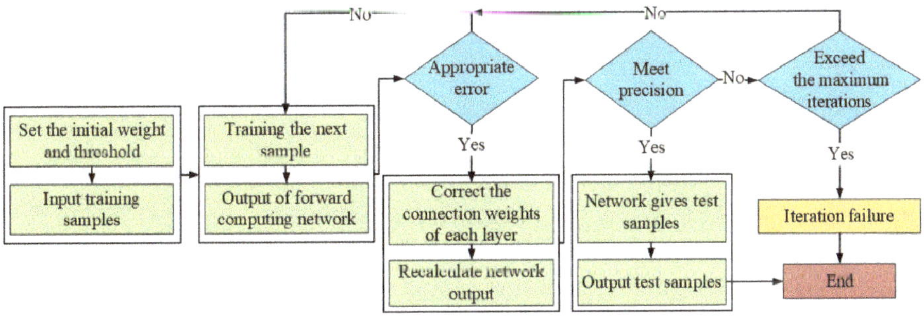

Figure 13. BP neural network training flowchart.

4.3. Prediction Results of Peak Shear Strength

Root mean square error (RMSE) and coefficient of determination (R^2) are used as model performance evaluation indicators [45]. Among them, RMSE is an error evaluation indicator. The closer its value is to 0, the better the model performance. R^2 is a trend evaluation indicator. The closer R^2 is to 1, the closer the model prediction result is to the true value and the higher the prediction accuracy. The calculation formulas Equations (3) and (4) for model evaluation indicators are as follows:

$$RMSE = \sqrt{\frac{1}{N}\sum_{i=1}^{N}\left(m_{\exp} - m_{\text{pred}}\right)^2} \quad (3)$$

$$R^2 = 1 - \frac{\sum_{i=1}^{N}\left(m_{\exp} - m_{\text{pred}}\right)^2}{\sum_{i=1}^{N}\left(m_{\exp} - m_{\text{ea}}\right)^2} \quad (4)$$

where $m_{\exp}$ is the actual value; m_{pred} is the predicted value; m_{ea} is the mean of the experimental test value, and N is the number of samples.

To obtain better generalization performance, the optimal number of training times should be determined when the number of hidden layer nodes is fixed. Therefore, the neural network is trained and tested alternately, and a training mean square error is recorded for each iteration. Meanwhile, the network weights should be unchanged, the network should be run forward with the test data, and the test mean square error should be recorded. As a result, two curves can be generated showing the error changing with the number of training times. Before a certain point in training, as the number of training iterations increases, both the training error and the test error decrease simultaneously. Once this point is surpassed, the training error continues to decrease while the test error starts to increase. This specific number of training iterations is considered the optimal amount. The number of steps for programming verification of the BP neural network is set to 6; that is, the network training verification stops after the error increases six consecutive times. It is indicated from Figure 14 that the convergence speed of the BP neural network prediction model is fast, and the optimal number of training steps is 12.

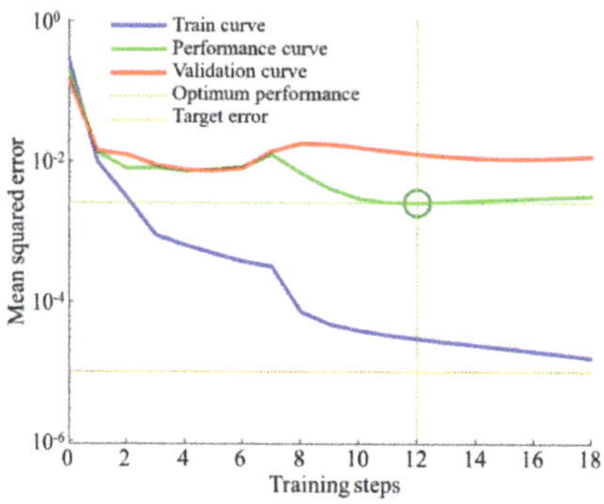

Figure 14. Plot of BP neural network training performance data versus training steps.

Figure 15 compares the actual and predicted values for the peak shear strength of the structural plane in both the training and test sets. The RMSE value of the neural network model is below 0.5 (Figure 15). This indicates a small prediction error, and it shows that the predicted shear strength of the structural plane obtained from the neural network closely aligns with the actual value.

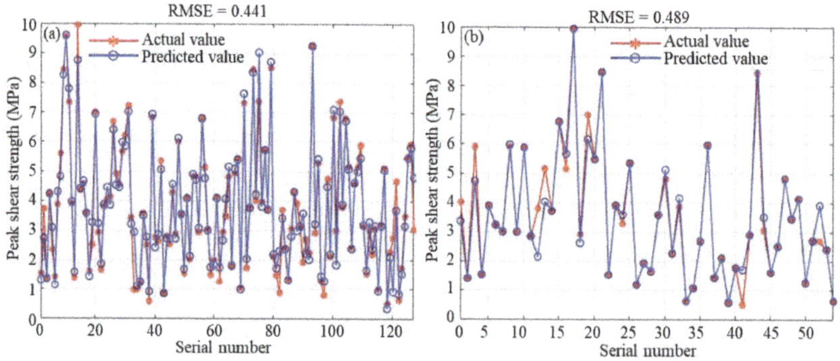

Figure 15. Comparison of actual and predicted values of peak shear strength: (**a**) Training set; (**b**) Test set.

Figure 16 illustrates the linear regression relationship between the actual peak shear strength values of both the training and test samples. The R^2 for the training and test sets are 0.957 and 0.947 (Figure 16), which signifies that the BP neural network model exhibits a strong prediction performance.

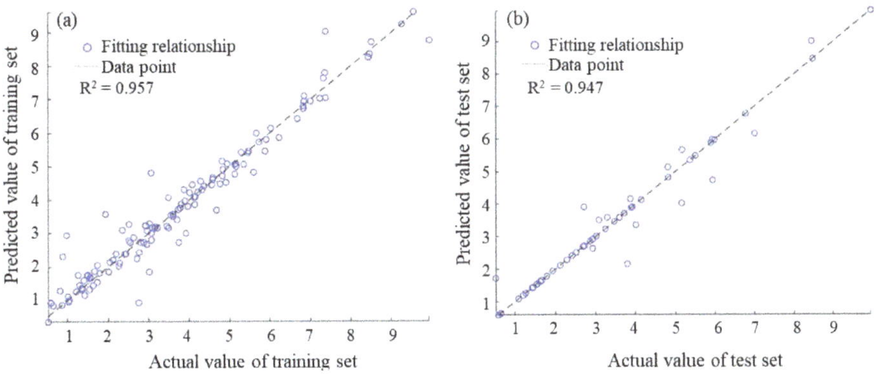

Figure 16. Regression plot between the actual value and the predicted value of the network output: (**a**) Training set; (**b**) Test set.

Combining the results of Figures 15 and 16, it can be seen that the overall regression curve of the predicted value and the actual value shows a high degree of regression, indicating that the predicted value and the actual value of the shear strength of the structural plane are very close. The comparison results indicate that the BP neural network provides a high level of accuracy in predicting the peak shear strength of structural planes. This method can effectively predict the shear strength of anisotropic structural planes.

5. Conclusions

Based on the discrete element method, in this paper, numerical models of anisotropic structural plane were established. After model verification, the direct shear numerical tests of anisotropic structural planes were carried out to evaluate the shear characteristics of structural planes with different sizes, normal stresses, and roughnesses. Then, a BP neural network model for predicting the peak shear strength of the structural plane was developed with numerical test data. The main conclusions are as follows:

(1) A direct shear test model for rock mass with anisotropic structural planes is established using 3DEC. By comparing the numerical and laboratory test results, it has been found that the error in peak shear strength was generally within 10%. Additionally, the shear displacement–stress curve obtained from the numerical tests closely resembles the curve from the laboratory tests, effectively replicating the shear process of the structural plane sample;

(2) The peak shear strength of anisotropic structural planes is negatively correlated with size. When the JRC is 4 to 6, the peak shear strength decreases by 0.022 to 0.045 MPa for every additional 1 cm in size with different normal stresses. When the JRC is from 8 to 10, the peak shear strength decreases by 0.025 to 0.053 MPa for every additional 1 cm in size with different normal stresses. When the JRC is 14 to 16, the peak shear strength decreases by 0.027 to 0.053 MPa for every additional 1 cm in size with different normal stresses. When the JRC is 18 to 20, the peak shear strength decreases by 0.031 to 0.066 MPa for every additional 1 cm in size with different normal stresses;

(3) The peak shear strength of anisotropic structural planes is positively correlated with normal stress and roughness. When the JRC is 4 to 6, the peak shear strength increases

by 0.253 to 0.904 MPa for each 1 MPa increase in normal stress with different sizes. When the JRC is 8 to 10, the peak shear strength increases by 0.473 to 1.049 MPa for each 1 MPa increase in normal stress with different sizes. When the JRC is from 14 to 16, the peak shear strength increases by 0.571 to 1.093 MPa for each 1 MPa increase in normal stress with different sizes. When the JRC is from 18 to 20, the peak shear strength increases by 0.580 to 1.287 MPa for each 1 MPa increase in normal stress with different sizes. Under identical size and normal stress conditions, the shear strength of the large structural plane with a high JRC is consistently the highest;

(4) The optimal number of training steps of the BP neural network model is 12. The RMSE of the BP neural network model is less than 0.5, and the R^2 value is greater than 0.94, which indicates that the predicted results align closely with the actual test results. Therefore, this model can effectively predict the peak shear strength of the anisotropic structural planes within rock masses.

Author Contributions: Conceptualization, W.-B.M. and G.L.; methodology, J.-L.Z. and G.L.; software, W.-B.M. and G.L.; validation, W.-B.M. and G.L.; formal analysis, W.-H.Z. and J.-L.Z.; investigation, W.-H.Z. and G.L.; resources, W.-H.Z.; data curation, W.-B.M.; writing—original draft preparation, W.-B.M.; writing—review and editing, W.-H.Z., J.-L.Z. and G.L.; visualization, W.-H.Z.; supervision, J.-L.Z.; project administration, G.L.; funding acquisition, W.-B.M. and G.L. All authors have read and agreed to the published version of this manuscript.

Funding: This work was supported by the Open Fund of State Key Laboratory of High-speed Railway Track Technology (2022YJ127-1), the National Natural Science Foundation of China (Grant No. 52304092), Zhejiang Provincial Natural Science Foundation (LQ24D020001).

Data Availability Statement: Data are contained within this article.

Acknowledgments: Thanks to the help of anonymous reviewers and journal editors.

Conflicts of Interest: Author Wei-Bin Ma, Wen-Hao Zou, Jin-Long Zhang was employed by the company China Academy of Railway Sciences Corporation Limited. The remaining authors declare that the research was conducted in the absence of any commercial or financial relationships that could be construed as a potential conflict of interest.

References

1. Zhang, L.; Yang, Q.; Liu, Y. Long-term stability analysis of the left bank abutment slope at Jinping I hydropower station. *J. Rock Mech. Geotech.* **2016**, *8*, 398–404. [CrossRef]
2. Zhong, J.H.; Yang, X. Pseudo-dynamic stability of rock slope considering Hoek-Brown strength criterion. *Acta Geotech.* **2022**, *17*, 2481–2494. [CrossRef]
3. Miao, C.; Liu, J.; Gao, M.; Li, J.; Pang, D.; Yuan, K. Insight into the overload failure mechanism of anchored slope with weak structural planes. *Bull. Eng. Geol. Environ.* **2024**, *83*, 390. [CrossRef]
4. Zhang, Z.R.; Wu, S.C.; Cheng, H.Y.; Han, L.; Chang, X.; Fu, X. Experimental study on the influence of structural planes on rockbursts in deep-buried hard-rock tunnels. *Rock Mech. Rock Eng.* **2024**, *57*, 8057–8080. [CrossRef]
5. Vick, L.M.; Böhme, M.; Rouyet, L.; Bergh, S.G.; Corner, G.D.; Lauknes, T.R. Structurally controlled rock slope deformation in Northern Norway. *Landslides* **2020**, *17*, 1745–1776. [CrossRef]
6. Zhao, L.H.; Yu, C.H.; Cheng, X.; Zuo, S.; Jiao, K. A method for seismic stability analysis of jointed rock slopes using Barton-Bandis failure criterion. *Int. J. Rock Mech. Min.* **2020**, *136*, 104487. [CrossRef]
7. Tan, X.; Ren, Y.K.; Li, T.L.; Zhou, S.H.; Zhang, J.C.; Zhou, S.K. In-situ direct shear test and numerical simulation of slate structural planes with thick muddy interlayer along bedding slope. *Int. J. Rock Mech. Min.* **2021**, *143*, 104791. [CrossRef]
8. Deng, D.P. Limit equilibrium analysis on the stability of rock wedges with linear and nonlinear strength criteria. *Int. J. Rock Mech. Min.* **2021**, *148*, 104967. [CrossRef]
9. Zhuang, Y.; Xing, A.; Leng, Y.; Bilal, M.; Zhang, Y.; Jin, K.; He, J. Investigation of characteristics of long runout landslides based on the multi-source data collaboration: A case study of the Shuicheng basalt landslide in Guizhou, China. *Rock Mech. Rock Eng.* **2021**, *54*, 3783–3798. [CrossRef]

10. He, K.; Liu, B.; Hu, X.W.; Zhou, R.; Xi, C.; Ma, G.; Han, M.; Li, Y.; Luo, G. Rapid characterization of landslide-debris flow chains of geologic hazards using multi-method investigation: Case study of the Tiejiangwan LDC. *Rock Mech. Rock Eng.* **2022**, *55*, 5183–5208. [CrossRef]
11. Wu, K.; Yi, X.B.; Fu, X.D.; Zhang, L.; Kang, J.Y.; Yuan, Q.; Shao, J. Geohazard development rules and post disaster reconstruction strategies of Hailuogou scenic road in the epicenter of Luding earthquake. *Chin. J. Rock Mech. Eng.* **2024**, *43*, 1909–1922. (In Chinese)
12. Huang, M.; Luo, Z.Y.; Du, S.G. Experimental study of sampling representativeness of structural plane of rock model. *Chin. J. Rock Mech. Eng.* **2013**, *32*, 2008–2014. (In Chinese)
13. Ming, H.J.; Liu, C.Y.; Wang, Y.X. Numerical simulation of size effect of rock structural plane. *Chin. J. Under Space Eng.* **2016**, *12*, 387–391. (In Chinese)
14. Barton, N. Review of a new shear strength criterion for rock joint. *Eng. Geol.* **1973**, *7*, 287–332. [CrossRef]
15. Tse, R.; Cruden, D.M. Estimating joint roughness coefficients. *Int. J. Rock Mech. Min. Geol. Abs.* **1979**, *16*, 303–307. [CrossRef]
16. Barton, N.; Choubey, V. The shear strength of rock joints in theory and practice. *Rock Mech. Rock Eng.* **1977**, *10*, 1–54. [CrossRef]
17. Ünlüsoy, D.; Süzen, M.L. A new method for automated estimation of joint roughness coefficient for 2D surface profiles using power spectral density. *Int. J. Rock Mech. Min.* **2020**, *125*, 104156. [CrossRef]
18. Barton, N.; Bandis, S. Effects of block size on the shear behavior of jointed rock. In Proceedings of the 23rd US Symposium on Rock Mechanics, Berkeley, CA, USA, 25–27 August 1982; pp. 739–760.
19. Du, X.F.; Du, S.G. Concise formula of Barton's straight edge method for joint roughness coefficient. *J. Eng. Geol.* **2008**, *16*, 196–200. (In Chinese)
20. Stigsson, M.; Ivars, D.M. A novel conceptual approach to objectively determine JRC using fractal dimension and asperity distribution of mapped fracture traces. *Rock Mech. Rock Eng.* **2019**, *52*, 1041–1054. [CrossRef]
21. Wang, G.; Zhang, Y.; Jiang, Y.; Liu, P.; Guo, Y.; Liu, J.; Ma, M.; Wang, K.; Wang, S. Shear behaviour and acoustic emission characteristics of bolted rock joints with different roughnesses. *Rock Mech. Rock Eng.* **2018**, *51*, 1885–1906. [CrossRef]
22. Feng, W.; Zou, D.; Wang, T.; Qiao, C.; Xu, S. Study on a nonlinear shear damage constitutive of structural plane and application of discrete element. *Comput. Geotech.* **2023**, *155*, 105190. [CrossRef]
23. Meng, F.; Zhou, H.; Wang, Z.; Zhang, L.; Kong, L.; Li, S.; Zhang, C. Experimental study on the prediction of rockburst hazards induced by dynamic structural plane shearing in deeply buried hard rock tunnels. *Int. J. Rock Mech. Min.* **2016**, *86*, 210–223. [CrossRef]
24. Du, S.G.; Lyu, Y.J.; Luo, Z.Y.; Huang, M. Combined test system for size effect of rock joint shear strength and its primary application research. *Chin. J. Rock Mech. Eng.* **2021**, *40*, 1337–1349. (In Chinese)
25. Mashhadiali, N.; Molaei, F. Theoretical and experimental investigation of a shear failure model for anisotropic rocks using direct shear test. *Int. J. Rock Mech. Min.* **2023**, *170*, 105561. [CrossRef]
26. Yin, Q.; Zhu, C.; Wu, J.; Pu, H.; Wang, Q.; Zhang, Y.; Jing, H.; Deng, T. Shear sliding of rough-walled fracture surfaces under unloading normal stress. *J. Rock Mech. Geotech.* **2023**, *15*, 2658–2675. [CrossRef]
27. Cao, X.; Wu, M.; Liu, S.; Fang, L.; Wang, B. Deformation simulation of steep over-dip stratified surrounding rock for small angle between tunnel and rock strike. *J. Mt. Sci. Eng.* **2013**, *31*, 385–390.
28. Yin, D.W.; Chen, S.J.; Chen, B.; Liu, X. Strength and failure characteristics of the rock-coal combined body with single joint in coal. *Geomech. Eng.* **2018**, *15*, 1113–1124. [CrossRef]
29. Le, H.L.; Wei, J.H.; Sun, S.R.; Zhang, C.C. Cracking behaviors and strength of rock-like samples with steps of different geometries under shear stress. *J. Mt. Sci. Eng.* **2021**, *18*, 1352–1370. [CrossRef]
30. Renaud, S.; Bouaanani, N.; Miquel, B. Experimental, analytical, and finite element assessment of the shear strength of concrete-rock interfaces at different scales. *Int. J. Numer. Anal. Met.* **2021**, *45*, 1238–1259. [CrossRef]
31. Greco, D.O.; Ferrero, A.M.; Oggeri, C. Experimental and analytical interpretation of the behaviour of laboratory tests on composite specimens. *Int. J. Rock Mech. Min.* **1993**, *30*, 1539–1543. [CrossRef]
32. Ban, L.R.; Zhu, C.; Hou, Y.H.; Du, W.S.; Qi, C.Z.; Lu, C.S. A method to predict the peak shear strength of rock joints based on machine learning. *J. Mt. Sci. Eng.* **2023**, *20*, 3718–3731. [CrossRef]
33. Xu, Z.H.; Guo, G.; Sun, Q.C.; Wang, Q.; Zhang, G.D.; Ye, R.Q. Structural plane recognition from three-dimensional laser scanning points using an improved region-growing algorithm based on the robust randomized Hough transform. *J. Mt. Sci. Eng.* **2023**, *20*, 3376–3391. [CrossRef]
34. Zhou, S.; Lei, Y.; Zhang, Z.-X.; Luo, X.; Aladejare, A.; Ozoji, T. Estimating dynamic compressive strength of rock subjected to freeze-thaw weathering by data-driven models and non-destructive rock properties. *Nondestruct. Test. Eva.* **2024**, *40*, 116–139. [CrossRef]
35. Zhou, S.; Zhang, Z.-X.; Luo, X.; Huang, Y.; Yu, Z.; Yang, X. Predicting dynamic compressive strength of frozen-thawed rocks by characteristic impedance and data-driven methods. *J. Rock Mech. Geotech.* **2024**, *16*, 2591–2606. [CrossRef]

36. Mahendra, S.; Bhawani, S. Laboratory and Numerical Modelling of a Jointed Rock Mass. In Proceedings of the 12th International Conference of International Association for Computer Methods and Advances in Geomechanics (IACMAG), Goa, India, 1–6 October 2008; ISBN 978-162276176-0.
37. Chen, S.; Qiao, C.S.; Yang, Z.; Qiao, C. Application of stochastic joint network simulation to composite strata of shallow-buried long-span metro tunnels. *Bull. Eng. Geol. Environ.* **2020**, *79*, 2085–2107. [CrossRef]
38. Yuan, W.; Cheng, Y.; Min, M.; Wang, X. Study on acoustic emission characteristics during shear deformation of rock structural planes based on particle flow code. *Comput. Part Mech.* **2024**, *11*, 105–118. [CrossRef]
39. Wang, T.T.; Huang, T.H. Anisotropic deformation of a circular tunnel excavated in a rock mass containing sets of ubiquitous Joints: Theory analysis and numerical modeling. *Rock Mech. Rock Eng.* **2014**, *47*, 643–657. [CrossRef]
40. Guo, W.; Zhang, C.; Wang, C.; Wang, M. Macroscopic and microscopic analysis of particle flow in direct shear mechanical characteristics of rock structural surface. *Chin. J. Comput. Mech.* **2023**, *40*, 237–248. (In Chinese)
41. Zhou, J.L. Study on Shear Characteristics of Rock-Mortar Heterogeneous Structural Plane. Ph.D. Thesis, Nanchang University, Nanchang, China, 2022. (In Chinese).
42. Kang, Y.S.; Hou, C.C.; Liu, B.; Bai, W. Influence of water content on the shear strength of rock joints with clay-rich fillings. *Rock Mech. Rock Eng.* **2023**, *56*, 1437–1449. [CrossRef]
43. Cao, Y.; Zhang, Y.Y.; Liu, Z.; Bao, M. An experimental study of the effects of filling degree on shear behavior and acoustic emission characteristics of kaolinite-filled granite joints. *Ksce. J. Civ. Eng.* **2024**, *28*, 2966–2981. [CrossRef]
44. Krahn, J.; Morgenstern, N.R. The ultimate frictional resistance of rock discontinuities. *Int. J. Rock Mech. Min.* **1979**, *16*, 127–133. [CrossRef]
45. He, Z.G.; Wilson, S.B.; Monjezi, M.; Tran, T.T. Estimating brazilian tensile strength of granite rocks using metaheuristic algorithms-based self-organizing neural networks. *Rock Mech. Rock Eng.* **2024**, *57*, 4653–4668. [CrossRef]

designs

MDPI

Article

Evaluating Seismic Isolation Design: Simplified Linear Methods vs. Nonlinear Time-History Analysis

Elias Yaacoub [1,2], Roberto Nascimbene [2,3,*], Marco Furinghetti [1,3] and Alberto Pavese [1,3]

1 Faculty of Engineering, University of Pavia, 27100 Pavia, Italy; elias.yaacoub01@universitadipavia.it (E.Y.); marco.furingheti@eucentre.it (M.F.); alberto.pavese@eucentre.it (A.P.)
2 STS Department, IUSS—Scuola Universitaria Superiore Pavia, 27100 Pavia, Italy
3 EUCENTRE—European Centre for Training and Research in Earthquake Engineering Pavia, 27100 Pavia, Italy
* Correspondence: roberto.nascimbene@iusspavia.it

Abstract: Seismic isolation is a vital strategy for improving the earthquake resilience of structures, utilizing flexible components such as lead–rubber bearings (LRBs) and curved surface sliders (CSSs) to attenuate ground motion effects. This paper presents a comprehensive comparative analysis of seismic isolation design methodologies prescribed in the U.S. code (ASCE 7-22) and the European code (EC8). The focus is on the equivalent lateral force method, also known as the simplified linear method, renowned for its simplicity and efficiency in seismic design applications. A six-story steel building serves as a case study to examine the discrepancies between the two codes. The structure was modeled and subjected to nonlinear time-history analysis (NTHA) using 20 ground motion records, selected and scaled to match a conditional mean spectrum (CMS). Key performance indicators—including displacement at the isolation level, base shear forces, story shear forces, and story drifts—were compared to assess the reliability and effectiveness of each code's design approach. The findings reveal notable differences between ASCE 7-22 and EC8, particularly in seismic hazard characterization and the calculation of design displacements. ASCE 7-22 generally adopts a more conservative stance, especially for CSSs, resulting in overestimations of design displacements and lateral seismic forces. In contrast, EC8's simplified method aligns more closely with observed performance for LRBs. However, when applied to CSSs, simplified methods prove less reliable, underscoring the need for more precise analytical techniques.

Keywords: seismic isolation; nonlinear time-history analysis; equivalent lateral force method; lead–rubber bearings; curved surface slider

1. Introduction

Seismic isolation strategies have evolved significantly over the years as a highly effective means of protecting the structural integrity of buildings and infrastructure during earthquakes [1]. The success of seismic isolation was convincingly demonstrated by the impressive performance of isolated structures during major seismic events such as the 1994 Northridge and 1995 Hyogoken-Nanbu earthquakes. These real-world examples reinforced the reliability of seismic isolation systems and validated the analytical models used to predict their behavior. As a result, their adoption has been increasingly encouraged in both new construction and retrofitting projects. Seismic isolation systems typically incorporate flexible components, such as rubber bearings, sliding, or rolling mechanisms, strategically placed between the primary mass of a structure and the source of ground motion. These components are often combined with energy-dissipating devices, such as dampers, to not only elongate the structure's natural periods but also increase its damping capacity. This prevents resonance with the shorter-period components of ground motions, which are often the most destructive during an earthquake. Furthermore, recent developments have expanded the application of seismic isolation to more flexible structures, offering a promising method to significantly reduce acceleration and displacement responses. This advancement allows designers to minimize the size of structural members, reduce potential damage, and improve the post-earthquake functionality of buildings, particularly those housing essential services or sensitive equipment. The primary benefits of implementing seismic isolation in structures, whether newly constructed or retrofitted, can be summarized as follows:

1. Increased Flexibility and Period Elongation: Seismic isolation introduces additional flexibility to the structural system, which results in an elongated fundamental period. By shifting the building's natural period away from the dominant frequencies typically associated with seismic events, this approach leads to a substantial reduction in the design forces for structures with shorter natural periods. Conversely, the impact on longer-period structures tends to be less significant. This added flexibility is particularly beneficial for nonstructural components [2], such as delicate equipment and mechanical systems, that are highly sensitive to seismic accelerations and could otherwise suffer damage during an earthquake.
2. Localized Inelastic Deformation: While the increased flexibility introduced by seismic isolation systems can result in larger system displacements, inelastic deformations are generally confined to the isolation system itself. This localized deformation allows for an elastic design approach for the remainder of the structure, ensuring that critical structural components remain undamaged even during strong seismic events. As a result, this strategy significantly improves the overall resilience of the building and reduces repair costs after an earthquake.
3. Energy Dissipation and Enhanced Damping: In addition to flexibility, seismic isolation systems play a crucial role in dissipating seismic energy, which is key to reducing the forces acting on the structure. Some systems inherently provide damping through the viscous or hysteretic properties of their materials. Hybrid isolation systems, which separate the functions of flexibility and damping, capitalize on the large movements at the isolation interface to integrate additional damping devices. By increasing the damping of the system, these devices further reduce seismic forces and help minimize the maximum displacement demands on the isolation bearings, thereby enhancing the performance and safety of the structure during an earthquake.

The market offers a wide variety of isolation devices, each designed to meet the three key requirements for an effective seismic isolator: vertical stiffness, lateral flexibility, and energy dissipation capacity. The selection of the optimal device for a particular project depends not only on factors such as availability and cost but also on the specific technical requirements dictated by the structure's location, design, and function. In this study, we focus on two of the most commonly used isolation systems, lead–rubber bearings (LRBs) and curved surface sliders (CSSs), and evaluate their capabilities in meeting these critical requirements.

LRBs are composite devices that combine the flexibility of rubber with the damping properties of a lead core [3]. These bearings are specifically designed to provide both horizontal flexibility and effective energy dissipation, which makes them particularly effective in reducing the seismic forces transmitted to the structure. The vertical stiffness of these devices is achieved through alternating layers of rubber and steel plates, which provide the necessary support to carry vertical loads without significant deformation. Lateral flexibility, on the other hand, is provided by the rubber layers, which accommodate horizontal displacements during seismic events, allowing the structure to move freely without transferring excessive forces to the building.

Friction-based isolation systems, such as CSSs, operate on the principle of controlled sliding. In this approach, the building is allowed to slide on a curved surface, with the weight of the structure transferred to the foundations while dissipating energy through friction. A simple flat sliding surface, however, is generally not suitable because it fails to control lateral sliding effectively, potentially leading to large and uncontrollable movements during an earthquake [4]. In some cases, flat sliders with low friction coefficients are used alongside additional energy dissipation devices to transfer gravity loads with minimal lateral resistance or damping. The force–displacement response of a CSS bearing is characterized by rigidity until the static friction threshold is exceeded, after which post-slip stiffness is determined by the supported weight and the radius of the curvature of the bearing.

This study explores the design requirements for seismically isolated structures, highlighting the differences between the U.S. code (ASCE 7-22) [5] and the European code (Eurocode 8) [6]. It also summarizes and compares the simplified design procedures for seismic isolation as prescribed by both codes and evaluates their effectiveness using nonlinear time-history analysis (NTHA). The case study focuses on a six-story steel building, modeled with both LRBs and CSSs, under each code's provisions. Key performance metrics—such as displacement at the isolation level, base shear, story shear forces, and story drifts—are analyzed and compared. Ultimately, this study aims to contribute to the ongoing efforts to refine seismic isolation design strategies and provide valuable guidance for engineers and policymakers working in earthquake-prone areas.

2. Case Study Building and Modeling Assumptions

The selected structure for this research is a six-story steel building with a well-documented history in structural studies. Initially analyzed by Tsai and Popov in [7], it was subsequently modified by Hall in [8] and later utilized by Filiatrault et al. in [9]. The six-story steel building was selected for this study due to its well documented structural properties and its relevance as a representative mid-rise seismically isolated structure. Its moderate height and moment-resisting frame configuration align with common applications of base isolation, particularly in critical infrastructure buildings where seismic resilience is paramount. Additionally, its prior use in seismic research ensures compatibility with both ASCE 7-22 and EC8 frameworks, making it an ideal candidate for evaluating the differences between these design methodologies. For the purposes of this study, minor adjustments were made to the building's design to ensure compliance with the specific constraints imposed by the simplified method of analysis stipulated in the relevant building codes. The building features a rectangular floor plan (Figure 1a) and is braced in both principal directions using moment-resisting frames, providing robust lateral stability. The design adheres to the provisions of the 1994 Uniform Building Code (UBC) [10] for structures located in Seismic Zone 4, constructed on soil type S2. Seismic Zone 4 represents the highest seismic hazard classification, covering areas prone to severe earthquakes (e.g., California). Soil Type S2 corresponds to stiff soil conditions with moderate seismic amplification, similar to EC8 Class C. The design parameters used in this study include a peak ground acceleration (PGA) of 0.4 g, a seismic coefficient of 0.44, and a response modification factor of 8.5, ensuring that the structure is designed to withstand high-intensity seismic loading conditions. Key dimensions, floor heights, and section details are presented in Figure 1b,c. The design gravity loads applied to the structure are as follows: a roof dead load of 3.8 kPa, a floor dead load of 4.5 kPa, a roof live load of 1.0 kPa, a floor live load of 3.8 kPa, and an exterior cladding weight of 1.7 kPa. The retrofit strategy includes the installation of seismic isolators at the building's base, as illustrated in Figure 2. The structure is assumed to rest on a large foundation mat with link frames connecting the base columns. The isolators are strategically positioned between the link frames and the top surface of the foundation mat to achieve optimal seismic performance. The structure is supported by a rigid reinforced concrete mat foundation (1.5 m thick), which provides a uniform and stable support surface for the isolation system. Since base-isolated structures experience significantly reduced force transmission to the foundation, the soil–structure interaction (SSI) is assumed negligible, consistent with prior research and seismic code assumptions. The large thickness and high stiffness of the mat foundation ensure that boundary conditions below the isolators remain fully fixed, preventing differential settlements or tilting effects that could influence isolator behavior.

This study adopts California as the hypothetical construction site for the building, with coordinates latitude 39.23 and longitude −123.584 and an elevation of 659.943 m. This location was chosen for several compelling reasons:

- California is classified as Seismic Zone 4 under the UBC framework [10], aligning with the seismic design requirements established during the building's original development.
- According to both American [5] and European [6] seismic design codes, the site is not considered near-fault, a classification further corroborated by disaggregation analysis.
- Comprehensive seismic data for this site are readily available from USGS (2024) [11], including critical parameters such as peak ground acceleration (PGA), spectral acceleration, and disaggregation data for various return periods. Specific data include seismic hazard values corresponding to a 2% probability of exceedance in 50 years (2475-year return period) and a 10% probability of exceedance in 50 years (475-year return period).

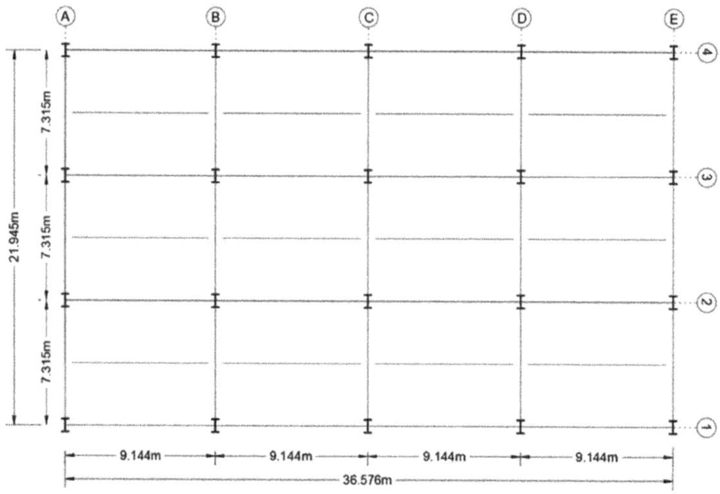

(a)

Figure 1. *Cont.*

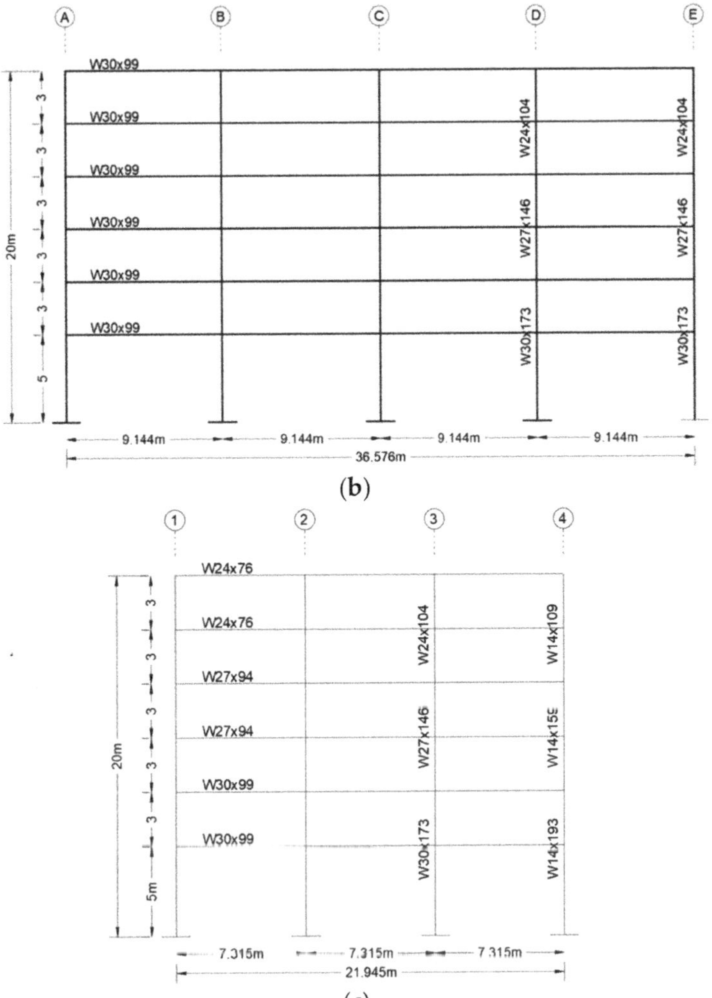

Figure 1. Plan view of the building (**a**); frame elevation in the long (**b**) and short (**c**) direction of the building.

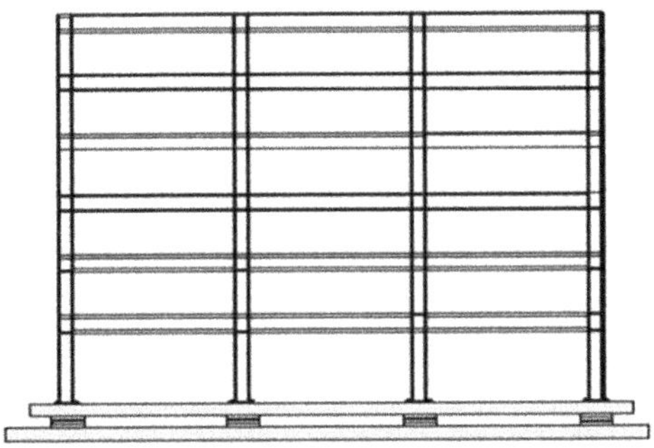

Figure 2. Schematic representation of the case building with isolator system at the base.

Structural modeling and analysis were conducted using SeismoStruct ver. 2024 [12–14], following the framework outlined by Christopoulos and Filatrault [3]. The model represents the bare steel frames, excluding any composite action from the slab. A bilinear stress–strain relationship with kinematic strain hardening is used to characterize the material behavior. The inelastic force-based plastic hinge element (infrmFBPH) was employed, with 150 section fibers specified to accurately capture the cross-sectional response. The hinges' plastic resistance is based on an expected yield strength of 290 MPa. Rigid-end offsets were included at the member joints to account for the actual physical dimensions of the elements. Gravity loads applied to the frame during seismic events comprise the roof and floor dead loads, exterior wall weights, and 20% of the floor live loads. P-Δ effects were incorporated into the analyses to capture second-order effects. A Rayleigh damping ratio of 3%, calibrated using the first two elastic vibration modes, was assigned to the model. Rayleigh damping was assigned at 3%, calibrated to the first two modes, as these modes dominate the response of the six-story base-isolated structure. While higher-mode damping effects are expected to have minimal influence due to the rigid-body behavior of the superstructure, they could become more significant in taller buildings where modal participation factors are higher [15,16]. For such cases, alternative damping models may be necessary to ensure accurate representation of dynamic behavior. The dynamic time-history analyses employed a time-step increment of 0.005 s to ensure numerical accuracy. The modeling of the isolation system reflects the spatial arrangement of isolator units, ensuring accurate representation of horizontal translations, overturning effects, and vertical axis rotations. Experimental studies have highlighted the significance of biaxial interactions on the force–displacement behavior of isolators, necessitating consideration of in-plane force coupling in the X and Y directions. However, the torsional moment contribution from individual isolators is considered negligible based on the findings by Nagarajaiah, Reinhorn, and Constantinou [17] and is excluded from the analysis. The assumption of negligible torsional moments is justified by the symmetric configuration of the six-story building and the uniform properties of the isolation system, which minimize rotational effects. Since base isolators primarily allow translational motion, the system's response is predominantly biaxial rather than torsional. However, in irregular structures with significant mass eccentricities or isolator stiffness variations, torsional effects should be explicitly analyzed to ensure accurate response predictions.

Both ASCE 7-22 [5] and EC8 [6] emphasize that the properties of the isolation system used in analysis should represent the most unfavorable values expected over the structure's service life. This requirement applies universally, irrespective of whether the analysis approach involves simplified linear methods, modal linear analysis, or nonlinear time-history simulations.

3. Design Procedure

The design forces acting on the superstructure are calculated by evaluating the forces transmitted through the isolators when the system reaches its design displacement. This design displacement is a critical parameter derived from the demand imposed on the structure during seismic events and the mechanical properties of the isolation system. These key properties include the effective stiffness and effective damping of the isolation system, both of which are intrinsic to the physical and mechanical characteristics of the isolator units. The design process aims to establish a logical and practical starting point by employing realistic initial values for the isolators' properties. This approach ensures that the preliminary design aligns with the anticipated behavior of the isolators under seismic loads, providing a foundation for iterative refinement as the analysis progresses.

By incorporating these initial parameters, the design ensures a balanced response of the superstructure, minimizing seismic forces while adhering to performance requirements.

Sections 3.1.1 and 3.2.1 introduce the simplified methodology employed to estimate the preliminary properties of the isolators. This procedure serves as the basis for initiating the design process, offering a systematic approach to determining effective stiffness and damping values that meet the desired performance criteria and requirements as indicated by ASCE7-22 [5] (Sections 3.1.2 and 3.2.2) and EC8 [6] (Sections 3.1.3 and 3.2.3).

3.1. Lead–Rubber Isolators

3.1.1. Preliminary Design

Static analysis of the building in Figure 1 shows that the pre-design requirements for isolators are the following:

1. Axial capacity of at least 3785.8 KN under zero lateral displacement (considering 100% DL + 100% LL).
2. Axial capacity of at least 2560 KN at design displacement (applying 100% DL + 20% LL).

As a result of the detailed analysis conducted on the building in Figure 1, the characteristics of the isolation device, selected for its suitability in meeting the structural and performance requirements, are summarized in Table 1.

Table 1. LRB product: main properties and geometrical dimension characteristics.

Outer Diameter (D_0) [mm]	800
Effective plan area (A_r) [mm^2]	491,345
Total rubber thickness (H) [mm]	2
Compressive stiffness (K_v) [kN/mm]	2960
Lead plug diameter (D_i) [mm]	120
Lead area (A_p) [mm^2]	11,309
Rubber code G0.4 with shear modulus (G) [N/mm^2]	0.385
Nominal long-term column load [kN]	5180
Initial stiffness (K_1) [kN/mm]	11.7
Post-yield stiffness (K_2) [kN/mm]	0.9
Characteristic strength (Q) [kN]	90.1
Shear stiffness of laminated rubber (K_r) [N/mm]	945.8
Additional shear stiffness by lead plug (K_p) [N/mm]	32.97

3.1.2. ASCE7-22 Requirements for Seismic Isolation Design

The fixed-base period of the building, representing the natural vibration period without seismic isolation, is calculated using the expression $T_f = 0.0488(20)^{0.75} = 0.462$ s. This calculation provides a baseline for comparing the dynamic behavior of the building with and without isolation. The LRB devices, which are installed beneath all the columns, possess the properties detailed in Table 1. These properties are essential for evaluating the isolators' ability to mitigate seismic forces and accommodate the expected displacements.

In compliance with ASCE 7-22 provisions, the equivalent lateral force (ELF) method leverages the fact that, in a seismically isolated structure, the majority of displacements are concentrated at the isolation level. As a result, the superstructure behaves nearly as a rigid body, simplifying the design process. This methodology focuses on a single mode of vibration, enabling the calculation of design forces acting on the superstructure based on the forces transmitted through the isolators at the design displacement. By following the procedural steps illustrated in Figure 3, derived from the guidelines set forth in ASCE 7-22 [5], and iteratively applying the equations governing this methodology, the necessary design parameters were derived. These calculations form the foundation for optimizing the seismic performance of the structure and ensuring compliance with code requirements.

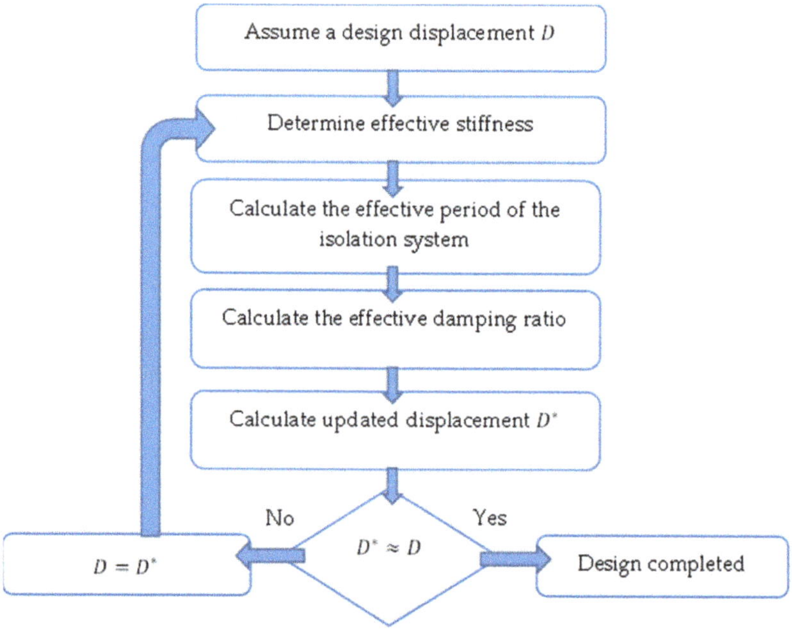

Figure 3. Flowchart of the iterative procedure implemented for the evaluation of the American requirements for seismic isolation design.

The isolation system is required to be designed to accommodate the maximum displacement (D_M) anticipated during seismic events. This maximum displacement is determined by evaluating the system's behavior using both upper-bound and lower-bound mechanical properties of the isolators, ensuring that the design captures the most critical conditions for horizontal response. This approach accounts for potential variability in isolator performance over the structure's lifetime and ensures the system's reliability under the most demanding seismic scenarios [5]:

$$D_M = \frac{gS_{M1}T_M}{4\pi^2 B_M} = 471.6 \text{ mm} \quad (1)$$

where g represents the acceleration due to gravity [mm/s^2]; S_{M1} is the spectral acceleration parameter corresponding to the maximum considered earthquake response (MCER) for a 5% damped system in a 1 s period; and B_M = 1.23 is a numerical coefficient related to the effective damping (β_M) of the isolation system, calculated using Equation (4), at the maximum displacement (D_M). Additionally, T_M denotes the effective period of the seismically isolated structure at the displacement D_M in the specific horizontal direction under analysis:

$$T_M = 2\pi\sqrt{\frac{W}{gk_M}} = 2.574 \text{ s} \quad (2)$$

where W represents the effective seismic weight of the structure situated above the isolation interface, while k_M denotes the effective stiffness of the isolation system [kN/mm] at the maximum displacement:

$$k_M = \frac{\sum|F_M^+| + \sum|F_M^-|}{2D_M} = 1.1 \text{ kN/mm} \quad (3)$$

where the term $\sum|F_M^+|$ represents the summation of the absolute values of the forces exerted by all isolator units when subjected to a positive displacement equal to D_M; similarly, $\sum|F_M^-|$ denotes the summation of the absolute values of the forces exerted by all isolator units when subjected to a negative displacement.

The effective damping of the isolation system at the maximum displacement D_M is the following:

$$\beta_M = \frac{\sum E_M}{2\pi k_M D_M^2} = 11\% \quad (4)$$

where $\sum E_M$ is the total energy dissipated in the isolation system during a full cycle of response at the displacement D_M.

Since both the effective stiffness (k_M) and the effective damping (β_M) of the isolation system are functions of the maximum displacement (D_M), the calculation of the effective period (T_M) and the damping coefficient (B_M) becomes an iterative process. This iterative approach is illustrated in Figure 3, where successive approximations refine these parameters until convergence is achieved. Following this, ASCE 7-22 mandates that designers calculate the total maximum displacement (D_{TM}) of the isolation system [5]:

$$D_{TM} = D_M\left[1 + \frac{y}{P_T^2}\left(\frac{12e}{b^2 + d^2}\right)\right] \quad (5)$$

The total maximum displacement, D_{TM}, must account for additional displacements caused by both actual and accidental torsion. These torsional effects arise from the spatial distribution of the isolation system's lateral stiffness and the most unfavorable positioning of the eccentric mass within the structure. ASCE 7-22 [5] further specifies that the value of D_{TM} shall not be taken as less than 1.15 times the maximum displacement (D_M), ensuring a conservative estimate that accommodates potential torsional amplifications and enhances the reliability of the isolation system under seismic loading. In Equation (5), y represents the distance, measured perpendicular to the direction of seismic loading under consideration, between the center of rigidity of the isolation system and the element of interest; the term e refers to the total eccentricity, which includes both the actual eccentricity and the accidental eccentricity. The actual eccentricity is the distance in plan between the center of mass of the structure above the isolation interface and the center of rigidity of the isolation system. The accidental eccentricity is conservatively taken as 5% of the longest plan dimension of the structure, measured perpendicular to the direction of the seismic force under consideration. The parameter b denotes the shortest plan dimension of the structure, measured perpendicular to d, where d represents the longest plan dimension of the structure. Lastly, P_T is defined as the ratio of the effective translational period of the isolation system to its effective torsional period. This ratio is a critical factor in assessing the dynamic behavior of the isolation system, particularly with respect to its torsional response under seismic excitation.

Finally, the minimum design force V_b acting beneath the isolation plane and the lateral seismic force V_s to be distributed along the height of the superstructure above the base level are presented in the corresponding table (Table 2) and in the following equations:

$$V_b = K_M D_M = 10{,}292 \text{ kN} \tag{6}$$

$$V_s = \frac{V_b}{R} = 5146 \text{ kN} \tag{7}$$

Table 2. Lateral seismic force, V_s, distributed across the height of the structure above the base level.

Level	w_i [kN]	h [m]	$w_i h^k$	F_i [kN]
6	3522.084	3	7665.811	598.77
5	4868.213	3	10,595.66	827.62
4	4868.213	3	10,595.66	827.62
3	4868.213	3	10,595.66	827.62
2	4868.213	3	10,595.66	827.62
1	5067.159	5	15,833.31	1236.72
Σ	28062.09	20	65,881.74	5145.95

The interstory drift ratio (IDR—in Table 3) should be based on elastic analysis of the fixed-base structure under design seismic force multiplied by R.

Table 3. Interstory drift ratio (IDR) based on elastic analysis of fixed-base structure under design seismic force multiplied by R.

Level	h [m]	F [kN]	D_x [mm]	D_y [mm]	IDR_x	IDR_y
6	3	598.8	52.59	186.08	0.11%	0.21%
5	3	827.6	49.23	179.91	0.18%	0.39%
4	3	827.6	43.84	168.36	0.22%	0.43%
3	3	827.6	37.19	155.51	0.28%	0.54%
2	3	827.6	28.92	139.27	0.33%	0.69%
1	5	1236.7	18.99	118.52	0.38%	2.37%

3.1.3. EC8 Requirements for Seismic Isolation Design

Considering the building described in Section 2, which is situated in California on a site classified as Class CD according to ASCE [5] (with a shear wave velocity $V_{s30} = 360\,\text{m/s}$), this classification aligns with Class C as per EC8 [6]. The reference peak ground acceleration (a_{gr}) on Type A ground for this site, corresponding to a 475-year return period (representing a 10% probability of exceedance in 50 years), is 0.32 g. This baseline value provides a starting point for seismic hazard characterization consistent with both ASCE [5] and EC8 [6] frameworks. To ensure a meaningful comparison between the design methodologies outlined in ASCE 7-22 [5] and EC8 [6], it is crucial to adopt a consistent reference point for seismic input. In this study, we extend the reference to align with a 2475-year return period, representing a 2% probability of exceedance in 50 years. This return period is consistent with the MCER in ASCE 7-22 [5] and provides a more conservative basis for evaluating the performance of the isolation system under extreme seismic events. By adopting this unified approach, this study facilitates a direct and reliable comparison of the design parameters and outcomes derived from the two codes. Consequently, the design ground acceleration on Type A ground is calculated as follows:

$$a_g = a_{gr}\left(\frac{P_L}{P_{LR}}\right)^{-\frac{1}{3}} = 0.32\left(\frac{2\%}{10\%}\right)^{-\frac{1}{3}} = 0.547\ \text{g} \tag{8}$$

The fixed-base period of the building, representing the natural vibration period without seismic isolation, is calculated using the expression $T_f = 0.085(20)^{0.75} = 0.8$ s. The parameters describing response spectrum for Type C ground are S = 1.15, T_B = 0.2 s, T_C = 0.6 s, and T_D = 2 s.

The simplified approach presented in EC8 [6] closely aligns with the methodology described in ASCE 7-22 [5] (and described in Figure 3), as both standards explicitly integrate displacement-based design principles into their frameworks. According to EC8 [6], the isolation system must be designed to withstand the design displacement (d_{dc}), which represents the maximum expected displacement under design-level seismic conditions:

$$d_{dc} = \frac{MS_e(T_{\text{eff}}, \xi_{\text{eff}})}{K_{\text{eff}}} = 338.82\ \text{mm} \tag{9}$$

where M is the mass of the superstructure; $S_e(T_{\text{eff}}, \xi_{\text{eff}})$ is the spectral acceleration; $T_{\text{eff}} = 2.49$ s is the effective period of the isolation system; K_{eff} is the effective horizontal stiffness of the isolation system when the design displacement d_{dc} takes place; $\xi_{\text{eff}} = 14.2\%$ is the equivalent viscous damping (effective damping) of the isolation system. EC8 does not provide detailed guidance on calculating K_{eff} and ξ_{eff}. Instead, practitioners often rely on the methods described in Equations (3) and (4). Since EC8 does not provide explicit equations for calculating effective stiffness and damping, this study adopts the formulations from ASCE 7-22 to ensure consistency in evaluating isolation system behavior across both standards. While this approach introduces some potential conservatism from ASCE provisions, it enables a more structured comparison of displacement and force demands without introducing arbitrary assumptions.

Finally, the minimum design force f_b acting beneath the isolation plane and the minimum seismic force f_s to be distributed along the height of the superstructure above the base level are presented in the corresponding table (Table 4) and in the following equations:

$$f_b = MS_e(T_{\text{eff}}, \xi_{\text{eff}}) = 7902\ \text{kN} \tag{10}$$

$$f_s = \frac{f_b}{q = 1.5} = 5268\ \text{kN} \tag{11}$$

Table 4. Lateral seismic force, f_s, distributed across the height of the structure above the base level.

Level	w_i [kN]	h [m]	F_i [kN]
6	3522.084	3	661.17
5	4868.213	3	913.87
4	4868.213	3	913.87
3	4868.213	3	913.87
2	4868.213	3	913.87
1	5067.159	5	951.22
$\sum$	28,062.09	20	5267.8

The IDR, in Table 5, should be based on elastic analysis of the fixed-base structure under design seismic force multiplied by q.

Table 5. IDR based on elastic analysis of fixed-base structure under design seismic force multiplied by q.

Level	h [m]	F [kN]	D_x [mm]	D_y [mm]	IDR_x	IDR_y
6	3	598.8	52.59	186.08	0.11%	0.21%
5	3	827.6	49.23	179.91	0.18%	0.39%
4	3	827.6	43.84	168.36	0.22%	0.43%
3	3	827.6	37.19	155.51	0.28%	0.54%
2	3	827.6	28.92	139.27	0.33%	0.69%
1	5	1236.7	18.99	118.52	0.38%	2.37%

3.2. Friction Pendulum Isolators

3.2.1. Preliminary Design

Figure 4 illustrates the relationship between the dynamic friction coefficient μ and the vertical load applied to the isolator, specifically focusing on the ratio of the vertical load N_{Sd} to the maximum vertical load N_{Ed}. The vertical load N_{Sd} is typically assumed to be constant and represents the quasi-permanent load, which is the average load experienced by the isolator during an earthquake. In contrast, N_{Ed} corresponds to the maximum vertical load under ultimate limit state (ULS) load combinations, which include the effects of seismic action. This ratio is a critical parameter for understanding how variations in vertical load influence the isolator's frictional behavior during dynamic conditions. For the present study, the devices under consideration can be equipped with two sliding materials, with distinct frictional properties: the former is characterized by a low friction coefficient, which is able to limit the base shear of the system with a significant displacement demand; the latter represents an intermediate case, which leads to a balance between the reduction in the building response and the limitation of the isolation displacement demand. In both cases, the friction coefficient is considered as a function of the applied vertical load. Such frictional properties are essential for evaluating the isolator's performance under varying vertical loads and seismic demands, ensuring that the device operates within the expected parameters to achieve optimal isolation efficiency. Graphical results are provided as friction coefficient values, as a function of the vertical load ratio N_{Sd}/N_{Ed}, normalized with respect to the low friction coefficient at the design vertical load (μ_{Ld}).

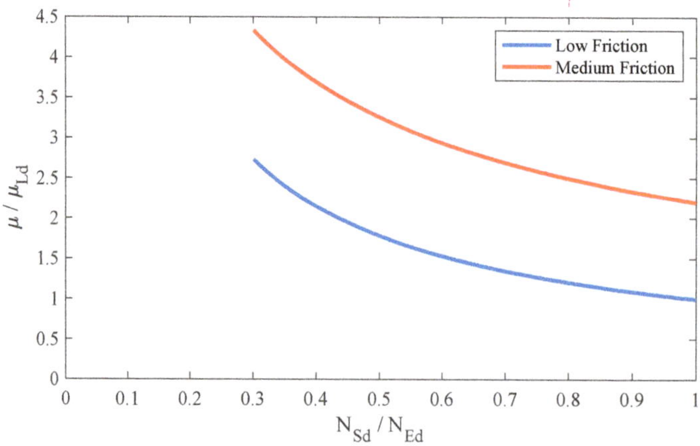

Figure 4. Graphical representation of the variation in friction coefficient with the vertical load.

The physical dimensions and mechanical properties of the device, chosen for its suitability in addressing the design requirements of the structure, are comprehensively detailed in Table 6.

Table 6. Friction pendulum product: main properties and geometrical dimension characteristics.

Diameter (D) [mm]	650
Height (H) [mm]	126
Maximum vertical load at ULS load combinations (N_{Ed}) [kN]	4000
Total displacement [mm]	500
Friction coefficient (μ) [%]	4.87
Initial stiffness (K_1) [kN/mm]	689.42
Post-yield stiffness (K_2) [kN/mm]	0.817
Characteristic strength (F_0 or Q) [kN]	87.56

3.2.2. ASCE7-22 Requirements for Seismic Isolation Design

By employing the same methodology illustrated in Figure 3 and utilizing the equations outlined in Section 3.1.2, we systematically derive key parameters for the CSS isolation system. These parameters include the maximum displacement (D_M = 487.7 mm), the effective period (T_M = 2.694 s), the effective stiffness (k_M = 1.0 kN/mm), and the effective damping ratio (β_M = 11.46%). Most importantly, this iterative process enables us to determine the total maximum displacement (D_{TM} = 557.3 mm), which accounts for additional effects such as torsion and spatial variability in stiffness. Finally, the minimum design force V_b acting beneath the isolation plane and the lateral seismic force V_s to be distributed along the height of the superstructure above the base level are presented in the corresponding table (Table 7) and in the following equations:

$$V_b = K_M D_M = 9713 \text{ kN} \quad (12)$$

$$V_s = \frac{V_b}{R} = 4856 \text{ kN} \quad (13)$$

Table 7. Lateral seismic force, V_s, distributed across the height of the structure above the base level.

Level	w_i [kN]	h [m]	$w_i h^k$	F_i [kN]
6	3522.084	3	7945.66	562.79
5	4868.213	3	10,982.5	777.89
4	4868.213	3	10,982.5	777.89
3	4868.213	3	10,982.5	777.89
2	4868.213	3	10,982.5	777.89
1	5067.159	5	16,687.2	1181.96
Σ	28062.09	20	68,562.7	4856.3

The IDR, in Table 8, should be based on elastic analysis of fixed-base structure under design seismic force multiplied by R.

Table 8. IDR based on elastic analysis of fixed-base structure under design seismic force multiplied by R.

Level	h [m]	F [kN]	D_x [mm]	D_y [mm]	IDR_x	IDR_y
6	3	598.8	52.59	186.08	0.11%	0.21%
5	3	827.6	49.23	179.91	0.18%	0.39%
4	3	827.6	43.84	168.36	0.22%	0.43%

Table 8. *Cont.*

Level	h [m]	F [kN]	D_x [mm]	D_y [mm]	IDR$_x$	IDR$_y$
3	3	827.6	37.19	155.51	0.28%	0.54%
2	3	827.6	28.92	139.27	0.33%	0.69%
1	5	1236.7	18.99	118.52	0.38%	2.37%

3.2.3. EC8 Requirements for Seismic Isolation Design

By employing the same methodology illustrated in Figure 3 and utilizing the equations presented in Section 3.1.3, we systematically derive the key parameters governing the performance of the CSS isolation system within the framework of EC8 [6] requirements. Specifically, the analysis yields the design displacement (d_{dc} = 325.66 mm), which represents the maximum expected horizontal movement of the isolator under seismic action. Furthermore, we determine the effective period of the isolated structure (T_{eff} = 2.58 s), which reflects the elongation of the structural vibration period introduced by the isolation system. Lastly, the analysis provides the effective damping ratio (ξ_{eff} = 15.7%), a critical parameter that accounts for the energy dissipation capacity of the CSS system during seismic excitation.

Finally, the minimum design force f_b acting beneath the isolation plane, as well as the minimum seismic force f_s to be distributed along the height of the superstructure above the base level, are presented in the corresponding Table 9 and in the following equations:

$$f_b = MS_e(T_{\text{eff}}, \xi_{\text{eff}}) = 7067 \text{ kN} \quad (14)$$

$$f_s = \frac{f_b}{q = 1.5} = 4711 \text{ kN} \quad (15)$$

Table 9. Lateral seismic force, f_s, distributed across the height of the structure above the base level.

Level	w_i [kN]	h [m]	F_i [kN]
6	3522.084	3	591.32
5	4868.213	3	817.33
4	4868.213	3	817.33
3	4868.213	3	817.33
2	4868.213	3	817.33
1	5067.159	5	850.73
Σ	28,062.09	20	4711.3

The IDR, in Table 10, should be based on elastic analysis of the fixed-base structure under design seismic force multiplied by q.

Table 10. IDR based on elastic analysis of fixed-base structure under design seismic force multiplied by q.

Level	h [m]	F [kN]	D_x [mm]	D_y [mm]	IDR$_x$	IDR$_y$
6	3	598.8	52.59	186.08	0.11%	0.21%
5	3	827.6	49.23	179.91	0.18%	0.39%
4	3	827.6	43.84	168.36	0.22%	0.43%
3	3	827.6	37.19	155.51	0.28%	0.54%
2	3	827.6	28.92	139.27	0.33%	0.69%
1	5	1236.7	18.99	118.52	0.38%	2.37%

4. Comparison of Design Assumptions and Procedures

In Sections 3.1 and 3.2, the specific requirements for the equivalent lateral force (ELF) method, also referred to as the simplified linear analysis method, for seismically isolated structures were presented. These sections provided a comprehensive overview of the equations, parameters, and procedural steps required for the design of isolated systems in accordance with both ASCE 7-22 [5] and EC8 [6]. Additionally, a practical application of the methodology was demonstrated for each code, focusing on two distinct scenarios: one involving a structure isolated using LRB and the other utilizing a CSS.

The comparative analysis of the two design codes reveals that ASCE 7-22 [5] consistently produces higher design values for both LRB and CSS isolation systems. This trend is reflected in several critical design parameters, including longer effective periods, larger design displacements, and higher minimum design forces acting beneath the isolation plane (Figure 5). Figure 5a–d represent the IDR profile, and this comparison shows the IDR for each story level in the x and y directions, demonstrating how the different methodologies estimate structural deformation under seismic loading. Figure 5e–f depict the shear comparison. This plot presents the total shear forces derived from each method, illustrating the variations in force demands imposed by different seismic design approaches.

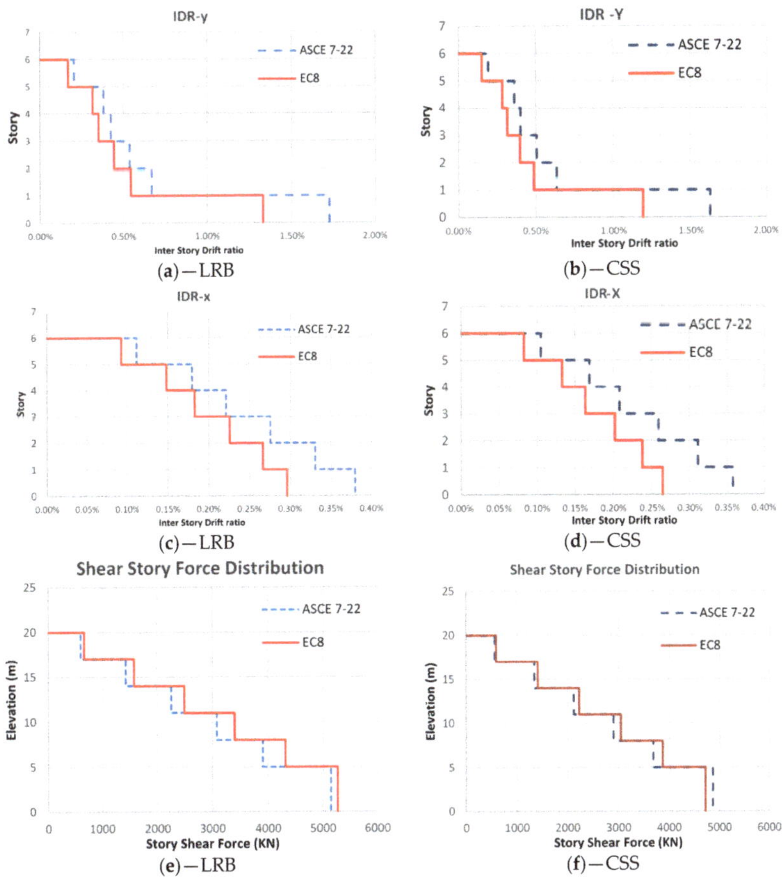

Figure 5. Comparative analysis of the two design codes: (**a**,**c**,**e**) are related to LRBs, and (**b**,**d**,**f**) are related to CSSs; (**a**–**d**) represent the comparison in terms of IDR while (**e**,**f**) refer to story shear force.

Although both ASCE 7-22 [5] and EC8 [6] adopt similar displacement-based design concepts and procedures, the key difference arises from the displacement spectra defined by each code. Specifically, the corner period—the point where the displacement spectrum flattens—is reached significantly earlier in EC8 [6], occurring at approximately 2 s in this study. In contrast, the corner period defined by ASCE 7-22 [5] occurs much later, at 8 s. This discrepancy becomes particularly relevant for structures with effective periods exceeding 2 s, where the displacement demands calculated using ASCE 7-22 [5] become notably larger. On the other hand, EC8 [6] offers slightly higher effective damping values, which can play a crucial role in reducing both displacement demands and seismic forces under specific conditions. The combination of higher damping and earlier corner periods may lead to more favorable results for certain isolation systems, particularly in scenarios where reducing forces and displacements is critical for performance optimization.

5. Nonlinear Time-History Analysis

SeismoStruct 2024 [12–14] was employed for the structural modeling and analysis of the building. The model exclusively includes the bare steel frames, meaning the contribution of the slab as a composite beam is intentionally omitted to focus on the steel frame behavior [18–22]. To represent the material behavior, a bilinear stress–strain model with kinematic strain hardening is adopted, ensuring accurate simulation of inelastic responses under seismic loads. The structural elements are modeled using an inelastic force-based plastic hinge frame element type (infrmFBPH) [23,24], with each section represented by 150 fibers to provide a detailed cross-sectional response. These plastic hinges exhibit bilinear hysteretic behavior, characterized by a curvature strain-hardening ratio of 0.02, and their length is specified as 1/6 of the respective member's span, ensuring a realistic representation of plasticity. The plastic resistance at the hinges is derived from an expected yield strength of 290 MPa, reflecting material properties typical of ductile steel. Additionally, rigid-end offsets are included at the ends of frame members to account for the actual physical dimensions and stiffness of joints. The dynamic behavior of the structure is further refined by incorporating Rayleigh damping at a 3% critical damping ratio, calibrated using the first two elastic vibration modes. The time-step increment for the time-history dynamic analysis is set at 0.005 s, ensuring precise temporal resolution for capturing seismic responses. The failure criterion applied to all steel beams and columns is defined by a plastic hinge rotation limit of 0.03 radians, consistent with the AISC design provisions [25] introduced following the 1994 Northridge earthquake [26,27]. This limit reflects the ductility demands expected in moment-resisting steel frames under severe seismic loading. To address the potential for brittle failure at welded beam-to-column connections, the model incorporates flexural strength degradation at the ends of beams and columns. This degradation simulates the reduction in strength to 1% of the yield moment upon reaching the ultimate curvature, accurately capturing the behavior of connections as they transition from ductile to brittle failure modes. This approach provides a comprehensive framework for evaluating the seismic performance of the structure, considering both global and local failure mechanisms.

Experimental investigations have consistently highlighted the substantial influence of biaxial effects on the force–displacement behavior of isolators. These biaxial effects arise from the simultaneous application of forces in two perpendicular directions, a scenario commonly encountered during seismic events when structures are subjected to complex, multi-directional ground motions. Accurate modeling of these interactions is essential for capturing the realistic behavior of isolators under such conditions. In this study, the interaction between in-plane forces acting along the X and Y directions is explicitly considered to ensure that the isolators' responses under biaxial loading are properly characterized.

For modeling LRBs, a biaxial hysteretic model has been employed. This model incorporates coupled plasticity properties for the two shear deformation modes, while linear stiffness properties are assumed for the remaining four deformation modes. The plasticity model is based on the hysteretic behavior proposed in [28], which extends the Wen model [29] for uniaxial behavior. This approach has been widely recommended for base-isolation analysis, as documented in [30]. The SeismoStruct 2024 software [12–14] implements this model under the name "Isolator1_Elastomeric Isolator Element", a 3D zero-length element specifically designed to replicate the described behavior. This element is capable of accurately simulating the nonlinear, coupled response of LRBs under biaxial loading, ensuring that both shear and axial deformation characteristics are faithfully reproduced.

For CSSs, the friction model established in [31] has been adopted. This model describes the friction coefficient as a velocity-dependent parameter governed by an exponential analytical law, which captures the complex behavior of CSS isolators under dynamic loading conditions. SeismoStruct 2024 [12–14] provides this model through the "Isolator2_Friction Pendulum Isolator" element, which, like the LRB element, is a 3D zero-length element. This element incorporates coupled plasticity properties for the two shear directions (axes 2 and 3 in the local coordinate system) while maintaining linear elastic behavior for the other four deformation types. The model is designed to accurately simulate the dynamic behavior of CSS isolators, including the interaction between shear forces and the isolator's frictional response.

5.1. Selection of Ground Motion Records

The conditional mean spectrum (CMS) is classically an appropriate target response spectrum and a valuable tool for selecting ground motions for dynamic analysis [32–35]. This approach ensures that the selected ground motions match the expected spectral shape at a given target period, providing a realistic representation of seismic demands. Baker in [36] presented a step-by-step procedure for computing the ground motion selection process to match CMS:

1. Step 1: Specify the Target Spectrum: The CMS is used as the target RotD50 spectrum (50th percentile). First, identify the target spectral acceleration (S_a) at a period of interest T^*. From previous calculations, the effective period of the isolated structure is found to be T_{ASCE} = 2.57 s and T_{EC8} = 2.5 s using lead–rubber bearings and T_{ASCE} = 2.7 s and T_{EC8} = 2.6 s using curved surface sliders. Approximately, we assume T^* = 2.5 s. The corresponding spectral acceleration (2% probability of exceedance in 50 years) is $S_a(2.5)$ = 0.4 g. Figure 6 represents the conditional spectrum associated with magnitude = 7.93, distance = 39.9 km, ε = 1.5, V_{s30} = 365 m/s, a California strike-slip rupture, and $S_a(2.5)$ = 0.4 g.
2. Step 2: Statistically Simulate Spectra: Monte Carlo simulation is used to probabilistically generate multiple response spectra from a distribution parameterized by the target means and variances of CMS (Figure 7).
3. Step 3: Specify Ground Motion Database: Here, we will consider the NGA-West2 database [37] of recorded ground motions and screen for suitable ground motions. Given that the target spectrum is associated with a magnitude 7.93, distance 39.9 km event, we restricted the selection to ground motions within 50 km of an earthquake with a magnitude between 6 and 9. With these criteria, the NGA West2 database has 1225 ground motions satisfying the initial screening. Scaling of ground motions was allowed but limited to a maximum of four. The selected motions were matched within a period range of $0.1T^*$ to $2T^*$ (0.25 to 5 s).

4. Step 4: From the database, an initial set of 20 ground motions was selected to match the simulated spectra. These motions were chosen based on their compatibility with the CMS, ensuring an appropriate distribution of spectral accelerations over the target period range.
5. Step 5: To ensure statistical accuracy, a maximum error tolerance of 10% was specified for the mean and standard deviation of spectral accelerations across the selected ground motions. This tolerance guarantees that the selected motions closely align with the target CMS.
6. Step 6: An optimization process was performed to further refine the selection of ground motions. Prior to optimization, the maximum errors in the mean and standard deviation of spectral accelerations were 11% and 33%, respectively. After optimization, these errors were significantly reduced to 6.5% and 9.4%, demonstrating the effectiveness of the optimization in achieving closer alignment with the CMS.
7. Step 7: The final set of selected ground motions was outputted, with their response spectra illustrated in Figure 6. The characteristics of these selected records, including their magnitude, distance, and scaling factors, are summarized in Table 11. These motions form the basis for dynamic analysis, ensuring that the selected inputs accurately represent the seismic demands expected at the site.

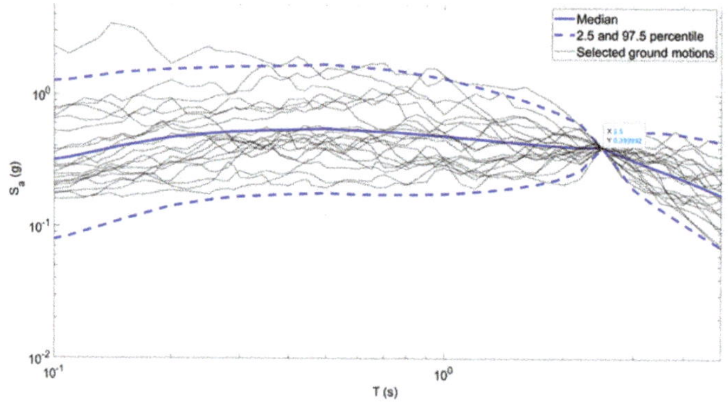

Figure 6. The conditional spectrum is associated with the rupture parameters, while the response spectra are derived from NGA-West2 ground motions selected to match this target spectrum.

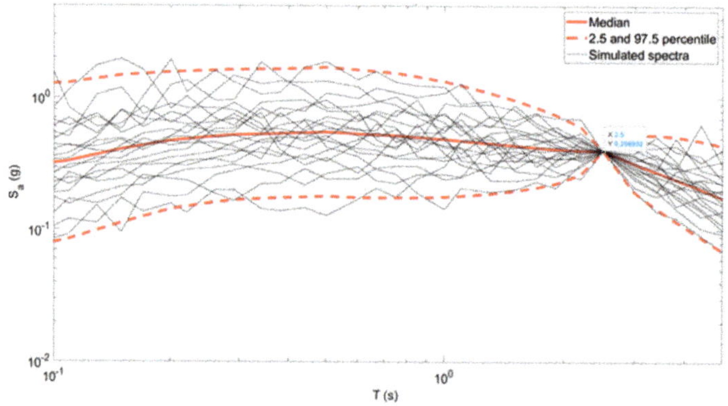

Figure 7. Statistically simulated spectra generated using Monte Carlo simulation, which probabilistically produces multiple response spectra based on a distribution parameterized by the target means and variances of the CMS.

Table 11. Final set of selected ground motions.

ID	Sequence Number	Earthquake	Scale Factor	Year	Station Name	*M*	Mechanism
1	1204	Chi-Chi_ Taiwan	3.06	1999	CHY039	7.62	Reverse Oblique
2	1227	Chi-Chi_ Taiwan	2.91	1999	CHY074	7.62	Reverse Oblique
3	1238	Chi-Chi_ Taiwan	2.32	1999	CHY092	7.62	Reverse Oblique
4	1504	Chi-Chi_ Taiwan	1.18	1999	TCU067	7.62	Reverse Oblique
5	1533	Chi-Chi_ Taiwan	1.62	1999	TCU106	7.62	Reverse Oblique
6	1534	Chi-Chi_ Taiwan	1.34	1999	TCU107	7.62	Reverse Oblique
7	1536	Chi-Chi_ Taiwan	0.92	1999	TCU110	7.62	Reverse Oblique
8	1538	Chi-Chi_ Taiwan	2.12	1999	TCU112	7.62	Reverse Oblique
9	1540	Chi-Chi_ Taiwan	1.82	1999	TCU115	7.62	Reverse Oblique
10	1542	Chi-Chi_ Taiwan	1.64	1999	TCU117	7.62	Reverse Oblique
11	1547	Chi-Chi_ Taiwan	1.62	1999	TCU123	7.62	Reverse Oblique
12	1550	Chi-Chi_ Taiwan	2.65	1999	TCU136	7.62	Reverse Oblique
13	1552	Chi-Chi_ Taiwan	2.74	1999	TCU140	7.62	Reverse Oblique
14	2459	Chi-Chi_ Taiwan-03	3.63	1999	CHY026	6.2	Reverse
15	3843	Chi-Chi_ Taiwan-03	2.85	1999	CHY002	6.2	Reverse
16	4483	L'Aquila_ Italy	3.33	2009	L'Aquila—Parking	6.3	Normal
17	4875	Chuetsu-oki_ Japan	0.34	2007	Kariwa	6.8	Reverse
18	5810	Iwate_ Japan	1.99	2008	Machimukai Town	6.9	Reverse
19	6889	Darfield_ New Zealand	0.96	2010	Christchurch Hospital	7	Strike slip
20	6969	Darfield_ New Zealand	1.96	2010	Styx Mill Transfer Station	7	Strike slip

5.2. Analysis Results and Comparisons

Using the SeismoStruct 2024 software [12–14], a comprehensive set of 40 nonlinear time-history analyses was performed—20 for the structures isolated with LRBs and 20 for those using CSSs. These analyses utilized the models and ground motion records previously described, ensuring consistency with the selected parameters and input conditions. For each analysis, the maximum displacement of the center of rigidity was extracted along both the *X* and *Y* directions. The maximum displacements were then averaged across the 20 analyses for both the LRB and CSS isolation systems, providing insight into the relative performance of each system. The resulting average displacements are summarized in Table 12, highlighting the effectiveness of each isolator type in controlling seismic displacements.

Table 12. Results of NTHA displacement and base shear force for each isolator type.

Quantity	LRB		CSS	
	NTHA-X	NTHA-Y	NTHA-X	NTHA-Y
Max displacement at the center of rigidity of the isolation system [mm]	310.7	298.0	142.9	157.3
Base shear [kN]	7087.7	5432.7	6329.5	5717.1
α	0.197	0.15	0.176	0.16

To evaluate the lateral forces acting on each floor and determine the base shear, the nodal accelerations at every floor were extracted for every time step of the analysis. The lateral force at each node was calculated by multiplying the nodal acceleration by the corresponding mass. These nodal forces were summed for each floor to determine the total lateral force at each time step. From these data, the maximum lateral forces experienced by each floor during the analyses were identified, and the average maximum lateral forces were computed across all 20 ground motions for both LRB and CSS systems. The results are presented in Table 13, which provides a detailed comparison of the lateral force distribution across the height of the structure.

Table 13. Results of NTHA story shear forces [kN] for each isolator type.

Level	LRB		CSS	
	NTHA-X	NTHA-Y	NTHA-X	NTHA-Y
6	1037.0	864.2	1027.3	928.1
5	1293.7	990.4	1191.8	1056.8
4	1199.5	877.3	1080.2	903.9
3	1127.8	863.0	1030.5	893.6
2	1159.6	888.1	991.7	930.3
1	1270.1	949.7	1008.0	1004.4
$\sum$	7087.7	5432.7	6329.5	5717.1

For the IDR, the displacement of the center of rigidity at each floor was extracted in both the X and Y directions for every time step. The IDR for each floor was calculated at every time step as the ratio of the relative displacement between adjacent floors to the story height. For each ground motion record, the maximum IDR was identified for every floor. Subsequently, the average maximum IDR was computed across all 20 ground motions for both LRB and CSS isolation systems. These results are summarized in Table 14, providing a clear comparison of the deformation behavior of the two isolator types.

Table 14. Results of NTHA interstory drift ratio for each isolator type.

Story	LRB		CSS	
	NTHA-X	NTHA-Y	NTHA-X	NTHA-Y
6	0.10%	0.15%	0.10%	0.17%
5	0.15%	0.26%	0.15%	0.29%
4	0.18%	0.26%	0.16%	0.28%
3	0.21%	0.31%	0.19%	0.32%
2	0.25%	0.36%	0.21%	0.36%
1	0.43%	1.76%	0.31%	1.54%

This rigorous analysis framework allows for a detailed evaluation of the seismic performance of the isolated structures, comparing key metrics such as displacement, lateral force, base shear, and interstory drift ratio. The selection of displacement, lateral force, base shear, and interstory drift ratio as primary evaluation metrics is justified by their direct impact on the design and performance assessment of seismically isolated structures. These metrics comprehensively capture both the isolator behavior (displacement and base shear) and the superstructure response (lateral force and IDR). By comparing these parameters across ASCE 7-22, EC8, and NTHA, this study provides a holistic evaluation of the reliability, accuracy, and conservatism of the simplified linear methods relative to a more advanced nonlinear approach. The results serve as a basis for assessing the relative effectiveness of the LRB and CSS systems in reducing seismic demands and enhancing structural resilience.

Tables 15 and 16 compare the predictions of the American code (ASCE 7-22), the European code (EC8), and NTHA for buildings using LRBs and CSSs, respectively. The comparison focuses on two primary parameters: the maximum displacement at the center of rigidity of the isolation system and the actual lateral seismic force above the base level, calculated without reductions from R-factors or q-factors. The percentage differences reported in Tables 15 and 16 are relative to the nonlinear time history analysis (NTHA) results, serving as a reference for comparison. These differences were calculated as the ratio of the deviation from NTHA to the NTHA value itself, ensuring a clear assessment of whether ASCE 7-22 and EC8 overestimate or underestimate key seismic response parameters.

Table 15. LRB: comparison between ASCE7-22, EC8, and NTHA.

Quantity	Direction NTHA		ASCE7-22			EC8		
	NTHA-X	NTHA-Y		Δx	Δy		Δx	Δy
D_M/d_{dc} [mm] [1]	310.7	298.0	471.6	52%	58%	338.8	9%	14%
V_s [2]	7087.7	5432.7	10,291.9	45%	89%	7901.8	11%	45%
α [3]	0.197	0.15	0.29	45%	89%	0.22	11%	45%

[1] Ratio between the maximum displacement at the center of rigidity of the isolation system in the direction under consideration and the design displacement. [2] The total unreduced lateral seismic design force above the base level. [3] Base shear coefficient.

Table 16. CSS: comparison between ASCE7-22, EC8, and NTHA.

Quantity	Direction NTHA		ASCE7-22			EC8		
	NTHA-X	NTHA-Y		Δx	Δy		Δx	Δy
D_M/d_{dc} [mm]	142.9	157.3	487.7	241%	210%	325.7	128%	107%
V_s [kN]	6329.5	5717.1	9712.6	53%	70%	7067.0	12%	24%
α	0.176	0.16	0.27	53%	70%	0.197	12%	24%

For buildings equipped with lead–rubber bearings, the American code (ASCE 7-22) significantly overestimates the maximum displacement when compared to NTHA results, with deviations reaching up to 58% in the Y-direction and 52% in the X-direction. In contrast, the European code (EC8) provides displacement estimates that are closer to the actual values, differing by only 14% in the Y-direction and 9% in the X-direction. Similarly, the lateral seismic forces predicted by ASCE 7-22 are considerably higher than those derived from NTHA, with differences of up to 89% in the Y-direction and 45% in the X-direction. The European code, while still overestimating the forces, performs better, with differences of 45% in the Y-direction and 11% in the X-direction.

These findings indicate that the European code aligns more closely with the actual behavior of the structure as determined by NTHA, particularly for both displacements and lateral forces. However, it still demonstrates some degree of conservatism. In contrast, the American code exhibits a much higher level of conservatism, particularly for displacement and force estimates, which may lead to over-designed and unnecessarily costly structures.

For structures employing CSSs, as shown in Table 16, both the American code (ASCE 7-22) and the European code (EC8) overestimate the maximum displacement by a considerable margin, with ASCE demonstrating much higher deviations. Specifically, ASCE overestimates the displacement by 241% in the X-direction and 210% in the Y-direction, while EC8 overestimates by 128% in the X-direction and 107% in the Y-direction. Similarly, for lateral seismic forces, ASCE continues to show significant overestimation, with deviations of 53% in the X-direction and 70% in the Y-direction. In contrast, EC8 presents more moderate overestimations for lateral forces, with deviations of 12% in the X-direction and 24% in the Y-direction. These findings highlight that, although both codes adopt conservative approaches, ASCE consistently yields higher and more conservative values, particularly for displacement estimates, which may lead to excessively robust and costly designs. EC8, while still conservative, provides results that are comparatively closer to actual performance as derived from NTHA.

These findings emphasize the practicality of applying the simplified method outlined in EC8 for LRB systems, as its predictions align more closely with actual performance. Alternatively, ASCE 7-22, while yielding more conservative results, prioritizes safety and can also be utilized effectively for LRB systems. However, for CSSs, the simplified method is less suitable due to the significant discrepancies it produces when compared to actual performance. In such cases, more precise methodologies, such as NTHA, should be employed for the final design to ensure accuracy and reliability. This conclusion is consistent with provisions in the European code, specifically Section 10.9.2. (1) of EC8. The code explicitly allows the use of equivalent linear models for isolation systems incorporating laminated elastomeric bearings, where the behavior can be adequately captured through linear assumptions. However, for elasto-plastic devices, such as friction-based systems, EC8 mandates the application of a bilinear hysteretic model to accurately represent the nonlinear behavior of these isolators. As a result, simplified linear analysis methods are not applicable for such systems, reinforcing the necessity of nonlinear time-history analysis to achieve a reliable and accurate design.

Tables 17 and 18 present a comparative analysis of the story shear forces calculated using three methods, EC8, ASCE 7-22, and NTHA, for structures employing LRBs and CSSs, respectively. For the LRB system, EC8 exhibits relative differences compared to NTHA. In the X-direction, which represents the stiffer direction of the structure, the differences range from −4% to 22%, indicating a relatively close match with NTHA in many cases. In the Y-direction, which is less stiff, EC8 results show increases of up to 59% compared to NTHA. On the other hand, ASCE 7-22 consistently predicts higher story shear forces than NTHA, with increases reaching up to 95% in the X-direction and as high as 160% in the Y-direction. These findings highlight the more conservative nature of ASCE, especially in the less stiff direction. For CSSs, EC8 also shows differences from NTHA. In the X-direction, the differences range from −14% to 27%, while in the Y-direction, the values vary from −4% to 37%, demonstrating a wider variability compared to the LRB case. ASCE 7-22 again yields higher values, with the most significant increase being 135%, reflecting its consistently conservative approach for shear force predictions across both directions.

Table 17. LRB: story shear forces.

Level	NTHA [kN]		ASCE7-22 [kN]			EC8 [kN]		
	NTHA-X	NTHA-Y		Δx	Δy		Δx	Δy
6	1037.0	864.2	1197.5	15%	39%	991.8	−4%	15%
5	1293.7	990.4	1655.2	28%	67%	1370.8	6%	38%
4	1199.5	877.3	1655.2	38%	89%	1370.8	14%	56%
3	1127.8	863.0	1655.2	47%	92%	1370.8	22%	59%
2	1159.6	888.1	1655.2	43%	86%	1370.8	18%	54%
1	1270.1	949.7	2473.4	95%	160%	1426.8	12%	50%
Σ	7087.7	5432.7	10,291.9			7901.8		

Table 18. CSS: story shear forces.

Level	NTHA [kN]		ASCE7-22 [kN]			EC8 [kN]		
	NTHA-X	NTHA-Y		Δx	Δy		Δx	Δy
6	1027.3	928.1	1125.6	10%	21%	887.0	−14%	−4%
5	1191.8	1056.8	1555.8	31%	47%	1226.0	3%	16%
4	1080.2	903.9	1555.8	44%	72%	1226.0	13%	36%
3	1030.5	893.6	1555.8	51%	74%	1226.0	19%	37%
2	991.7	930.3	1555.8	57%	67%	1226.0	24%	32%
1	1008.0	1004.4	2363.9	135%	135%	1276.1	27%	27%
Σ	6329.5	5717.1	9712.6			7067.04		

These results underscore the variability in performance predictions among the methods. While EC8 aligns more closely with NTHA in many cases, ASCE 7-22 consistently provides more conservative estimates, particularly for shear forces, which may lead to over-design in some scenarios.

Regarding the IDR, as presented in Tables 19 and 20, both ASCE 7-22 and EC8 consistently overestimate the IDR across all stories for structures equipped with LRBs. However, EC8 provides results that are closer to the NTHA values compared to ASCE. Notably, for the first story, both standards underestimate the IDR, with EC8 showing a deviation of up to −31% and ASCE underestimating by −11%. This indicates that while EC8 generally aligns better with NTHA, it significantly underestimates first-story deformations for LRB systems. For structures with CSSs, EC8 also underestimates the IDR for the first story, demonstrating a recurring limitation of both codes in accurately predicting first-story drift for isolator types. These findings suggest that while both standards adopt conservative approaches for upper-story IDRs, they may require additional considerations or adjustments to better capture the behavior of the first story, particularly for CSS systems.

Table 19. LRB: interstory drift ratio.

Level	NTHA		ASCE7-22				EC8			
	NTHA-X	NTHA-Y	X	Δx	Y	Δy	X	Δx	Y	Δy
6	0.10%	0.15%	0.11%	15%	0.21%	39%	0.09%	−5%	0.17%	15%
5	0.15%	0.26%	0.18%	20%	0.38%	47%	0.15%	0%	0.32%	22%
4	0.18%	0.26%	0.22%	26%	0.43%	63%	0.18%	4%	0.35%	35%
3	0.21%	0.31%	0.28%	30%	0.54%	74%	0.23%	7%	0.45%	44%
2	0.25%	0.36%	0.33%	31%	0.67%	87%	0.27%	6%	0.55%	52%
1	0.43%	1.76%	0.38%	−11%	1.73%	−2%	0.30%	−31%	1.33%	−24%

Table 20. CSS: interstory drift ratio.

Level	NTHA		ASCE7-22				EC8			
	NTHA-X	NTHA-Y	X	Δx	Y	Δy	X	Δx	Y	Δy
6	0.10%	0.17%	0.11%	6%	0.19%	15%	0.08%	−17%	0.15%	−9%
5	0.15%	0.29%	0.17%	15%	0.36%	24%	0.13%	−9%	0.28%	−2%
4	0.16%	0.28%	0.21%	27%	0.40%	43%	0.16%	−1%	0.32%	12%
3	0.19%	0.32%	0.26%	37%	0.51%	59%	0.20%	7%	0.40%	25%
2	0.21%	0.36%	0.31%	45%	0.64%	78%	0.24%	11%	0.49%	37%
1	0.31%	1.54%	0.36%	14%	1.63%	6%	0.26%	−15%	1.19%	−22%

6. Conclusions

This study outlines the design requirements for seismically isolated structures, emphasizing the distinctions between the U.S. code (ASCE 7-22) and the European code (EC8). It provides a detailed comparison of the simplified design procedures for seismic isolation as prescribed by both codes and evaluates their effectiveness through NTHA. A case study was conducted on a six-story steel building, which was analyzed under the provisions of each code using two types of isolation systems: LRBs and CSSs. The building was subjected to 20 recorded ground motions selected and scaled to match a CMS. Key performance metrics, including displacement at the isolation level, story shear forces, and interstory drift ratios, were compared. The main findings of the study are summarized below:

1. Seismic Hazard Definitions: ASCE 7-22 and EC8 differ significantly in their definitions of seismic hazard for the design of isolated structures. ASCE 7-22 considers a 2475-year return period for earthquakes (2% probability of exceedance in 50 years),

whereas EC8 uses a 245-year return period (10% probability of exceedance in 50 years) as the basis for design. This fundamental difference leads to varying levels of conservatism between the two codes.

2. Displacement-Based Design: Both codes use displacement-based design principles in their ELF or simplified linear analysis methods. This approach assumes that the superstructure behaves nearly as a rigid body, concentrating deformations at the isolation level.
3. Displacement Spectra Differences: A key difference between ASCE and EC8 lies in their displacement spectra [38–40]. EC8 defines an earlier corner period (e.g., 2 s in this study), resulting in lower design displacements for effective periods exceeding 2 s. In contrast, ASCE 7-22 uses a later corner period (e.g., 8 s), leading to higher displacements and force estimates.
4. Requirements for Simplified Methods: Both codes specify conditions under which the simplified method is permissible. These include considerations for the effective stiffness and damping, the structural regularity, and the restoring force provided by the isolation system. However, EC8 imposes more stringent requirements for these criteria, reflecting its greater emphasis on isolator performance and structural regularity.
5. Bounding Analysis and Device Testing: both codes mandate a bounding analysis to account for variability in isolator properties and require extensive testing of the isolation devices to ensure reliability under seismic conditions.
6. Amplification Factor in EC8: EC8 introduces an amplification factor (γ_x) to enhance the reliability of the isolators. This factor, however, does not influence the design displacement (d_{dc}) or the lateral seismic forces of the structure. The misconception that this factor equates EC8's requirements to a 2475-year return period earthquake, as in ASCE, is incorrect.
7. Conservatism of Simplified Methods: Both codes' simplified methods provide conservative estimates compared to NTHA, particularly for the design displacement of the isolation system. While conservatism enhances safety, excessive conservatism can lead to over-designed and cost-prohibitive structures.
8. Performance for LRBs: For LRB systems, both codes produce results that align well with NTHA. The design displacement (d_{dc}) and lateral seismic forces (V_s) calculated using EC8 show close agreement with NTHA results, making the simplified method reliable for LRB systems.
9. Performance for CSSs: Discrepancies arise for CSS systems, where EC8's design displacement deviates by up to 128% from NTHA results. ASCE 7-22, however, significantly overestimates design displacement and forces, with deviations of up to 241% compared to NTHA. The underestimation of first-story IDR for CSS isolators in ASCE 7-22 and EC8 likely results from a combination of equivalent linearization assumptions and isolator modeling limitations. Standard methods neglect higher-mode contributions, velocity-dependent friction effects, and bidirectional coupling, leading to lower drift predictions than those observed in nonlinear time-history analysis. To better capture first-story drift behavior, refined nonlinear modeling and site-specific analysis may be necessary, particularly for structures with flexible superstructures or significant torsional effects.
10. Reliability and Precision: EC8's simplified method is generally more reliable for LRB systems, producing results closer to observed performance. In contrast, ASCE 7-22, though more conservative, may be preferable where safety is prioritized. For CSS systems, however, reliance on simplified methods is less advisable, and more precise analytical techniques should be used in the final design stages.

This study highlights that while ASCE 7-22 and EC8 provide solid frameworks for the design of seismically isolated structures, their inherent conservatism should be carefully evaluated. Designers must balance safety and cost-efficiency, especially when the simplified methods produce overly conservative results. For LRB systems, EC8 appears to offer a more practical and reliable design approach. Conversely, for CSS systems, reliance on more detailed analyses, such as NTHA, is essential to ensure both performance and cost-effectiveness.

Author Contributions: Conceptualization, E.Y. and M.F.; methodology, E.Y.; software, E.Y.; validation, E.Y.; resources, M.F. and R.N.; writing—original draft preparation, R.N.; writing—review and editing, R.N. and A.P.; supervision, A.P. All authors have read and agreed to the published version of the manuscript.

Funding: This research received no external funding.

Data Availability Statement: Data are contained within the article.

Conflicts of Interest: The authors declare no conflicts of interest.

References

1. Feng, D.; Liu, W.; Masuda, K.; Wang, S.; Huan, S. A comparative study of seismic isolation codes worldwide, part 1 Design Spectrum. In Proceedings of the First European Conference on Earthquake Engineering and Seismology, Geneva, Switzerland, 3–8 September 2006.
2. Merino, R.J.; Perrone, D.; Nascimbene, R.; Filiatrault, A. Performance-based seismic classification of acceleration-sensitive non-structural elements. *Earthq. Eng. Struct. Dyn.* **2023**, *52*, 4222–4244. [CrossRef]
3. Christopoulos, C.; Filatrault, A. *Principles of Passive Supplemental Damping and Seismic Isolation*, 1st ed.; IUSS Press: Pavia, Italy, 2006.
4. Grant, D. Seismic isolation and supplemental damping. In *Seismic Design of Buildings*, 2nd ed.; CRC Press: London, UK, 2016.
5. *ASCE 7-22*; Minimum Design Loads and Associated Criteria for Buildings and Other Structures (ASCE/SEI 7-22). American Society of Civil Engineers: Reston, VA, USA, 2022.
6. *EN 1998-1*; Eurocode 8: Design of Structures for Earthquake Resistance—Part 1: General Rules, Seismic Actions and Rules for Buildings. European Committee for Standardization: Brussels, Belgium, 2004.
7. Tsai, K.-C.; Popov, E.P. *Steel Beam-Column Joints in Seismic Moment Resisting Frames*; Earthquake Engineering Research Center, University of California: Berkeley, CA, USA, 1988.
8. Hall, J. Seismic response of steel frame buildings to near-source ground motions. *Earthq. Eng. Struct. Dyn.* **1998**, *27*, 1445–1464. [CrossRef]
9. Filiatrault, A.; Tremblay, R.; Wanitkorkul, A. Performance evaluation of passive damping systems for the seismic retrofit of steel moment-resisting frames subjected to near-field ground motions. *Earthq. Spectra* **2001**, *17*, 427–456. [CrossRef]
10. International Conference of Building Officials. *Uniform Building Code (UBC)*; International Conference of Building Officials: Lansing, MI, USA, 1994.
11. USGS Earthquake Hazard Toolbox. Available online: https://earthquake.usgs.gov (accessed on 20 December 2024).
12. SeismoStruct2024. A Computer Program for Static and Dynamic Nonlinear Analysis of Framed Structures. Available online: https://seismosoft.com/ (accessed on 20 December 2024).
13. Nascimbene, R. Numerical model of a reinforced concrete building: Earthquake analysis and experimental validation. *Period. Polytech. Civ. Eng.* **2015**, *59*, 521–530. [CrossRef]
14. Almeida, J.P.; Correia, A.A.; Pinho, R. Force-based higher-order beam element with flexural-shear-torsional interaction in 3D frames. Part II: Applications. *Eng. Struct.* **2015**, *89*, 218–235. [CrossRef]
15. Mazzotta, V.; Brunesi, E.; Nascimbene, R. Numerical modeling and seismic analysis of tall steel buildings with braced frame systems. *Period. Polytech. Civ. Eng.* **2017**, *61*, 196–208. [CrossRef]
16. Fan, H.; Li, Q.S.; Tuan, A.Y.; Xu, L. Seismic analysis of the world's tallest building. *J. Constr. Steel Res.* **2009**, *65*, 1206–1215. [CrossRef]
17. Nagarajaiah, S.; Reinhorn, A.M.; Constantinou, M.C. *Nonlinear Dynamic Analysis of Three-Dimensional Base Isolated Structures (3D-Basis)*; National Center for Earthquake Engineering Research: Buffalo, NY, USA, 1989.
18. Song, B.I.; Giriunas, K.A.; Sezen, H. Progressive collapse testing and analysis of a steel frame building. *J. Constr. Steel Res.* **2014**, *94*, 76–83. [CrossRef]

19. Rodrigues, D.; Brunesi, E.; Nascimbene, R. Fragility and sensitivity analysis of steel frames with bolted-angle connections under progressive collapse. *Eng. Struct.* **2021**, *228*, 111508. [CrossRef]
20. Yang, B.; Tan, K.H. Experimental tests of different types of bolted steel beam-column joints under a central-column-removal scenario. *Eng. Struct.* **2013**, *54*, 112–130. [CrossRef]
21. Gonzalez, A.; Spacone, E.; Nascimbene, R. Performance-based seismic design framework for RC floor diaphragms in dual systems. *Procedia Eng.* **2017**, *199*, 3546–3551. [CrossRef]
22. Sarkis, A.; Sullivan, T.; Brunesi, E.; Nascimbene, R. Investigating the Effect of Bending on the Seismic Performance of Hollow-Core Flooring. *Int. J. Concr. Struct. M.* **2023**, *17*, 18. [CrossRef]
23. Spacone, E.; Filippou, F.C.; Taucer, F.F. Fibre beam–column model for non-linear analysis of R/C frames: Part I. Formulation. *Earthq. Eng. Struct. Dyn.* **1996**, *25*, 711–725. [CrossRef]
24. Pavese, A.; Lanese, I.; Nascimbene, R. Seismic Vulnerability Assessment of an Infilled Reinforced Concrete Frame Structure Designed for Gravity Loads. *J. Earthq. Eng.* **2017**, *21*, 267–289. [CrossRef]
25. American Institute of Steel Construction (AISC). Available online: https://www.aisc.org (accessed on 23 December 2024).
26. Hough, S.; Graves, R.W.; Cochran, E.S.; Yoon, C.E.; Blair, L.; Haefner, S.; Wald, D.J.; Quitoriano, V. The 17 January 1994 Northridge, California, Earthquake: A Retrospective Analysis. *Seism. Rec.* **2024**, *4*, 151–160. [CrossRef]
27. Richards, P.W. Finite Element Evaluation of New Retrofit Schemes for Pre-Northridge Connections. *Lect. Notes Civ. Eng.* **2024**, *520*, 552–563.
28. Park, Y.J.; Wen, Y.K.; Ang, H.S. Random vibration of hysteretic systems under bi-directional ground motions. *Earthq. Eng. Struct Dyn.* **1986**, *14*, 543–557. [CrossRef]
29. Wen, Y.K. Method for Random Vibration of Hysteretic Systems. *J. Eng. Mech. ASCE Div.* **1967**, *102*, 249–263. [CrossRef]
30. Nagarajaiah, S.; Reinhorn, A.M.; Constantinou, M.C. *3D-BASIS—Nonlinear Dynamic Analysis of Three-Dimensional Base Isolated Structures: Part II.*; National Center for Earthquake Engineering Research: Buffalo, NY, USA, 1991.
31. Constantinou, M.C.; Tsopelas, P.; Kasalanati, A.; Wolff, E.D. *Property Modification Factors for Seismic Isolation Bearings*; National Center for Earthquake Engineering Research: Buffalo, NY, USA, 1999.
32. Baker, J.W. CS_Selection. Available online: https://github.com/bakerjw/CS_Selection (accessed on 23 December 2024).
33. Baker, J.W.; Cornell, C.A. Correlation of response spectral values for multicomponent ground motions. *BSSA* **2006**, *96*, 215–227. [CrossRef]
34. Baker, J.W.; Lee, C. An Improved Algorithm for Selecting Ground Motions to Match a Conditional Spectrum. *J. Earthq. Eng.* **2018**, *22*, 708–723. [CrossRef]
35. Baker, J.W.; Jayaram, N. Correlation of Spectral Acceleration Values from NGA Ground Motion Models. *Earthq. Spectra* **2008**, *24*, 299–317. [CrossRef]
36. Baker, J.W. Conditional Mean Spectrum: Tool for ground motion selection. *J. Struct. Eng.* **2011**, *137*, 322–331. [CrossRef]
37. NGA-West2 Database. Available online: https://ngawest2.berkeley.edu (accessed on 23 December 2024).
38. Merino Vela, R.J.; Brunesi, E.; Nascimbene, R. Derivation of floor acceleration spectra for an industrial liquid tank supporting structure with braced frame systems. *Eng. Struct.* **2018**, *171*, 105–122. [CrossRef]
39. Ibrahim, A.; El-Araby, I.; Saleh, A.; Shaaban, M. Comparison between the Egyptian and international codes based on seismic response of mid- to high-rise moment resisting framed buildings. *Struct. Eng. Mech.* **2023**, *87*, 347–361.
40. Assi, R.; Mazloom, S.; Abouda, A. Effect of nonlinearity on seismic response of nonstructural components in a code-conforming RC shear wall building. *Can. J. Civ. Eng.* **2024**, *52*, 89–104. [CrossRef]

designs

MDPI

Article

Artificial-Intelligence-Based Prediction of Crack and Shrinkage Intensity Factor in Clay Soils During Desiccation

Abolfazl Baghbani [1], Tanveer Choudhury [2] and Susanga Costa [1,*]

[1] School of Engineering, Deakin University, Waurn Ponds, VIC 3216, Australia; abaghbani@deakin.edu.au
[2] The Institute of Innovation, Science and Sustainability, Federation University, Churchill, VIC 3842, Australia; t.choudhury@federation.edu.au
* Correspondence: susanga.costa@deakin.edu.au; Tel.: +613-5277-3698

Abstract: Desiccation-induced cracking in clay soils significantly affects the structural performance and durability of geotechnical systems. This study presents a data-driven approach to predict the Crack and Shrinkage Intensity Factor (CSIF), a comprehensive index quantifying both crack formation and shrinkage behavior in drying soils. A database of 100 controlled desiccation tests was developed using five clay mixtures with varying plasticity indices, which were subjected to a range of drying rates, soil thicknesses and initial conditions. Four predictive models—Multiple Linear Regression (MLR), Classification and Regression Random Forest (CRRF), Artificial Neural Network (ANN) and Genetic Programming (GP)—were evaluated. The ANN model using Bayesian Regularization demonstrated superior performance (R = 0.99, MAE = 5.44), followed by CRRF and symbolic GP equations. Sensitivity analysis identified drying rate and soil thickness as the most influential parameters, while initial moisture content and ambient conditions were found to be redundant when the drying rate was included. This study not only advances the predictive modeling of desiccation cracking but also introduces interpretable equations for practical engineering uses. The developed models offer valuable tools for crack risk assessment in clay liners, soil covers and expansive soil foundations.

Keywords: desiccation cracking; clay soils; CRRF; ANN; GP; artificial intelligence

1. Introduction

Desiccation cracking in soil is multi-physic in nature and is a highly non-linear phenomenon which involves hydraulic, mechanical and thermal processes. Numerical models developed in the past have been able to capture specific elements of the desiccation cracking process, such as the number of cracks and the depth of cracks (e.g., [1–4], but were not able to model the complete process under given soil properties and boundary conditions. One limitation of numerical models is that they are often limited by the size of the computational domain that can be modeled. This is because the number of grid elements required to accurately model the domain increases exponentially with the domain size. As a result, numerical models of desiccation cracking are often limited to small-scale laboratory experiments or to simplified geometries. Another limitation of numerical models is that they require accurate input parameters, such as material properties and boundary conditions, which can be difficult to measure experimentally. Inaccurate input parameters can lead to inaccurate predictions, and it can be challenging to identify the sources of error in the model.

Analytical models, on the other hand, often make simplifying assumptions to handle complicated equations and to make them solvable in a closed form. For example, some

https://doi.org/10.3390/designs9030054

analytical models assume that the soil is homogeneous and isotropic, and that the cracks are straight and parallel [5–9]. While these assumptions may be reasonable in some cases, they can lead to inaccuracies in the predictions. Another limitation of analytical models is that they often assume linear behavior, which is not often accurate for materials undergoing desiccation cracking [10]. This can limit their usefulness in predicting the behavior of real-world materials.

The above-mentioned limitations of numerical and analytical models can be overcome by artificial intelligence (AI) and machine learning (ML) tools. AI algorithms, such as Artificial Neural Networks (ANNs), random forest (RD) and deep learning, can learn from data and make predictions without explicitly modeling the underlying physical and chemical processes. This allows for more accurate and efficient predictions of desiccation cracking. This can lead to new insights and a deeper understanding of the underlying processes that drive desiccation cracking.

AI has already been successfully applied in various geotechnical fields [11], such as slope stability [12,13], tunneling [14,15], rock mechanics [16,17], foundations [18], etc. For example, Luo et al. [13] applied particle swarm optimization combined with cellular automata (PSO-CA) for slope stability prediction, while Jamhiri et al. [19] used probabilistic machine learning methods to predict desiccation cracking under uncertainty. Sharma et al. [17] employed machine learning models to predict rock fragmentation in coal seams. These studies demonstrate the growing utility of AI in handling geotechnical datasets characterized by complexity and non-linearity. Building upon these advances, the current study aims to refine the prediction of Crack and Shrinkage Intensity Factor (CSIF) and offers both improved accuracy and deeper insight into influential parameters.

Choudhury and Costa [20] used ANN to predict desiccation cracks in thin, long clay layers. The output of the model was the number of cracks, which was predicted with a high accuracy (R) of 99% and a mean absolute error (MAE) of 0.11. However, the dataset used in this study was relatively small—16 samples. Baghbani et al. [21] used a classification and regression tree (CART) model to examine a 31-set database of desiccation cracking. The results indicated that CART performed satisfactorily. In both studies ([20,21]), four inputs were considered: initial moisture content, specimen thickness, specimen length and specimen width, and the output was the number of cracks. However, not all influential parameters were considered in these studies. For example, drying rate and dry density were not among the inputs. The output used—number of cracks—is not a true representation of the severity of cracking and cannot be used for comparison as it does not describe the morphology of cracks. Another limitation is the size of the database, which affects prediction intervals.

This study contributes to the field by developing and comparing four mathematical models—MLR, CRRF, ANN and GP—for accurately predicting the Crack and Shrinkage Intensity Factor (CSIF) during clay soil desiccation. By generating a comprehensive experimental database of 100 drying tests across a range of soil types and environmental conditions, this research identifies the most influential parameters governing desiccation behavior. This study not only demonstrates the superior performance of AI-based models, particularly ANN with Bayesian Regularization, but also introduces a symbolic GP-based equation for practical use. Importantly, this is the first known study to successfully apply AI to predict the CSIF with such high accuracy, offering a scalable and interpretable framework for future field and laboratory applications in geotechnical engineering.

2. Methodology

A large number of soil drying tests were conducted to generate a sizeable database for the model. The scope of this study was limited to parallel cracks in clay, which are usually observed in long soil layers. Hence, the soil samples for drying were prepared in thin, long molds similar to the work of Nahlawi and Kodikara [22] and Costa et al. [23]. This type of crack can be considered one-dimensional (1D).

2.1. Materials

Two types of clay, kaolin and bentonite, were used to prepare a range of soil mixtures with different plasticity characteristics. According to Table 1, the different weight percentages of these two types of clays were mixed. The Atterberg limit tests, including LL and PL, were performed on the 5 mixtures described in Table 1.

Table 1. Physical characteristics of the test mixtures.

Mixture No.	Kaolin (%)	Bentonite (%)	LL (%)	PL (%)	PI (%)
1	100	0	74	34	40
2	75	25	105	34	71
3	50	50	135	35	100
4	25	75	165	35	130
5	0	100	205	36	169

2.2. Shrinkage and Cracking Test

The Australian Standard of linear shrinkage [24] was followed for the shrinkage and crack testing of mixtures. According to Tables 1 and 2, different dry mixtures were prepared based on different percentages of kaolinite and bentonite. The LL and PI were used as input parameters to quantify the mixture type.

Table 2. Experimental test plan for 1D desiccation cracking: soil properties and environmental conditions.

<table>
<tr><th>Mixtures</th><th>LL (%)</th><th>PI (%)</th><th>Thickness (mm)</th><th>Initial Moisture Content (%)</th><th>Dry Density (kg/m^3)</th><th>Drying Rate (g/min.)</th><th>Relative Humidity (%)</th><th>Ambient Temperature (°)</th></tr>
<tr><td>1</td><td>74</td><td>40</td><td>5</td><td rowspan="5">Both LL and 1.5 LL</td><td rowspan="5">Depends on the mixtures</td><td rowspan="5">Both D and 0.75 D</td><td rowspan="5">Both 17 and 19</td><td rowspan="5">Both 38 and 50</td></tr>
<tr><td>2</td><td>105</td><td>71</td><td>10</td></tr>
<tr><td>3</td><td>135</td><td>100</td><td>15</td></tr>
<tr><td>4</td><td>165</td><td>130</td><td>20</td></tr>
<tr><td>5</td><td>205</td><td>169</td><td>25</td></tr>
</table>

The cracking test plan included five different soil layer thicknesses, 5, 10, 15, 20 and 25 mm, based on previous laboratory tests on desiccation cracking [25]. For achieving the desired thickness, rectangular acrylic molds with widths and lengths of 25 and 600 mm, respectively, and target heights were prepared (Figure 1).

Soil mixtures were prepared at LL and 1.5 LL moisture contents and were cured for 24 h at room temperature to attain moisture equilibrium. During filling, air bubbles were removed from the sample by tapping for 30 s.

Two different drying rates were used, controlled by four lights over the length of the sample: for a higher drying rate (*D*), all four lights were on, and for a lower drying rate (0.75 *D*), two lights were on. Figure 2 shows the schematic drawing of the setup used for the drying tests.

All drying tests were conducted in a closed indoor laboratory environment to maintain consistency across experiments. While ambient temperature and relative humidity were allowed to vary slightly to reflect realistic conditions, their range was narrow—between 39.8 °C and 50.0 °C for temperature, and 15.9% to 17.9% for relative humidity—as listed in Table 2. No airflow or wind sources were present during the tests. These conditions ensured minimal environmental noise and a high consistency in the drying process, providing a reliable basis for model training and validation.

Figure 1. Setting up molds for the parallel (1D) cracks.

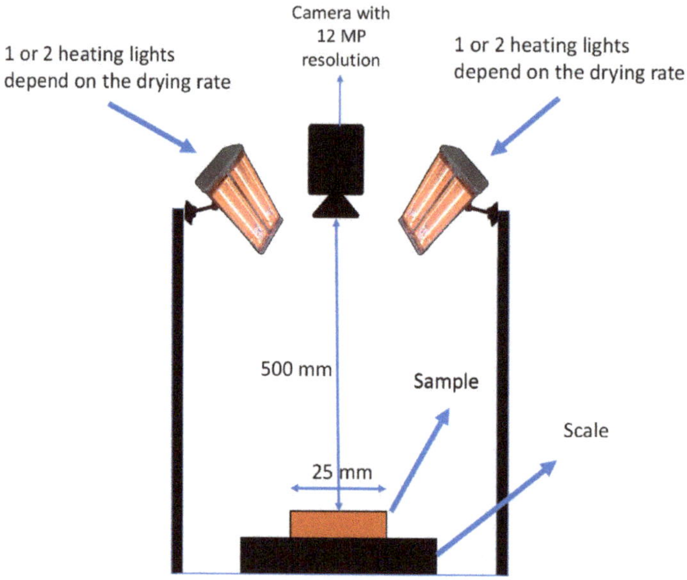

Figure 2. Schematic drawing of the setup used for the shrinkage quantification.

All tests were conducted for a duration of 270 min, and the drying rate was measured accordingly by using Equation (1) for each test. The drying rate (D) was assumed to be uniform during the drying period, as given by Equation (1).

$$D = \frac{m_f - m_i}{t} \quad (1)$$

where m_f is the final sample mass, m_i is the initial sample mass and t is the total test time. The surrounding ambient temperature and humidity were also recorded during the drying process.

2.3. Shrinkage and Crack Quantification

In order to quantify shrinkage and cracks, images were taken every 30 min with a 12-megapixel camera. The final Crack and Shrinkage Intensity Factor (*CSIF*) was calculated using the final image taken after 270 min at the end of the test. The image analysis software ImageJ (Version 1.51) [26] was used to analyze the images and to calculate *CSIF* according to Equation (2).

$$CSIF = (Area\ of\ cracks\ and\ shrinkage)/(Initial\ sample\ area\ before\ drying) \quad (2)$$

2.4. Modeling Methods

2.4.1. Multiple Linear Regression (MLR)

Multiple Linear Regression (MLR) is a fundamental statistical technique used to model the relationship between a dependent variable and multiple independent variables through a linear equation. It estimates regression coefficients by minimizing the sum of the squared differences between observed and predicted values, allowing for the straightforward interpretation of each input's contribution to the output. While MLR is computationally efficient and easy to implement, it inherently assumes linearity, independence and homoscedasticity among variables. These assumptions may limit its predictive performance when modeling highly non-linear processes such as soil desiccation cracking. Despite this, MLR serves as a useful baseline model for comparison with the more advanced machine learning techniques explored in this study.

2.4.2. Classification and Regression Random Forest (CRRF)

Breiman [27] combined decision trees with bootstrap aggregation or bagging to reduce classification and regression errors. Classification and Regression Random Forest (CRRF) is an ensemble learning method that combines multiple decision trees to improve predictive performance and reduce overfitting. Developed by Breiman [27], the random forest algorithm constructs numerous decision trees using bootstrap samples of the training data and random subsets of input variables at each split. For regression tasks, such as predicting the Crack and Shrinkage Intensity Factor (CSIF), the model aggregates the outputs of individual trees by averaging their predictions. This method is particularly robust when handling non-linear relationships, noisy data and complex interactions between variables. CRRF also provides insights into variable importance by evaluating the contribution of each input to the overall reduction in prediction error. Due to its simplicity, high accuracy and resistance to overfitting, CRRF is widely adopted in geotechnical engineering applications and serves as a reliable benchmark for evaluating model performance in this study.

2.4.3. Artificial Neural Network (ANN)

The ANN is a mathematical model composed of artificial neurons capable of finding complex, non-linear relationships between inputs and outputs. The ANN first identifies the relationship between inputs and outputs based on the weights that are the connecting power between two neurons [28]. Second, the network seeks to optimize the obtained weights, which is called the paradigm. This study used a feedforward paradigm [29], one of the strongest and most widely used paradigms. The feedforward paradigm has different algorithms. Bayesian Regularization (BR) [30] and Levenberg–Marquardt (LM) [31] algorithms with a strong ability in testing non-linear relationships were used in this study.

Input layers, hidden layers and output layers make up the ANN architecture. A total of five architecture types were considered, including one, two, three, four and five hidden layers, for each algorithm, LM and BR. The number of 60 neurons was selected after several trials. Each neuron was re-trained five times in every network. Figure 3 shows the ANN schematic architecture used in this study.

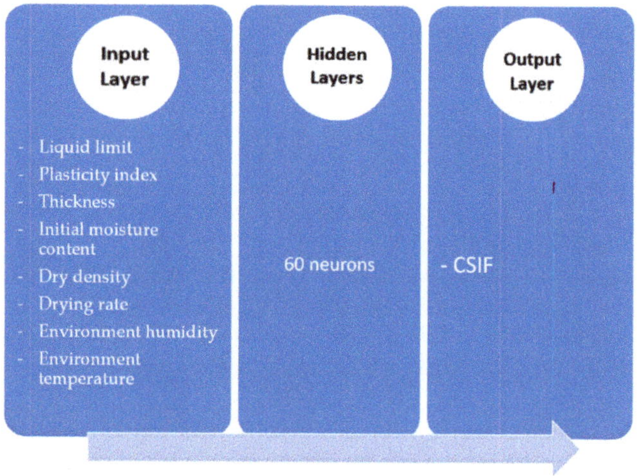

Figure 3. The architecture of the ANN in this study.

2.4.4. Genetic Programming (GP)

Genetic Programming (GP) is an evolutionary algorithm-based method that automatically evolves mathematical models or symbolic expressions to solve complex regression or classification problems. Inspired by the principles of natural selection and genetics, GP iteratively refines a population of candidate solutions—represented as computer programs or symbolic equations—through operations such as selection, crossover and mutation. In this study, GP was employed to derive explicit mathematical expressions for predicting the Crack and Shrinkage Intensity Factor (CSIF) using normalized input variables. Unlike traditional black-box models, GP offers interpretable and deployable formulas that can be directly used in practical engineering design without requiring specialized software. While GP may require significant computational effort and careful tuning to avoid overfitting, its ability to discover non-linear relationships and generate human-readable equations makes it a powerful tool for knowledge discovery and predictive modeling in geotechnical applications.

2.5. Data Processing

Drying tests resulted in 100 datasets with 8 effective inputs: dry density, initial water content, soil thickness, drying rate, LL, PI, ambient humidity and temperature; and one output parameter: the CSIF. The following steps were taken to prepare the database:

2.5.1. Normalization

In the existing database, each variable had its own unit. Data normalization reduces network error and increases the network training speed. Below is the normalization linear function which was used in this study.

$$X_{norm} = \frac{X - X_{min}}{X_{max} - X_{min}} \tag{3}$$

where X_{max}, X_{min}, X and X_{norm} are the maximum, minimum, actual and normalized values, respectively.

2.5.2. Testing and Training Databases

For all three mathematical models, the dataset was randomly divided into two subsets: 80% for training and 20% for testing. To ensure the consistency and reliability of the model performance, the statistical characteristics of both subsets, along with the minimum, maximum, mean and standard deviation, were examined. As shown in Tables 3 and 4, the training and testing datasets exhibited similar statistical distributions, indicating that the data segmentation was balanced and suitable for accurate model evaluation.

Table 3. Statistical information of training database.

Variable	Observations	Minimum	Maximum	Mean	Std. Deviation
CSIF	80	4.133	57.898	25.610	11.929
The drying rate (g/min)	80	0.142	0.252	0.207	0.031
Humidity (%RH)	80	15.900	17.900	16.850	1.005
Temperature (c)	80	39.800	50.000	44.645	5.126
Initial moisture (%)	80	71.097	312.500	175.452	71.677
LL	80	74.000	205.000	135.650	46.829
PI	80	40.000	169.000	100.709	46.114
Dry density	80	265.000	862.667	502.019	169.160
The thickness (mm)	80	5.000	25.000	14.625	7.016

Table 4. Statistical information of testing database.

Variable	Observations	Minimum	Maximum	Mean	Std. Deviation
CSIF	20	10.394	50.591	26.174	11.997
The drying rate (g/min)	20	0.148	0.250	0.213	0.034
Humidity (%RH)	20	15.900	17.900	17.100	1.005
Temperature (c)	20	39.800	50.000	45.920	5.127
Initial moisture (%)	20	74.439	249.300	158.787	48.480
LL	20	74.000	205.000	141.400	42.503
PI	20	40.000	169.000	106.371	41.854
Dry density	20	342.933	832.000	516.370	142.730
The thickness (mm)	20	5.000	25.000	16.500	7.452

For a more comprehensive assessment, k-fold cross-validation (e.g., 5-fold or 10-fold) could be implemented, especially in future studies involving larger and more diverse datasets. This would allow each data point to be used for both training and validation, thereby minimizing variance and better capturing the model's generalization capability. Although computational constraints limited its implementation in the current work—especially given the extensive hyperparameter tuning and re-training required for the ANN and GP models—we recognize its importance and recommend cross-validation as a key component in future extensions of this research.

3. Results

The experimental results obtained from the soil cracking tests and the results from the mathematical models are presented in the following sections.

3.1. Experimental Results from Cracking Tests

Based on the test plan in Table 2, 100 desiccation cracking tests were conducted under different test conditions. Figure 4 illustrates the CSIF data obtained from the cracking tests.

The change in CSIF with dry density and layer thickness followed the patterns identified by previous researchers [22,23].

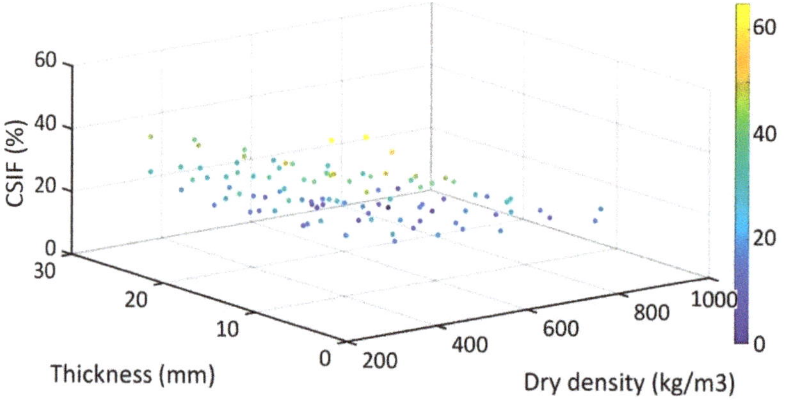

Figure 4. Crack and Shrinkage Intensity Factor (CSIF) of thin, long samples against dry density and soil layer thickness.

3.2. Effect of Input Variables

An important factor in network accuracy is the selection of input parameters. Five groups of MLR, CRRF and ANN models were evaluated for their impact on network accuracy (each ANN group contained ten ANNs, five networks with LM algorithms and five networks with BR algorithms). Table 5 shows the five input groups that were run with mathematical models.

Table 5. Different groups of ANN modeling with their input variables.

Group No.	Input Variables							
	Dry Density	Thickness	Drying Rate	Initial Water Content	Liquid Limit (LL)	Plasticity Index (PI)	Ambient Temperature	Ambient Humidity
1	√	√	√	√	√	√	√	√
2	√	√	√	√	√	√	×	×
3	√	√	√	×	√	√	×	×
4	√	√	×	√	√	√	×	×
5	√	√	×	×	√	√	×	×

3.3. Mathematical Modeling Results

The correlation coefficient (R) and mean absolute error (MAE) between the predicted and measured values were used to assess the model's performance. R and MAE are defined in Equations (4) and (5).

$$R = \frac{\sum_N \left(X_m - \overline{X_m}\right)\left(X_p - \overline{X_p}\right)}{\sqrt{\sum_N \left(X_m - \overline{X_m}\right)^2 \sum_N \left(X_p - \overline{X_p}\right)^2}} \quad (4)$$

$$MAE = \frac{\sum_N \left|\left(X_m - X_p\right)\right|}{N} \quad (5)$$

where X_m, X_p, $\overline{X_m}$ and $\overline{X_p}$ are the actual value, predicted value, the average of actual value and the average of predicted value, respectively, and N is the number of datasets. The ideal model is the one that has an R of 1 and an MAE equal to 0.

3.3.1. Multiple Linear Regression (MLR)

The Pearson correlation matrix of parameters is shown in Table 6. There is a strong correlation between ambient temperature and humidity and the drying rate, which is expected. The other inputs do not show any significant correlation, which indicates that multicollinearity does not affect them. More importantly, the correlation between inputs and the CSIF is less than 75%, which indicates that MLR cannot adequately predict CSIF. This affirms the non-linear nature of the desiccation cracking problem and the need for more sophisticated models to capture the cracking process accurately.

Table 6. Correlation matrix of parameters.

	Drying Rate	Ambient Humidity	Ambient Temperature	Initial Moisture	LL	PI	Dry Density	Thickness	CSIF
Drying rate	1	0.904	0.904	0.214	0.195	0.195	−0.231	0.290	0.335
Humidity	0.904	1	1.000	0.046	0.025	0.025	−0.050	−0.021	0.408
Temperature	0.904	1.000	1	0.046	0.025	0.025	−0.050	−0.021	0.408
Initial moisture	0.214	0.046	0.046	1	0.870	0.870	−0.947	0.010	0.692
LL	0.195	0.025	0.025	0.870	1	1.000	−0.888	−0.007	0.750
PI	0.195	0.025	0.025	0.870	1.000	1	−0.888	−0.007	0.750
Dry density	−0.231	−0.050	−0.050	−0.947	−0.888	−0.888	1	−0.015	−0.689
Thickness	0.290	−0.021	−0.021	0.010	−0.007	−0.007	−0.015	1	−0.452
CSIF	0.335	0.408	0.408	0.692	0.750	0.750	−0.689	−0.452	1

Figure 5 shows the actual and predicted values of CSIF for the training and testing databases for input groups 3 and 4. Based on the results of the MLR method, several points failed to meet 80% accuracy across all groups of databases.

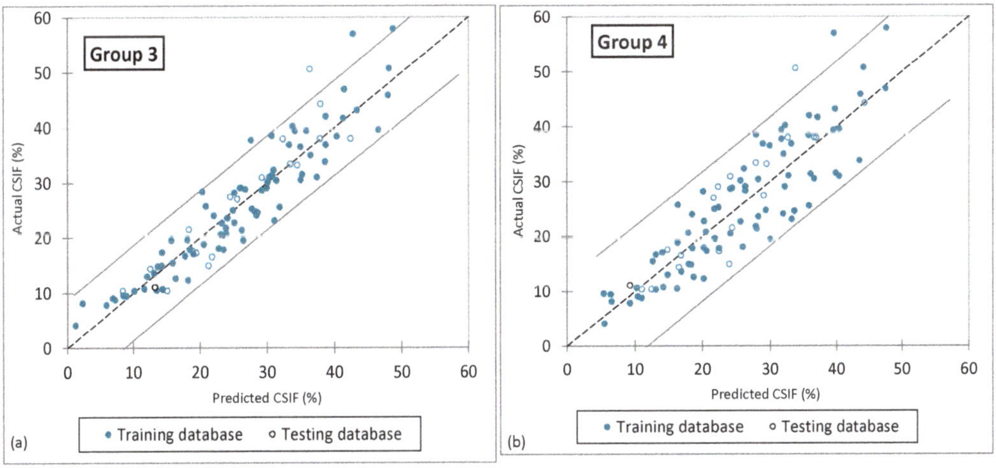

Figure 5. The results of the best MLR model to predict CSIF for the database groups (**a**) 3 and (**b**) 4.

Based on the R and MAE values shown in Table 7, input group 1 achieves the best MLR performance. Groups 1, 2 and 3 show small differences in R, while groups 4 and 5 show much lower R. The selection of input parameters in group 3 can be considered the optimum combination according to Table 7.

3.3.2. Classification and Regression Random Forest (CRRF)

Table 8 lists the tree and forest parameters used in the CRRF model. M_{try} is the number of random subsets of variables per node. Increasing the number of trees increased the analysis time. In addition, the accuracy of the tree increased with depth until it reached 10, then declined, so 10 was the optimal depth.

Table 7. Results of MLR models for different databases and groups.

Group Number	Number of Inputs	Inputs	Training Database		Testing Database	
			R	*MAE*	*R*	*MAE*
1	8	Dry density, thickness, drying rate, initial water content, liquid limit, plasticity index, ambient temperature, ambient humidity	0.853	10.412	0.847	10.447
2	6	Dry density, thickness, drying rate, initial water content, liquid limit, plasticity index	0.851	11.532	0.843	11.799
3	5	Dry density, thickness, drying rate, liquid limit, plasticity index	0.844	11.341	0.838	11.499
4	5	Dry density, thickness, initial water content, liquid limit, plasticity index	0.759	12.274	0.750	12.045
5	4	Dry density, thickness, liquid limit, plasticity index	0.734	15.376	0.721	15.674

Table 8. The specifications of the best CRRF.

Tree Parameters					Forest Parameters		
Min. Node Size	Min. Son Size	Max Depth	M_{try}	CP	Sampling	Sample Size	Number of Trees
2	1	10	2	0.00001	Random with replacement	100	1000

Figure 6 shows the actual and predicted values for CSIF for input groups 3 and 4, both for the training and test databases. Group 3 slightly outperforms group 4 in terms of the *R* and MAE values. The results for all five input groups are shown in Table 9. In the first three databases, namely groups 1, 2 and 3, the models have high *R*, while in the fourth and fifth databases, the models have lower *R*. However, group 3 can be considered the most desired database since it has fewer input parameters. The *R* value for group 3 is 0.97, which indicates good accuracy and performance.

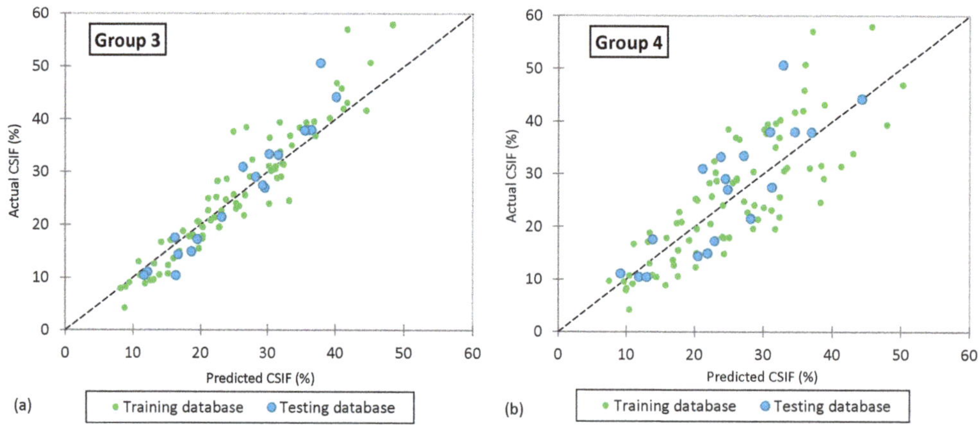

Figure 6. The results of the best CRRF model to predict CSIF for the input groups (**a**) 3 and (**b**) 4.

Table 9. The performance of the best CRRF model at predicting CSIF.

Group Number	Number of Inputs	Inputs	Training Database		Testing Database	
			R	MAE	*R*	MAE
1	8	Dry density, thickness, drying rate, initial water content, liquid limit, plasticity index, ambient temperature, ambient humidity	0.983	5.926	0.974	6.008
2	6	Dry density, thickness, drying rate, initial water content, liquid limit, plasticity index	0.978	6.062	0.965	6.329
3	5	Dry density, thickness, drying rate, liquid limit, plasticity index	0.981	6.492	0.967	6.978
4	5	Dry density, thickness, initial water content, liquid limit, plasticity index	0.867	7.914	0.853	8.066
5	4	Dry density, thickness, liquid limit, plasticity index	0.810	9.075	0.789	9.966

3.3.3. Artificial Neural Network (ANN)

Tables 10–14 present the ANN results for each input group for different algorithms and hidden layers. The ANN with five hidden layers using the LM algorithm was the best model for group 1. For the testing database, this model resulted in an R of 0.994 and MAE of 5.449. Hence, this network was sufficiently accurate.

The LM algorithm network with five hidden layers had the best accuracy with R and MAE values of 0.991 and 5.464, respectively, for the second input group as well (see Table 11). Despite removing ambient temperature and humidity, the network accuracy was still high, and as shown by the MLR analysis in Section 3.3.1, the effect of ambient temperature and humidity can be accounted for by the drying rate.

Table 10. The ANN results for group 1 with eight inputs.

Model	*R*		MAE	
	Testing Database	Training Database	Testing Database	Training Database
BR-1H	0.992	0.999	5.377	4.418
BR-2H	0.989	0.999	5.630	4.195
BR-3H	0.981	0.999	5.988	4.197
BR-4H	0.935	0.928	10.231	10.063
BR-5H	0.928	0.916	11.673	10.703
LM-1H	0.989	0.997	5.664	4.528
LM-2H	0.991	0.998	5.375	4.575
LM-3H	0.990	0.997	5.525	4.527
LM-4H	0.991	0.997	5.328	4.499
LM-5H	0.994	0.998	5.449	4.473

Table 11. The ANN results for input group 2 with six inputs.

Model	*R*		MAE	
	Testing Database	Training Database	Testing Database	Training Database
BR-1H	0.991	0.999	5.246	4.425
BR-2H	0.990	0.999	5.411	4.436
BR-3H	0.983	0.999	5.858	4.185
BR-4H	0.983	0.999	5.801	4.399
BR-5H	0.896	0.885	11.745	12.082
LM-1H	0.989	0.996	5.643	4.693
LM-2H	0.987	0.996	5.758	4.495
LM-3H	0.990	0.997	5.883	4.506
LM-4H	0.983	0.999	5.801	4.399
LM-5H	0.991	0.997	5.464	4.530

Table 12. The ANN results for input group 3 with five inputs.

	R-Test	*R*-Train	*MAE*-Test	*MAE*-Train
BR-1H	0.990	0.999	5.471	4.327
BR-2H	0.991	0.999	5.443	4.537
BR-3H	0.988	0.999	5.825	4.149
BR-4H	0.979	0.999	6.403	4.158
BR-5H	0.944	0.915	9.714	10.082
LM-1H	0.987	0.997	5.575	4.660
LM-2H	0.988	0.996	5.773	4.609
LM-3H	0.986	0.998	5.724	4.526
LM-4H	0.987	0.997	5.759	4.518
LM-5H	0.989	0.997	5.554	4.490

Table 13. The ANN results for group 4 with five inputs (excluding drying rate).

	R-Test	*R*-Train	MAE-Test	MAE-Train
BR-1H	0.904	0.904	8.092	8.388
BR-2H	0.904	0.904	8.092	8.357
BR-3H	0.898	0.904	9.146	8.377
BR-4H	0.896	0.908	9.209	8.307
BR-5H	0.898	0.876	10.464	10.405
LM-1H	0.910	0.928	7.922	7.562
LM-2H	0.910	0.933	8.017	7.800
LM-3H	0.899	0.920	8.330	7.676
LM-4H	0.933	0.924	8.329	7.502
LM-5H	0.920	0.926	8.196	7.693

Table 14. The ANN results for input group 5 with four inputs.

Model	*R*		*MAE*	
	Testing Database	Training Database	Testing Database	Training Database
BR-1H	0.905	0.912	8.012	8.245
BR-2H	0.905	0.910	8.004	8.231
BR-3H	0.875	0.905	9.162	8.362
BR-4H	0.898	0.905	9.189	8.360
BR-5H	0.901	0.907	9.435	8.324
LM-1H	0.920	0.920	8.057	7.552
LM-2H	0.911	0.916	7.758	7.797
LM-3H	0.911	0.920	8.122	7.408
LM-4H	0.905	0.921	8.196	7.667
LM-5H	0.913	0.922	8.370	7.699

The *R* and MAE values of the ANN for input group 3 are shown in Table 12. Using the BR algorithm (with two hidden layers), the best ANN model had an *R* of 0.991 and MAE of 5.443. In this model, the network accuracy remained high even after removing initial moisture content, ambient temperature and humidity. The low significance of initial moisture content was due to initial moisture content being prepared as a function of the liquid limit (LL and 1.5 LL), which reduced the importance of initial water content as an independent influential factor.

The results for input group 4 of the training and testing databases are shown in Table 13. Compared to group 3, the *R* value of the highest performing algorithm dropped to 0.93. This sharp drop in model accuracy suggests the importance of drying rate in crack prediction. The model with four hidden networks and the LM algorithm was the best for group 4.

The best ANN model for input group 5 had an *R* of 0.92 and MAE of 8.05, as shown in Table 14. This model was based on the LM algorithm with one hidden layer. Although the network's accuracy decreased slightly from 0.93 to 0.92, it was approximately the same as group 4. Thus, it was affirmed that the removal of initial soil water content did not significantly affect the accuracy of the predictions.

Comparing Tables 10–14, it can be concluded that input group 3 is the most appropriate combination of input parameters for desiccation crack prediction. Group 3 achieved almost the same accuracy as groups 1 and 2, but with fewer inputs. Hence, it was the most optimal of the five input groups. The CSIF results predicted by the best network for group 3 are shown in Figure 7. There is excellent agreement between the actual and predicted CSIF values.

The number of hidden layers (ranging from 1 to 5) and neurons per layer (fixed at 60) were selected based on extensive trial-and-error optimization. Each network configuration was evaluated five times, and the average *R* and MAE values were compared to identify the best-performing structure. The choice of 60 neurons per layer was found to offer a strong balance between training stability and model complexity, while deeper networks (e.g., 4 or 5 layers) tended to slightly overfit the data without significant gains in prediction accuracy.

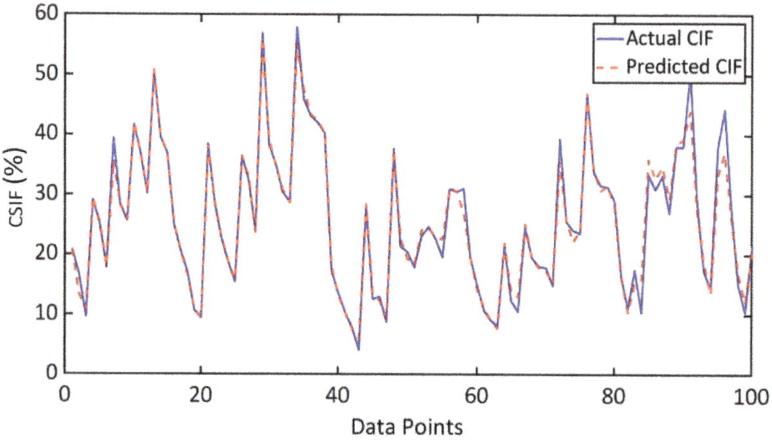

Figure 7. The result of the best ANN model based on input group 3.

3.3.4. Genetic Programming (GP)

The Genetic Programming method represents a technique capable of generating novel equations. This approach involves the introduction of diverse functions into the model, allowing the formulation of the desired equation based on these functions. In desiccation cracking, among the essential parameters considered, the drying rate stands out [32]. Previous studies have demonstrated that crack evolution, modeled by an exponential function, is a function of the drying rate. For example, El Youssoufi et al. [33] proposed a pragmatic model for soil drying kinetics that took into account the changes in the grain radius R over time during the drying process.

$$R = R_0 \exp\left(-\alpha \frac{t}{\tau}\right) \quad (6)$$

where R_0 is the radius at $t = 0$, α is a shrinkage parameter and τ is the total duration of the experiment.

The proposed GP model assumes the use of the exponential function in Equation (6). Notably, Equation (7), which comprises five inputs, is presented for the first time and the values in the equation are normalized based on Table 4. To use this equation, normalization must be performed linearly using the values provided in Table 4.

$$\begin{aligned} CSIF = {} & (R_1 \times (((((R_2 - X_5) \times (2 \times X_5)) \times ((X_2 - X_1) + (\exp(R_1)))) \times (((X_5 - R_1) \times (R_2 + X_3)) + ((X_4 \times R_1) - \\ & (R_1 + R_2)))) - ((X_5 - ((X_5{}^2) + (X_3 - X_5))) - ((R_2 - (X_1 \times X_4)) + ((\exp(X_1)) - (X_5 - X_2)))))) \end{aligned} \quad (7)$$

where X_1, X_2, X_3, X_4 and X_5 are drying rate, LL, PI, dry density and thickness. R_1 and R_2 are 0.170 and 0.963.

Equation (8) represents the second equation proposed to estimate CSIF. In this equation, the $x^{\frac{1}{3}}$ function is utilized.

$$CSIF = (R_1 \times (((R_1{}^{\frac{1}{3}} - ((X_4 \times R1) \times \left(X_1{}^2\right))) - ((X_4 \times (X_5 - X_4)) \times ((X_1 - X_2) \times (X_5 - R_2)))) + (X_1 + (X_3 - X_5{}^{\frac{1}{3}})))) \quad (8)$$

where X_1, X_2, X_3, X_4 and X_5 are drying rate, LL, PI, dry density and thickness. R_1 and R_2 are 0.374 and 0.460.

Table 15 reports the accuracy of both equations, and the results demonstrate that Equation (7) achieves a higher accuracy than Equation (8). These findings suggest that cracking modeled by a function of $x^{\frac{1}{3}}$ represents a function of the drying rate.

Table 15. The performance of two proposed GP equations for group 3 with five inputs.

Proposed Model	*R*		*MAE*	
	Testing Database	Training Database	Testing Database	Training Database
Equation (7)	0.947	0.970	2.527	2.039
Equation (8)	0.958	0.973	2.431	2.048

Comparing Equations (7) and (8), it can be seen that Equation (8) outperforms Equation (7) in both the *R*-Test and MAE-Test metrics. This indicates that Equation (8) has better predictive power on both the training and testing databases, and it makes fewer errors in predicting the target variable.

Figure 8 illustrates the anticipated CSIF values by the proposed Equations (7) and (8) against the actual values obtained during the experiments for training and testing. The outcomes presented in Figure 8 indicate that both proposed equations were able to predict the values of CSIF accurately. The close alignment between the model's anticipated values and the actual values obtained indicates the model's high accuracy in forecasting CSIF values.

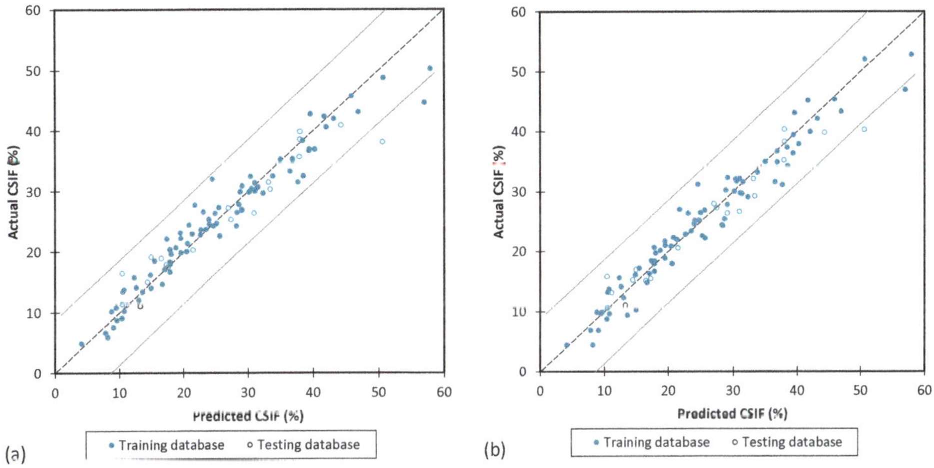

Figure 8. The actual CSIF versus the predicted CSIF values by (**a**) Equation (7) and (**b**) Equation (8).

4. Discussion

4.1. Comparison of Mathematical Models

The results of the four mathematical models used to predict CSIF using the five input groups are presented in Table 16. Based on the R values of the three mathematical models using the first three input groups, 1, 2 and 3, the initial moisture content, ambient temperature and humidity did not appear to have much impact on prediction when drying rate was included. When drying rate was removed, the accuracy of all three mathematical models was greatly reduced, indicating the importance of drying rate in predicting CSIF.

Among all five input groups, the Artificial Neural Network (ANN) consistently outperformed the other models, achieving the highest predictive accuracy with an R value of 0.99 and a mean absolute error (MAE) of approximately 5. The Classification and Regression Random Forest (CRRF) model ranked second in performance, particularly for input groups 1, 2 and 3, where it attained an R of 0.97 and an MAE close to 6. In contrast, the Multiple Linear Regression (MLR) model showed the lowest accuracy, with an R value around 0.84 and MAE value near 11 for the same input groups, highlighting its limitation in capturing the non-linear nature of this problem. Notably, the Genetic Programming (GP) model developed for input group 3 yielded an R of 0.958 and an exceptionally low MAE of 2.431, demonstrating strong predictive capability and enhanced interpretability.

Table 16. Comparison of model performances (MLR, CRRF, ANN and GP) at predicting CSIF across different input groups.

Group Number	Number of Inputs	Mathematical Methods						Proposed GP Equation	
		MLR		CRRF		ANN			
		R	*MAE*	*R*	*MAE*	*R*	*MAE*	*R*	*MAE*
1	8	0.847	10.447	0.97	6.01	0.99	5.45		
2	6	0.843	11.799	0.96	6.33	0.99	5.46		
3	5	0.838	11.499	0.97	6.98	0.99	5.44	0.958	2.431
4	5	0.750	12.045	0.85	8.07	0.93	8.33		
5	4	0.721	15.674	0.79	9.97	0.92	8.05		

Figure 9 presents the Taylor diagram for input group 3, identified as being the most optimal input combination. This diagram was used to compare the predictive performance of the mathematical models against the actual laboratory results for the testing dataset (denoted as REF in the figure). Among all models, the ANN demonstrated the closest agreement with the experimental data, confirming its superior accuracy and reliability in predicting the Crack and Shrinkage Intensity Factor (CSIF).

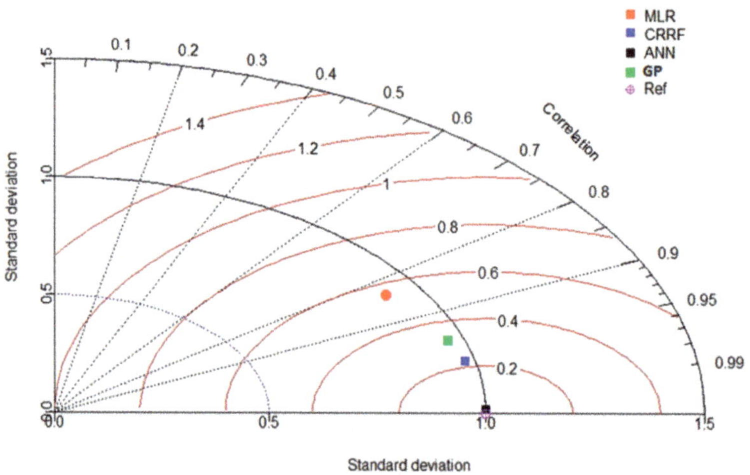

Figure 9. Taylor diagram of testing for prediction of CSIF.

The ANN model outperformed the other approaches across all input groups, particularly when trained using Bayesian Regularization. Unlike traditional algorithms that focus solely on minimizing error, BR introduces a Bayesian inference framework to optimize network weights while constraining overfitting. This allows the network to remain flexible enough to model complex, non-linear interactions (e.g., between LL, PI and drying rate) while avoiding noise fitting, which was a limitation in both the MLR and over-parameterized networks. These characteristics make BR especially effective for geotechnical problems involving coupled processes with limited datasets.

4.2. Sensitivity Analysis of Input Parameters in ANN Models

To assess the sensitivity, each input parameter was changed from −100% to +100% separately. The ANN model sensitivity analysis for the input parameters is shown in Figure 10.

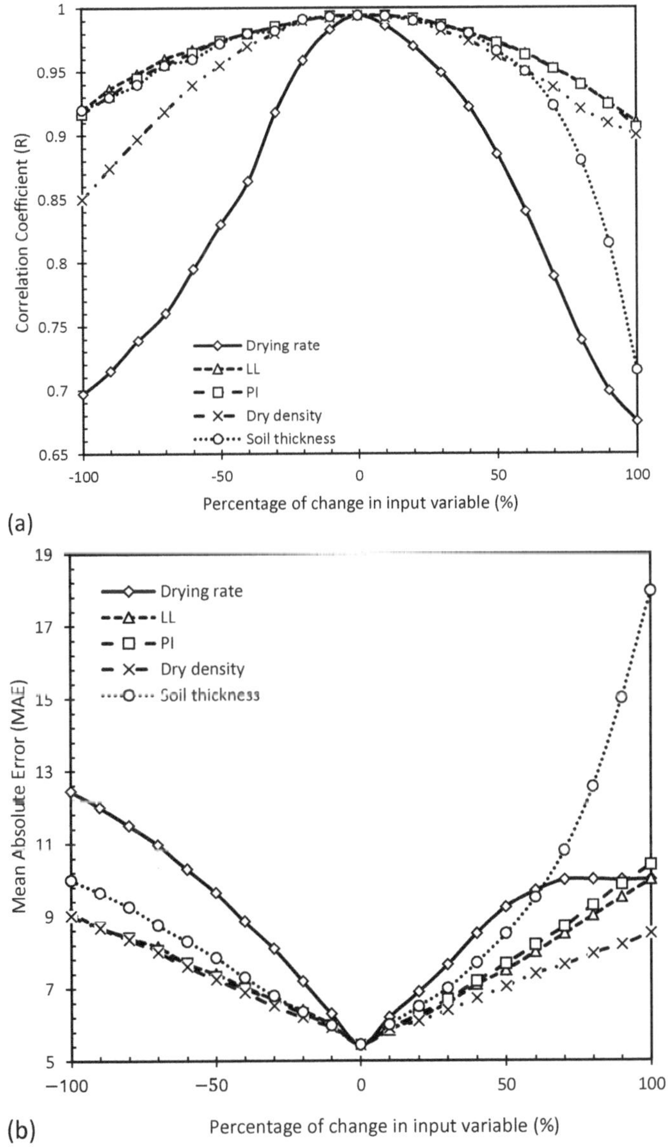

Figure 10. The ANN model sensitivity analysis for the input parameters based on (**a**) R and (**b**) MAE.

Increasing or decreasing the drying rate by 100% decreased the R of the model to about 0.675 and increased the MAE to about 12.5. For LL, MAE was almost symmetrical with LL changes. In the worst-case scenario, when LL increased by 100%, R was reduced to about 0.9 and MAE was increased to about 10. For PI, the MAE of the ANN increased most when PI increased by 100%, resulting in a change in MAE to 10.4. In response to changing the dry density by 100%, the model's R and MAE reached 0.85 and 9. Finally, for soil thickness, a 100% increase in soil thickness caused the largest network error (MAE). As soil thickness increased, R decreased to 0.715 and MAE increased to 17.95.

4.3. The Number of Neurons and Re-Training in the Best ANN Model

When modeling ANNs, layers, neurons and neuron re-training are important. Figure 11 illustrates the best ANN model with two hidden layers, using the BR algorithm, for various hidden neurons and network training numbers. Consequently, the ANN model had the greatest error (MAE) in the first and third network re-trainings, but the lowest error (MAE) in the second and fifth.

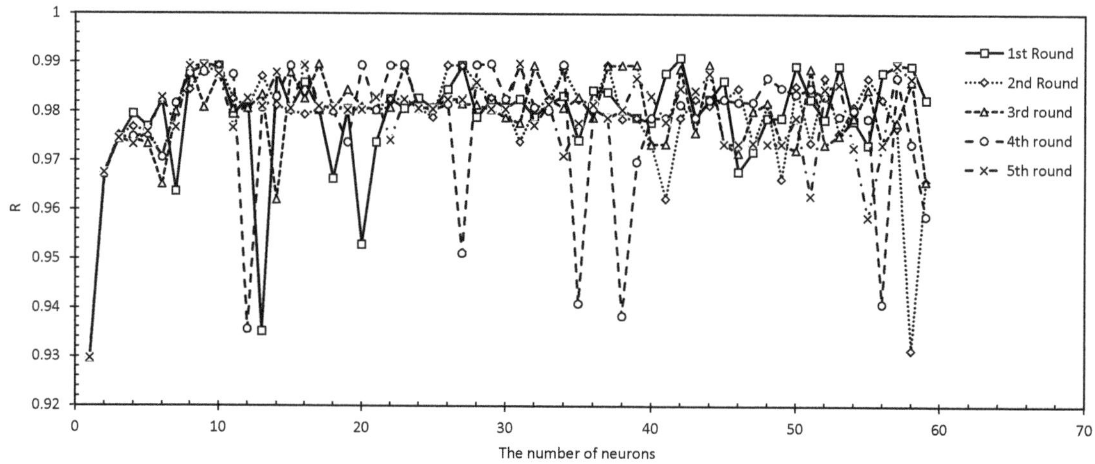

Figure 11. Variation in the correlation coefficient (R) with the number of neurons in the ANN model across five training rounds.

4.4. The Importance of the Input Parameters

The CRRF and ANN models with two hidden layers using the BR algorithm were investigated for the influence of input parameters on MAE. Figure 12 shows the sensitivity of the best CRRF model to input parameters. The highest sensitivity was related to soil thickness and drying rate, while the lowest sensitivity was related to ambient humidity and temperature. Jamhiri et al. [19] also ranked temperature and relative humidity among the least influential parameters for desiccation cracking.

Table 17 presents the ranking of the importance of input parameters based on the increase in mean absolute error (MAE) observed in three AI models: ANN, CRRF and GP. Across all models, drying rate and soil thickness consistently emerged as the most critical factors influencing the prediction of the Crack and Shrinkage Intensity Factor (CSIF), with the lowest MAE rankings (1 or 2). In contrast, ambient humidity and temperature showed the lowest influence, receiving the highest MAE rankings (7 or 8), confirming their redundancy when drying rate was included. Intermediate significance was observed for the liquid limit (LL), plasticity index (PI), dry density and initial moisture content, with moderate MAE impacts across the models. These results aligned with the sensitivity analysis findings and emphasized the dominant role of drying dynamics and soil geometry in desiccation crack development.

Regarding the importance of drying rate, the results of Zhang et al. [34] showed that the drying rate had a significant effect on the total pore volume. They found that samples dried at higher drying rates had larger total pore volumes than those at lower drying rates.

In addition, the experimental results of Krisdani et al. [35] confirmed that the changes in the proportion of empty space and the degree of soil saturation were strongly influenced by the rate of drying.

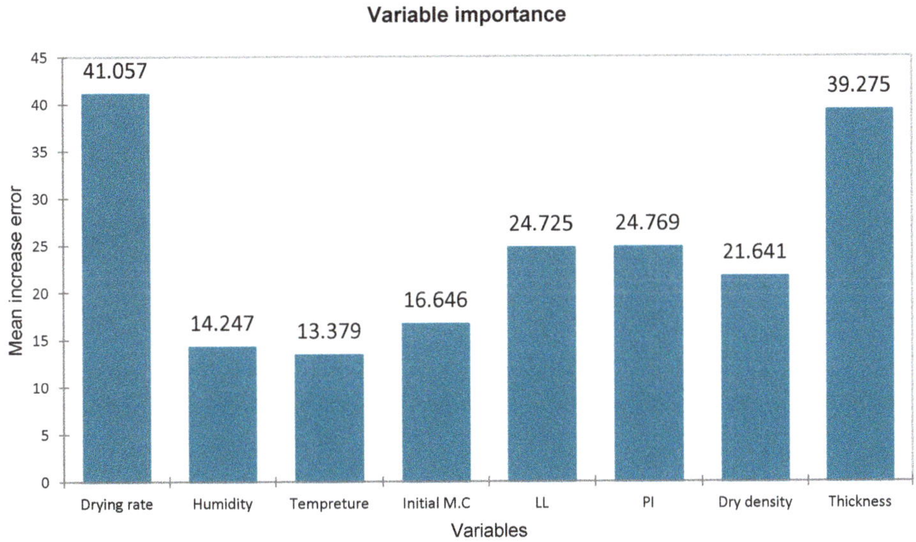

Figure 12. Variable importance based on the mean increase in prediction error from the CRRF model.

Table 17. The results of the importance of variables for three AI models.

	Input Parameters																							
	Drying Rate			Humidity			Temperature			Initial MC			LL			PI			Dry Density			Thickness		
	ANN	CRRF	GP	ANN	CRRF	GP	ANN	CRRF	GP	ANN	CRRF	GP	ANN	CRRF	GP	ANN	CRRF	GP	ANN	CRRF	GP	ANN	CRRF	GP
MAE	2	1	1	7	7	8	8	8	7	6	6	6	4	4	5	3	3	4	5	5	3	1	2	2

LL and PI were the third and fourth most significant and influential factors in predicting CSIF in three AI models, according to Table 17. LL and PI play an important role in affecting pore water volume and contraction strain. PI increases the pore water volume and therefore the volumetric shrinkage strain during drying, according to Rayhani et al. [36].

Dry density was the fifth most important parameter in both the ANN and CRRF and the third parameter in GP, followed by the initial moisture content, ambient humidity and temperature. Section 3.3.1, Section 3.3.2. and Section 3.3.3 discuss how the removal of these three parameters affects model accuracy.

The dominance of soil thickness and drying rate as the most important variables is consistent across all AI models tested. Thicker soil layers are more susceptible to crack propagation due to the greater energy required to overcome tensile strength and form multiple fracture planes. In contrast, thinner layers are more constrained and develop fewer cracks. The drying rate governs the rate of moisture loss, which directly influences suction generation, shrinkage strain accumulation and the initiation of desiccation cracking. Higher drying rates often result in steeper moisture gradients, leading to rapid crack formation and wider fracture networks. These behaviors are supported by the experimental findings from Costa et al. [37], Krisdani et al. [35] and Zhang et al. [34], all of whom observed significant changes in crack morphology with variations in thickness and drying conditions.

The effect of removing potentially redundant parameters, such as initial moisture content, was investigated through a comparison of different input groups. Notably, input group 3, which excluded initial moisture content, ambient temperature and humidity, achieved a nearly identical performance to groups 1 and 2, with an R of 0.991 and MAE of 5.44 for the testing dataset. This indicated that the removal of initial moisture content did not significantly affect the model's predictive capability. This can be attributed to the experimental setup, in which initial moisture was fixed at either LL or 1.5 LL for each test case. As a result, the variation in initial moisture content was already embedded in the LL parameter, making its explicit inclusion redundant. This finding supports the idea that carefully reducing the number of input parameters can streamline the model, improve computational efficiency and reduce overfitting risks without compromising accuracy.

The sensitivity analysis results provide a clear understanding of how each input parameter affects prediction error and thus offers valuable insight for simplifying models without compromising accuracy. The dominant influence of drying rate and soil thickness aligns with physical observations from past studies, highlighting their role in controlling moisture migration and crack energy release. Such insights are useful for prioritizing field measurements when data availability is limited.

4.5. Sustainability and Future Research Directions

The application of AI-based models in predicting soil desiccation cracking contributes to sustainability in geotechnical engineering by minimizing the need for extensive physical testing and enabling the more efficient use of resources. The accurate prediction of crack intensity factors can assist engineers in proactively designing infrastructure and soil covers, thereby reducing long-term maintenance costs and the environmental impacts associated with soil failure and water seepage.

Furthermore, the reduced reliance on trial-and-error approaches for crack control in clayey soils promotes a more sustainable design methodology, especially in expansive soil regions where desiccation-related damage is common.

Future research should explore the generalization of these models to more complex crack geometries (2D and 3D patterns), as well as their validation in field-scale conditions under natural environmental variability. Incorporating additional environmental factors (e.g., wind speed, solar radiation), developing hybrid physical–AI models and creating lightweight predictive tools for field use (e.g., mobile apps or edge computing devices) are promising directions to enhance both the sustainability and practicality of this approach in real-world applications.

4.6. Comparison with Previous Studies

An ANN model was used by Choudhury and Costa [20] to predict 1D cracks in thin, long layers. Table 18 compares the accuracy of predictions from this study to Choudhury and Costa [20]. The best performing model from each study has been selected for comparison. The current study achieved improved R and MAE values compared to Choudhury and Costa [20].

Table 18. The results of the proposed models using the training and testing databases by Choudhury and Costa [20] and the ANN (in this study).

Database	Proposed Models			
	Choudhury and Costa [20]		ANN	
	R	MAE	*R*	MAE
Training	0.92	7.55	0.99	4.53
Testing	0.93	7.25	0.99	5.44

Earlier studies in the domain of desiccation crack prediction primarily focused on limited datasets and fewer input parameters. For instance, Choudhury and Costa [20] applied an ANN model to predict the number of cracks in long, thin clay layers based on only four input variables and a relatively small dataset of 16 samples. While they achieved high predictive accuracy ($R = 0.93$), their output was limited to the number of cracks, which did not fully reflect the severity or morphology of desiccation cracking.

In contrast, the current study introduces the Crack and Shrinkage Intensity Factor (CSIF) as a more comprehensive and continuous output variable that better captures the extent of desiccation damage. By developing a larger experimental database (n = 100) with eight physical and environmental input parameters, and by testing multiple AI and regression models, we present a more robust framework for crack prediction.

The ANN model in this study achieves $R = 0.99$ and MAE = 5.44, which represents a notable improvement over previous studies—not just numerically, but in terms of the model generalizability, input diversity and output quality. Furthermore, unlike past research, this study performs a detailed input sensitivity analysis and explores parameter optimization, including the effect of removing redundant inputs.

Additionally, the introduction of a symbolic Genetic Programming (GP) model adds interpretability and usability by providing analytical equations that can be easily deployed in engineering calculations. The inclusion of residual plots and Taylor diagrams further strengthens the reliability of the findings, going beyond simple correlation metrics.

5. Conclusions

A series of drying tests were conducted on five types of clay (with different LLs and PIs) in different conditions, including initial moisture content, ambient temperature and humidity, dry density, drying rate and soil thickness. This experimental database was used to predict the Crack and Shrinkage Intensity Factor (CSIF) using four mathematical models, namely MLR, CRRF, ANN and GP.

- The ANN models produced the most accurate predictions for CSIF, with an R of 0.99 and MAE of 5.44 for the testing database. Predictions from CRRF and GP were also satisfactory. A higher MAE was noticed in the MLR predictions, indicating the limitations of linear models in encapsulating non-linear processes in desiccation crack formation.
- Critical input parameters for AI modeling of desiccation cracks were identified: liquid limit, plasticity index, dry density, soil layer thickness and drying rate. Ambient temperature and humidity were deemed redundant when drying rate was included. Similarly, initial moisture content could also be omitted when liquid limit was included.
- The most sensitive input parameters to predict CSIF were soil layer thickness and drying rate followed by plasticity index, liquid limit and dry density.

From a practical engineering standpoint, the developed models can be used to support predictive assessments of soil cover performance, crack mitigation in clay liners and risk evaluations for expansive soil foundations. For example, the ANN model could be integrated into early-stage design tools to estimate the risk of shrinkage cracking under varying environmental conditions, enabling the more efficient selection of soil types or thicknesses. The symbolic GP equations further offer opportunities for fast estimations in field or design office settings without the need for complex software.

While the models developed in this study demonstrated excellent predictive performance under controlled laboratory conditions, further work is needed to assess their generalizability to real-world scenarios. Future research will aim to incorporate field data, explore climate variability and extend the models to 2D and 3D crack patterns. Additionally, integrating uncertainty quantification and cross-site validation will be essential for establishing the models as reliable tools in practical geotechnical applications.

Author Contributions: Conceptualization, S.C.; Formal analysis, T.C. and S.C.; Investigation, A.B.; Methodology, A.B.; Software, A.B. and T.C.; Supervision, S.C.; Validation, A.B. and T.C.; Writing—original draft, A.B.; Writing—review and editing, T.C. and S.C. All authors have read and agreed to the published version of the manuscript.

Funding: This research received no external funding.

Data Availability Statement: Additional data are available upon request.

Acknowledgments: The authors acknowledge the scholarship provided by Deakin University to Abolfazl Baghbani under the Deakin University Postgraduate Scholarship (DUPR) scheme to undertake this research.

Conflicts of Interest: The authors declare no conflicts of interest.

References

1. Amarasiri, A.; Kodikara, J.; Costa, S. Numerical Modelling of Desiccation Cracking. *Int. J. Numer. Anal. Methods Geomech.* **2011**, *35*, 82–96. [CrossRef]
2. Amarasiri, A.; Kodikara, J. Use of material interfaces in DEM to simulate soil fracture propagation in mode I cracking. *Int. J. Geomech.* **2011**, *11*, 314–322. [CrossRef]
3. Gui, Y.L.; Zhao, Z.Y.; Kodikara, J.; Bui, H.H.; Yang, S.Q. Numerical modelling of laboratory soil desiccation cracking using UDEC with a mix-mode cohesive fracture model. *Eng. Geol.* **2016**, *202*, 14–23. [CrossRef]
4. Pouya, A.; Vo, T.D.; Hemmati, S.; Tang, A.M. Modeling soil desiccation cracking by analytical and numerical approaches. *Int. J. Numer. Anal. Methods Geomech.* **2019**, *43*, 738–763. [CrossRef]
5. Morris, P.H.; Graham, J.; Williams, D.J. Cracking in drying soils. *Can. Geotech. J.* **1992**, *29*, 263–277. [CrossRef]
6. Kodikara, J.K.; Choi, X. A simplified analytical model for desiccation cracking of clay layers in laboratory tests. In Proceedings of the Fourth International Conference on Unsaturated Soils, Carefree, Arizona, 2–6 April 2006; Volume 2, pp. 2558–2567.
7. Costa, S.; Kodikara, J.; Barbour, S.L.; Fredlund, D.G. Theoretical analysis of desiccation crack spacing of a thin, long soil layer. *Acta Geotech.* **2018**, *13*, 39–49. [CrossRef]
8. Teng, J.; Zhang, X.; Zhang, S.; Zhao, C.; Sheng, D. An analytical model for evaporation from unsaturated soil. *Comput. Geotech.* **2019**, *108*, 107–116. [CrossRef]
9. Zhu, L.; Shen, T.; Ma, R.; Fan, D.; Zhang, Y.; Zha, Y. Development of cracks in soil: An improved physical model. *Geoderma* **2020**, *366*, 114258. [CrossRef]
10. Costa, S.; Htike, W.Y.; Kodikara, J.; Xue, J. Determination of J-integral for clay during desiccation. *Environ. Geotech.* **2016**, *3*, 372–378. [CrossRef]
11. Baghbani, A.; Choudhury, T.; Costa, S.; Reiner, J. Application of artificial intelligence in geotechnical engineering: A state-of-the-art review. *Earth-Sci. Rev.* **2022**, *228*, 103991. [CrossRef]
12. Suman, S.; Khan, S.Z.; Das, S.K.; Chand, S.K. Slope stability analysis using artificial intelligence techniques. *Nat. Hazards* **2016**, *84*, 727–748. [CrossRef]
13. Luo, Z.; Bui, X.N.; Nguyen, H.; Moayedi, H. A novel artificial intelligence technique for analyzing slope stability using PSO-CA model. *Eng. Comput.* **2021**, *37*, 533–544. [CrossRef]
14. Baghbani, A.; Baghbani, H.; Shalchiyan, M.M.; Kiany, K. Utilizing artificial intelligence and finite element method to simulate the effects of new tunnels on existing tunnel deformation. *J. Comput. Cogn. Eng.* **2022**, *3*, 166–175. [CrossRef]
15. Mahdevari, S.; Torabi, S.R.; Monjezi, M. Application of artificial intelligence algorithms in predicting tunnel convergence to avoid TBM jamming phenomenon. *Int. J. Rock Mech. Min. Sci.* **2012**, *55*, 33–44. [CrossRef]
16. Tariq, Z.; Elkatatny, S.M.; Mahmoud, M.A.; Abdulraheem, A.; Abdelwahab, A.Z.; Woldeamanuel, M. Estimation of rock mechanical parameters using artificial intelligence tools. In Proceedings of the 51st US Rock Mechanics/Geomechanics Symposium, San Francisco, CA, USA, 25–28 June 2017; OnePetro: Richardson, TX, USA, 2017.
17. Sharma, M.; Choudhary, B.S.; Raina, A.K.; Khandelwal, M. Prediction of rock fragmentation in a fiery seam of an open-pit coal mine in India. *J. Rock Mech. Geotech. Eng.* **2024**, *16*, 2879–2893. [CrossRef]
18. Ebid, A.M.; Onyelowe, K.C.; Salah, M. Load-Settlement Curve and Subgrade Reaction of Strip Footing on Bi-Layered Soil Using Constitutive FEM-AI Coupled Techniques. *Designs* **2022**, *6*, 104. [CrossRef]

19. Jamhiri, B.; Xu, Y.; Shadabfar, M.; Costa, S. Probabilistic machine learning for predicting desiccation crack in clayey soils. *Bull. Eng. Geol. Environ.* **2023**, *82*, 355. [CrossRef]
20. Choudhury, T.; Costa, S. Prediction of Parallel Clay Cracks Using Neural Networks—A Feasibility Study. In *Contemporary Issues in Soil Mechanics*; Hemeda, S., Bouassida, M., Eds.; GeoMEast 2018; Sustainable Civil Infrastructures; Springer: Cham, Switzerland, 2019. [CrossRef]
21. Baghbani, A.; Costa, S.; Choundhury, T.; Shirani Faradonbeh, R. Prediction of Parallel Desiccation Cracks of Clays Using a Classification and Regression Tree (CART) Technique. In Proceedings of the 8th International Symposium on Geotechnical Safety and Risk (ISGSR 2022), Newcastle, Australia, 14–16 December 2022.
22. Nahlawi, H.; Kodikara, J.K. Laboratory experiments on desiccation cracking of thin soil layers. *Geotech. Geol. Eng.* **2006**, *24*, 1641–1664. [CrossRef]
23. Costa, S.; Kodikara, J.; Thusyanthan, N.I. Modelling of desiccation crack development in clay soils. In Proceedings of the 12th International Conference of IACMAG, Goa, India, 1–6 October 2008; pp. 1099–1107.
24. *AS 1289.3.4.1-2008*; Soil Classification Test—Determination of the Linear Shrinkage of a Soil Standard Method 2008. Standards Australia: Sydney, Australia, 2008.
25. Costa, S.; Kodikara, J.; Thusyanthan, N.I. Study of desiccation crack evolution using image analysis. In *Unsaturated Soils*; Advances in Geo-Engineering; CRC Press: Boca Raton, FL, USA, 2008; pp. 175–180.
26. Schneider, C.A.; Rasband, W.S.; Eliceiri, K.W. NIH Image to ImageJ: 25 years of image analysis. *Nat. Methods* **2012**, *9*, 671–675. [CrossRef]
27. Breiman, L. Random forests. *Mach. Learn.* **2001**, *45*, 5–32. [CrossRef]
28. McCord-Nelson, M.; Illingworth, W.T. A Practical Guide to Neural Nets. Addison-Wesley Longman Publishing Co., Inc.: Boston, MA, USA, 1991.
29. Fahlman, S.E. *An Empirical Study of Learning Speed in Back-Propagation Networks*; Carnegie Mellon University, Computer Science Department: Pittsburgh, PA, USA, 1988; pp. 35–36.
30. Burden, F.; Winkler, D. Bayesian regularization of neural networks. In *Artificial Neural Networks*; Humana Press: Totowa, NJ, USA, 2008; pp. 23–42.
31. Marquardt, D.W. An algorithm for least-squares estimation of nonlinear parameters. *J. Soc. Ind. Appl. Math.* **1963**, *11*, 431–441. [CrossRef]
32. Tang, C.-S.; Zhu, C.; Cheng, Q.; Zeng, H.; Xu, J.-J.; Tian, B.-G.; Shi, B. Desiccation cracking of soils: A review of investigation approaches, underlying mechanisms, and influencing factors. *Earth-Sci. Rev.* **2021**, *216*, 103586. [CrossRef]
33. El Youssoufi, M.S.; Delenne, J.Y.; Radjai, F. Self-stresses and crack formation by particle swelling in cohesive granular media. *Phys. Rev. E* **2005**, *71*, 051307. [CrossRef] [PubMed]
34. Zhang, J.Z.; Tang, C.S.; Zhu, C.; Zhou, Q.Y.; Xu, J.J.; Shi, B. Monitoring and mapping the evolution of clayey soil desiccation cracking using electrical resistivity tomography. *Bull. Eng. Geol. Environ.* **2023**, *82*, 430. [CrossRef]
35. Krisdani, H.; Rahardjo, H.; Leong, E.C. Effects of different drying rates on shrinkage characteristics of a residual soil and soil mixtures. *Eng. Geol.* **2008**, *102*, 31–37. [CrossRef]
36. Rayhani, M.H.T.; Yanful, E.K.; Fakher, A. Physical modeling of desiccation cracking in plastic soils. *Eng. Geol.* **2008**, *97*, 25–31. [CrossRef]
37. Costa, S.; Kodikara, J.; Shannon, B. Salient factors controlling desiccation cracking of clay in laboratory experiments. *Géotechnique* **2013**, *63*, 18–29. [CrossRef]

designs

MDPI

Article

Cautionary Note on the Current EN1998-4 Formula of the Additional Pressure in the Seismic Design of Circular Silos

Sulyman Mansour [1,2] and Stefano Silvestri [3,*]

[1] School of Engineering, University of Galway, H91 HX31 Galway, Ireland; sulyman.mansour@universityofgalway.ie
[2] Department of Architecture, Construction and Design—ArCoD, Polytechnic University of Bari, 70125 Bari, Italy
[3] Department of Civil, Chemical, Environmental and Materials Engineering—DICAM, University of Bologna, 40136 Bologna, Italy
* Correspondence: stefano.silvestri@unibo.it

Abstract: Silos are strategic structures widespread in the industrial sectors for post-harvest preservation purposes. Current standards on the seismic design of silos are understandably based on approximate and simplified assumptions, leading intentionally to conservative design-oriented formulae. However, unjustified over-estimation might lead to unnecessary economic losses. As part of the authors' analytical and experimental ongoing research on the complex seismic behavior of filled silo systems, in this short paper, an in-depth reading of the theoretical framework originally proposed during the 1970s and 1980s is provided to present a better understanding of the unexplained design-oriented formula of the seismic additional pressure in the European standard. A conceptual incongruence in the Eurocode EN1998-4:2006 is pointed out and discussed regarding the dynamic overpressure formula in the case of ground-supported flat-bottom circular silos subjected to seismic excitation. Specifically, a potential miscounting of the geometrical aspect in circular silos, with respect to rectangular ones, leads to an inconsistent amplification of the additional pressures in the range 1.65–2, depending on the filling aspect ratio of the silo. This inconsistency provides the reason for several unexplained results recently published in the scientific literature. A proposal for a physically based correction, retaining the current assumptions made by the EN1998-4, is finally given.

Keywords: silo; granular material; seismic design; dynamic overpressure; Eurocode EN1998-4

1. Introduction

Silos filled with granular solid are exposed to particular horizontal and vertical loading conditions during seismic events that are still not fully understood [1–3]. Two main open questions remain insufficiently addressed by the current state of knowledge in the field of the dynamic and seismic response of silos: (1) the fundamental period of the filled silo structure, as no current formula is available to even roughly estimate this value; and (2) the seismic effective mass (i.e., global response), along with the corresponding seismic additional pressure or dynamic overpressure (i.e., local response) [3]. Earthquake excitation induces additional, non-uniform dynamic pressures on the silo wall. In circular silos, these pressures result in radial and circumferential forces acting on the wall, as well as vertical forces in the stiffeners, if present. Unlike the static equilibrium problem well-established in the 19th century [4,5], the dynamic equilibrium problem was investigated for the first time during the 1970s and 1980s by Wood [6,7] and Trahair [8], resulting in a simple formula describing a uniform profile of the horizontal dynamic overpressure along the height of the silo wall. Different numerical and experimental investigations were devoted to understanding the suitability of this formula (e.g., Rotter and Hull, 1989 [9]), which is currently explicitly referred to by the American FEMA P-2082-1/2020 standard [10]. The European standard EN1998-4:2006 [11] gives a formula that, even without explicitly making reference to the Wood and Trahair one, can be traced back to it at least for the original rectangular silo case. It is worth noticing that all other international codes do not even provide formulas for the seismic overpressures on the silo wall [3]. This lack of a robust formulation, along with the reliance on simplified models originally proposed in earlier research, stems from the intrinsic difficulty of capturing the peculiar features of silo structures—particularly those related to the granular nature of the stored material—which cause their seismic behavior to differ significantly from that of other, seemingly similar structures, such as liquid storage tanks [12], and even more so from conventional building structures [13–15].

Significant attention has recently been given to the dynamic and seismic response of silos, particularly at numerical and experimental levels, with very few studies addressing it analytically. A comprehensive review of the main open questions on the dynamic and seismic response of ground-supported silos (mainly: the fundamental period of vibration and the damping ratio, as well as the effective mass and the corresponding seismic overpressures) can be found in [3] by the same authors. Another recent wide review regarding the structural behavior and failure modes of ground-supported and elevated silos (i.e., their response under ordinary and extreme loads, including the effects of imperfections and the interacting structures, discussing the stored material properties together with the risk assessment and mitigation strategies) is also available in [1] conducted by other researchers. The interested reader is referred to these two reviews [1,3] and further recent works [16–24] for a comprehensive understanding of the general framework and current level of knowledge within which this work is presented, with a direct focus on the dynamic overpressure problem.

As part of the authors' ongoing analytical [25,26] and multiscale experimental [27–30] efforts to investigate the complex seismic behavior of filled silo systems—and to provide a more consistent, physically based, and experimentally validated alternative formulation for additional pressure [25,26]—this technical note aims to shed light on the essential development of dynamic overpressure formulations for both rectangular and circular silos.

In light of this, an inconsistency between the provisions of the current European standard EN1998-4:2006 [11] and the original theoretical formulation by Wood and Trahair [6–8] on the seismic forces provoked by the stored material onto cylindrical silo walls is pointed out and discussed for the case of ground-supported silos, even if it can refer also to column-supported elevated silos.

In this short paper, we provide a methodological explanation of the incongruence in the design-oriented formulae presented in EN 1998-4:2006. This incongruence, on one hand, has different economic implications for practitioners, and, on the other hand, has been encountered by different researchers [31–35] in numerical studies without a comprehensive explanation for the unexpectedly low numerical values compared to Eurocode expectations. Furthermore, this work suggests the corrected formula for the circular silo case, adhering to the same Eurocode assumptions, as the absence of experimental verifications currently prevents the proposal of a new fully design-oriented formula.

2. Early Dynamic Pressure Models

The earliest investigations on the seismic response of silos can be traced back to the first half of the 1970s, when Wood [6,7] began studying the relatively similar phenomenon of earthquake-induced soil pressures on a rigid wall. This can be considered as a starting point for the analytical model specifically developed for ground-supported silos by Trahair et al. [8], which represented the first closed-form prediction formula that provides the so-called dynamic overpressures (i.e., additional pressures on the silo wall provoked by the inertial loads of the granular solid under seismic excitation). The work distinguishes between squat and slender ground-supported silos on the basis of the filling aspect ratio. The aspect ratio is defined as the ratio between the filling height of the content H and a reference base dimension D (namely: the base length L along the seismic input direction for rectangular silos, and the actual base diameter d_c for circular silos). Trahair identified the 0.5 value as a significant splitting value between two different seismic behavior cases: silos characterized by aspect ratio $H/D > 0.5$ were classified as slender, whilst silos characterized by aspect ratio $H/D < 0.5$ were classified as squat. The impact of the aspect ratio on the seismic response is primarily related to the continuity or discontinuity of the mobilized mass along the radial direction and is explained hereafter. Trahair started from the physical consideration that for squat silos only a fraction of the contained mass is mobilized during an earthquake, whilst for slender silos the full mass is mobilized. In both cases, one half of the mobilized mass is supposed to push on the "windward" side of the silo by increasing the static pressure, whilst the other half is supposed to apply a decompression force by reducing the static pressure on the "leeward" side.

The formula is simply derived by solving a plane strain problem wherein a homogenous and isotropic elastic soil mass is contained within two rigid vertical boundaries (i.e., rigid walls) of height H, overlying a rigid horizontal base of length L as represented in Figure 1, subjected to a uniform horizontal body acceleration equal to 1.0 g.

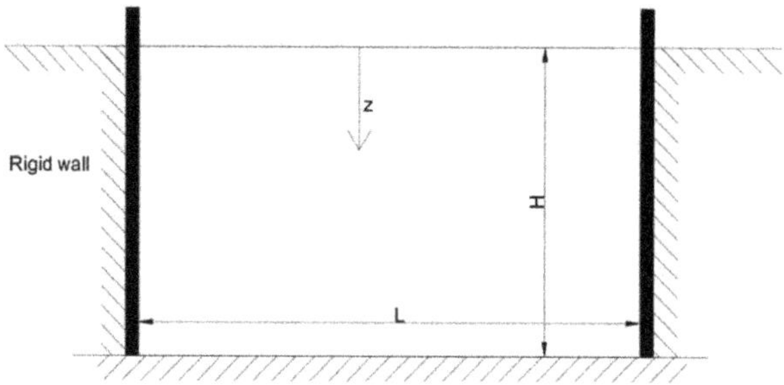

Figure 1. Rigid wall problem adopted in Wood [1,2] and Trahair [3].

This problem led to a solution depending on the aspect ratio. The main solution was originally obtained for rectangular silos of dimensions L and H and a unit dimension into the page.

For $H/L > 0.5$ (slender rectangular silos according to Trahair's definition),

$$p_{hf} = \gamma \cdot \frac{L}{2} \tag{1}$$

For $H/L < 0.5$ (squat rectangular silos according to Trahair's definition),

$$p_{hf} = \gamma \cdot H \tag{2}$$

where p_{hf} is the dynamic overpressure onto the silo wall due to a horizontal acceleration of 1.0 g, and γ is the specific weight of the granular material. Equations (1) and (2) indicate full and partial mass mobilization, respectively. Figure 2 graphically compares the difference in mobilized mass between squat and slender silos.

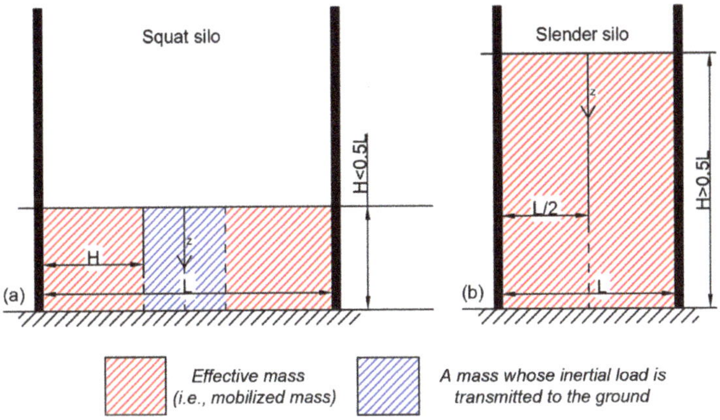

Figure 2. Mobilized mass: (**a**) squat rectangular silo (i.e., partial mass mobilization) and (**b**) slender rectangular silo (i.e., full mass mobilization).

The solution obtained for rectangular silos was then adapted to circular silos (defined by radius R or diameter $d_c = 2R$) on the basis of the same physical interpretation, i.e., one half of the mobilized mass in the cylindrical container is supposed to push on the "windward" side of the silo by increasing the static pressure, whilst the other half is supposed to apply a decompression force by reducing the static pressure on the "leeward" side. It can be demonstrated that this observation leads to the following formulae for the dynamic overpressure.

For $H/d_c > 0.5$ (slender circular silos according to Trahair's definition),

$$p_{hf} = \gamma \cdot \frac{d_c}{4} = \gamma \cdot \frac{R}{2} \quad (3)$$

For $H/d_c < 0.5$ (squat circular silos according to Trahair's definition),

$$p_{hf} = \gamma \cdot \frac{H}{2} \quad (4)$$

By considering the case of circular silos, the predefined reference dynamic overpressure in the rectangular silo case is to be subdivided into two components (radial $p_{hf,r}$ and circumferential $p_{hf,\theta}$) depending on the angle θ, as shown in Figure 3. The two components are evaluated as

$$\begin{cases} p_{hf,r} = p_{hf} \cdot \cos\theta \\ p_{hf,\theta} = p_{hf} \cdot \sin\theta \end{cases} \text{for } 0 \leq \theta \leq 2\pi \quad (5)$$

where θ is the angle formed between the direction of the horizontal component of the seismic input and the radial line connecting the silo section center with the silo wall at the circumferential point of interest.

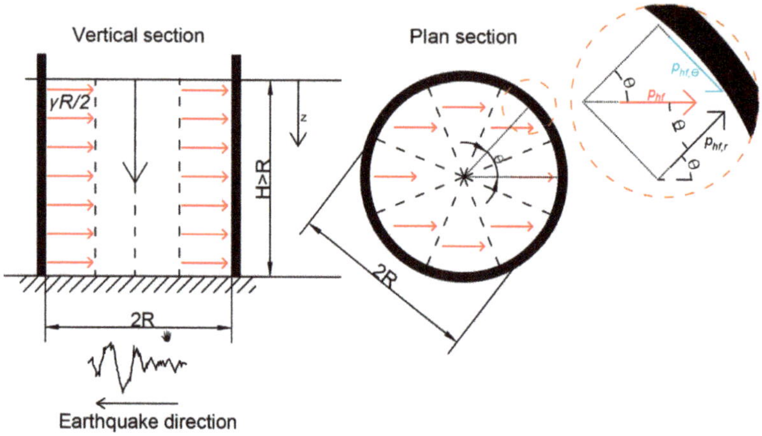

Figure 3. Dynamic overpressure profile and radial and circumferential components of the seismic overpressures provoked by the granular material onto the cylindrical silo wall.

If the horizontal acceleration input is different from 1.0 g and variable over the height of the silo, the dynamic overpressure proposed by Trahair becomes

$$\Delta_{ph,Trahair}(z) = \alpha(z) \cdot p_{hf} \tag{6}$$

where $\alpha(z)$ is the ratio between the acceleration at depth z from the equivalent surface of the stored content and the gravity acceleration (hereafter simply referred to as the "acceleration factor").

It is worth mentioning that Trahair's original formula for circular silos can be derived from alternative models (i.e., Silvestri et al., 2012 [25], which accounts for the system's mechanical properties, such as friction coefficients and pressure ratio), as the maximum pure dynamic overpressure that can be triggered by a unit slice under seismic excitation.

3. An Inconsistency in the Current European Standard

The EN1998-4:2006 [11] standard provides general principles and practical application rules for the earthquake design of elevated and ground-supported silos. In seismic design conditions, the additional pressure should be applied only over the part of the wall that is in contact with the stored content, i.e., up to the height corresponding to the equivalent surface of the stored content. For ground-supported circular silos, the additional normal pressure (i.e., radial component of the dynamic overpressure) can be evaluated as $\Delta_{ph,EN1998} \cdot \cos\theta$, with $\Delta_{ph,EN1998}$ being the reference pressure and θ the angle between 0 and 2π. The reference pressure is given by

$$\Delta_{ph,EN1998}(z) = \alpha(z) \cdot \gamma \cdot \min\left(H; \frac{d_c}{2}; 3x\right) \tag{7}$$

where $\alpha(z)$ is the "acceleration factor"; H is the overall height of the granular solid from the base level up to the equivalent surface; $x = H - z$ is the height of the studied point from the silo base; and d_c is the silo diameter.

A simple comparison between the original Trahair formulae (Equations (1)–(4) and (6)) and the current Eurocode provisions (Equation (7)) is sufficient to highlight a perfect matching between the two formulations for the following cases:

- Slender rectangular silo;
- Squat rectangular silo;

but an inconsistency associated with the following cases:

- Slender circular silo;
- Squat circular silo.

The incongruence is concerned with the magnitude of the additional pressure, with possible important consequences on the seismic design of silos. Indeed, for the circular silo cases, the dynamic overpressure provided by EN1998-4:2006 [11] is nearly twice the value suggested by the Trahair formula, whilst for the rectangular silo cases both references give the same prediction from an engineering point of view. The perfect agreement in the case of rectangular silos and the significant discrepancy in the case of circular silos between the two additional pressure formulations suggest that both references are based on the same assumptions and that the Eurocode formula for circular silos is most likely the result of an unintentional error—such as a miscalculation or misunderstanding—rather than a deliberate overestimation or conservative design choice.

Note that there is also a slight secondary difference regarding the dynamic overpressure profile, which is however understandable and of minor importance. For actual industrial silos (i.e., aspect ratio $H/d_c \geq 0.5$), the overpressure profile has a constant value from the free granular surface down to a certain point (according to EN1998-4:2006 [11]), defined by the $3x$ term that gives a height from the silo base equal to $R/3$, from which the dynamic overpressure starts to linearly decrease till reaching zero at the silo base level. The overpressure reduction near the base might be correlated with the possible interaction of the lower portion of the granular solid with the base, occurring directly through the potential friction surface at ground level (Figure 4a).

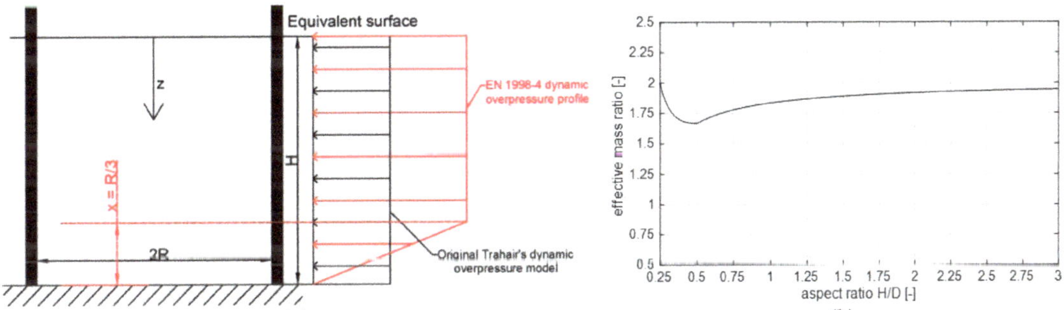

Figure 4. (**a**) A qualitative comparison between EN1998-4 and Trahair's dynamic overpressure profiles and (**b**) effective mass ratio (between Eurocode EN1998-4 and Trahair et al., 1983 [8]) for squat, intermediate, and slender silo categories as a function of the filling aspect ratio.

The main fundamental difference in the magnitude between the Trahair (and, consequently, the American standard) [8,10] and the EN1998-4:2006 [11] predictions can be addressed by introducing the Effective Mass Ratio (*EMR*). The *EMR* is a ratio between two seismic forces (i.e., effective mass multiplied by design acceleration). The numerator is the EN 1998-4 seismic force, obtained by integrating the seismic additional pressures over the internal surface (i.e., the contact area between the stored material and the silo wall). The denominator is the Trahair seismic force, calculated by integrating the Trahair pressures over the same contact area.

After a few mathematical steps, and assuming that $\alpha(z) = \alpha$ is constant over the silo height, the *EMR* can be expressed as a function of the filling aspect ratio H/d_c, as follows.

For $H/d_c > 0.5$ (slender circular silos according to Trahair's definition),

$$EMR = \frac{\int_0^H \int_0^{2\cdot\pi} \Delta_{ph,EN1998} \cdot Rd\theta \cdot dz}{\int_0^H \int_0^{2\cdot\pi} \Delta_{ph,Trahair} \cdot Rd\theta \cdot dz} = \frac{\int_0^{R/3} \int_0^{2\cdot\pi} (\alpha \cdot \gamma \cdot 3x) \cdot Rd\theta \cdot dx + \int_{R/3}^{H} \int_0^{2\cdot\pi} (\alpha \cdot \gamma \cdot R) \cdot Rd\theta \cdot dx}{\int_0^H \int_0^{2\cdot\pi} \left(\alpha \cdot \gamma \cdot \frac{R}{2}\right) \cdot Rd\theta \cdot dz} = 2 - \frac{1}{6}\left(\frac{d_c}{H}\right) \tag{8}$$

Note that, in the case of slender silos, the denominator corresponds to the full mass (density times volume of the whole silo content) multiplied by the acceleration factor $\left(\int_0^H \int_0^{2\cdot\pi} \Delta_{ph,Trahair} \cdot Rd\theta \cdot dz = \alpha \cdot \gamma \cdot V_{cylindrical} = \alpha \cdot \gamma \cdot \pi \cdot R^2 \cdot H\right)$.

For $1/6 < H/d_c < 0.5$ (squat circular silos according to Trahair's definition, which are not substantially affected by the modification introduced by EN1998-4:2006 [11] through the $3x$ term),

$$EMR = \frac{\int_0^H \int_0^{2\cdot\pi} \Delta_{ph,EN1998} \cdot Rd\theta \cdot dz}{\int_0^H \int_0^{2\cdot\pi} \Delta_{ph,Trahair} \cdot Rd\theta \cdot dz} = \frac{\int_0^{R/3} \int_0^{2\cdot\pi} (\alpha \cdot \gamma \cdot 3x) \cdot Rd\theta \cdot dx + \int_{R/3}^{H} \int_0^{2\cdot\pi} (\alpha \cdot \gamma \cdot H) \cdot Rd\theta \cdot dx}{\int_0^H \int_0^{2\cdot\pi} \left(\alpha \cdot \gamma \cdot \frac{H}{2}\right) \cdot Rd\theta \cdot dz} = 2 - \frac{1}{3}\left(\frac{d_c}{H}\right) + \frac{1}{12}\left(\frac{d_c}{H}\right)^2 \tag{9}$$

In the case of squat silos, instead, the denominator corresponds to a mass portion of the silo content multiplied by the acceleration factor $\left(\int_0^H \int_0^{2\cdot\pi} \left(\alpha \cdot \gamma \cdot \frac{H}{2}\right) \cdot Rd\theta \cdot dz = \alpha \cdot \gamma \cdot V_{partial} = \alpha \cdot \gamma \cdot \pi \cdot \frac{H \cdot R}{2} \cdot H\right)$, involving a partial volume lower than the total volume of the silo content.

To better illustrate the issue numerically, a variation in the filling aspect ratio H/d_c was set from 0.25 to 3. It is worth noticing here that the classification of silos provided by the Eurocode is also different from the one provided by Trahair. The squat silos according to Trahair's definition correspond to the retaining silos as defined by EN1991-4:2006 [36]. The slender silos according to Trahair's definition include squat, intermediate, and slender silos as defined by EN1991-4:2006 [36]. Figure 4b clearly shows that the effective mass calculated from EN1998-4:2006 [11] ranges roughly between 165% and 200% of the total mass, which is very conservative and physically inconsistent.

The origin of the difference between EN1998-4 and the original Trahair formula for the specific case of circular silos might be associated with one of the following events:

1. A potential misleading consideration of the full inertial contribution of the total mobilized mass on the "windward" side of the silo wall, as a result of a suction condition development on the "leeward" side of the wall.
2. A potential miscounting of the geometrical aspects, different from rectangular to circular cross-sections.

On one hand, the former possibility is associated with the condition represented by the "almost rare" case in which the seismic overpressure is larger than the static pressure, so that the earthquake effects do not translate only into an increment of the static pressure on the "windward" side of the silo and into a reduction in the static pressure on the "leeward" side, but into a physically inconsistent (corresponding to twice the total mass of the granular solid) overpressure on the "windward" side of the silo and into a full detachment of the granular solid from the "leeward" side. It is clear that the total mass of the granular solid can push on the "windward" side of the silo, but it is also clear that it cannot correspond to twice the mass (assuming that $\alpha(z) = \alpha$ is constant over the silo height as in Figure 5).

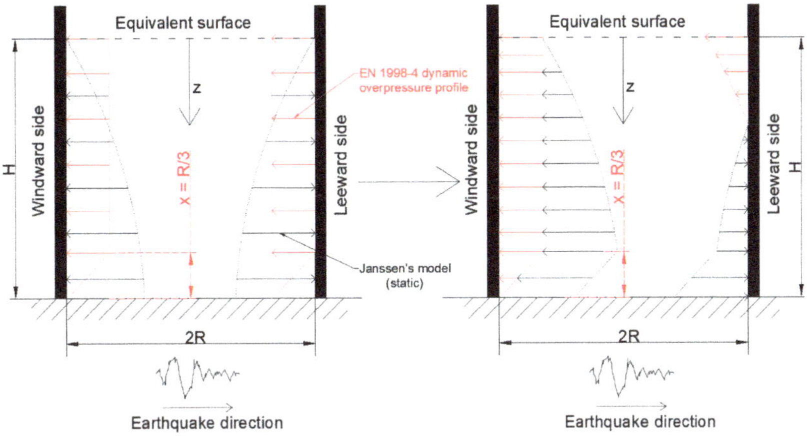

Figure 5. A schematic representation of the static pressure and dynamic overpressure profiles over the silo height according to the European standards.

On the other hand, the latter possibility is more plausible. Differently from the rectangular case for which the solution can be derived from the simplified plane strain problem (i.e., a vertical section along the input as in Figure 2), attention should be paid towards the case of circular silos since the vertical section is not sufficient to fully address the issue, but a plan section is also necessary to account for the fan-shaped slice in the dynamic overpressure estimation. To better illustrate this issue, Figure 6 presents the plan view of a rectangular-shaped slice in the rectangular silo and of a fan-shaped slice in the circular silo.

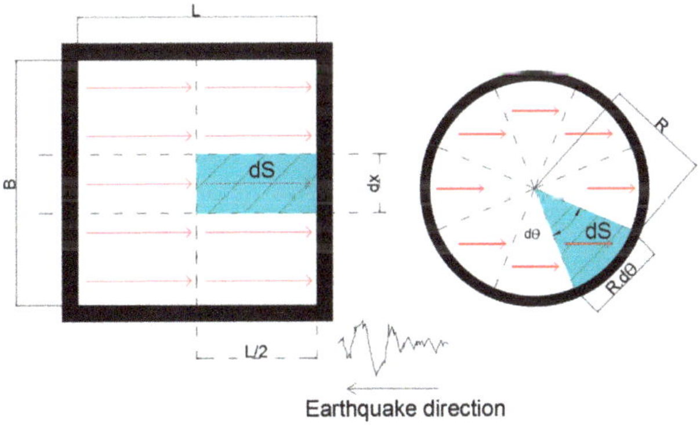

Figure 6. Plan view of a rectangular-shaped slice in the rectangular silo and of a fan-shaped slice in the circular silo.

Under a uniform acceleration profile and assuming, for instance, full mass mobilization (e.g., slender case), the seismic base shear V_{base} should correspond to the total volume of the stored material inside the silo. The total volume is equal to the integral of the unit slice area dS highlighted in Figure 6, along the plan perimeter and over the height of the silo, as reported hereafter for the rectangular and for the circular silo cases, respectively:

$$V_{base\_r} = \int_{silo} \alpha \cdot \gamma \cdot dV = \int_{silo} \alpha \cdot \gamma \cdot (dS) \cdot dz = 2 \cdot \int_0^H \int_0^B \alpha \cdot \gamma \cdot \left(\frac{L}{2} \cdot dx\right) \cdot dz = \alpha \cdot \gamma \cdot L \cdot B \cdot H \quad (10)$$

$$V_{base\_c} = \int_{silo} \alpha \cdot \gamma \cdot dV = \int_{silo} \alpha \cdot \gamma \cdot (dS) \cdot dz = \int_0^H \int_0^{2\cdot\pi} \alpha \cdot \gamma \cdot \left(\frac{R}{2} \cdot Rd\theta \right) \cdot dz = \alpha \cdot \gamma \cdot \pi \cdot R^2 \cdot H \quad (11)$$

Consequently, the unit dynamic overpressure exerted on the wall area dA (equal to $dxdz$ for rectangular silos and to $Rd\theta dz$ for circular silos) should lead to the same base shear. In this respect, Equations (12) and (13) provide the same seismic base shear as obtained starting from the correct dynamic overpressures for the rectangular and for the circular silo cases, respectively:

$$V_{base\_r} = \int_{silo\ wall} p_{hf} \cdot dA = 2 \cdot \int_0^H \int_0^B \left(\alpha \cdot \gamma \cdot \frac{L}{2} \right) \cdot dxdz = \alpha \cdot \gamma \cdot L \cdot B \cdot H \quad (12)$$

$$V_{base\_c} = \int_{silo\ wall} p_{hf} \cdot dA = \int_0^H \int_0^{2\cdot\pi} \left(\alpha \cdot \gamma \cdot \frac{R}{2} \right) \cdot Rd\theta dz = \alpha \cdot \gamma \cdot \pi \cdot R^2 \cdot H \quad (13)$$

It is clear that p_{hf} as written in Equation (12) is consistent with Equation (7), whilst p_{hf} as written in Equation (13) is not. Placing the Eurocode term $d_c/2$ instead of $R/2$ in Equation (13) would result in an amplified value as shown in Equation (8) and Figure 4b.

Moreover, it should be noted that assuming an amplified dynamic overpressure profile would not be consistent with the rationale behind the Eurocode profile whose overpressure values tend to be reduced at the base level (through the term $3x$).

4. Brief Discussion and Proposal

Different studies can be found in the literature, almost all of them numerical, where the highlighted difference (e.g., pressures predicted by the Eurocode are almost double those numerically obtained) was encountered and observed, but not sufficiently explained [31–35]. It should be noted that the calibration and validation of numerical models remains a challenge, primarily due to the lack of supporting experimental data. In this respect, particular focus can be placed on two very recent studies [34,35]. In the more recent one [35], the researchers calibrated a numerical model using the dynamic response of a tested silo specimen on a shaking table [27], which was performed by the second author of this work. The study found that the average numerical values of additional pressure were significantly lower than those predicted by the Eurocode (EC), approximately half of the expected values as shown in Figure 9 of [35]. The earlier study [34], on the other hand, combined both experimental and numerical approaches. It found that the additional pressure values—both experimental and numerical—averaged approximately half of the Eurocode prediction in Figure 10 and Figure 14 of [34].

In light of the previous discussions in Section 3, the values found by previous researchers are physically consistent with the maximum feasible base shear force that can be triggered by the solid material inside the silo. This differs from the EN1998-4 formula, which yields a higher value that exceeds full mass mobilization.

Moreover, as far as this inconsistency is not addressed, the following design-oriented consequences are expected:

1. In the case of steel silos, the relatively high level of design additional pressures (i.e., horizontal forces in lateral and circumferential components) would lead to a general overdesign of the silo wall sheets (i.e., thicker walls) over the silo height, and specifically in the upper part where the rationale of the EN1991-4:2006 [11] standard aims at reducing the forces in static conditions by accounting for Reimberts' modification [37] on Janssen's model for the squat silo category (as per the Eurocode definition).

2. Overdesign of the stiffener cross-sections in steel silos (i.e., larger sections), as the earthquake-induced vertical axial forces would be nearly twice as large as those predicted by Trahair's model.
3. Higher thickness of the reinforced concrete shell wall of concrete silos over the silo height.
4. Higher overall weight of concrete silos, resulting in slightly higher static vertical and seismic inertial forces in the design phases of the silo foundation. Such an issue is neglected in steel silos as their wall thicknesses are lower, and therefore, their weights are negligible with respect to the filling material weight.

The previous points clearly indicate that a physically consistent revision of the Eurocode formula might lead to economic savings in different countries classified as moderate-to-high-seismic areas (e.g., Italy, Greece, Portugal, Romania, and other EU and Extra-EU countries where EN1998-4 is applicable or commonly used).

Finally, current analytical studies and experimental works have recently shown that full mass mobilization is imprecise and even conservative for a broad category of industrial silos with a filling aspect ratio lower than 2, especially when characterized by a flat-wall section [25,27]. A similar trend is observed for corrugated-wall silos, albeit to a lesser extent. However, given the lack of sufficient available experimental results, it is prudent to retain the Eurocode assumptions. Nonetheless, the current formula should be modified to align more closely with the one suggested by Trahair as follows:

For rectangular silos,

$$\Delta_{ph,EN1998}(z) = \alpha(z) \cdot \gamma \cdot \min\left(H; \frac{L}{2}; 3x\right) \tag{14}$$

For circular silos,

$$\Delta_{ph,EN1998}(z) = \alpha(z) \cdot \gamma \cdot \min\left(\frac{H}{2}; \frac{R}{2}; 3x\right) \tag{15}$$

Equations (14) and (15) represent a physically based correction of the Eurocode formulae. If the Eurocode were to remain conservative, then it might consciously include a model safety factor (up to a certain desired value) for both circular and rectangular cases.

5. Concluding Remarks

This paper aims at achieving an in-depth interpretation and an improved understanding of the design-oriented formulae of the additional pressure explicitly prescribed by the EN1998-4:2006 European code and implicitly referenced by the American standard, for the seismic design of silo walls.

A critical pursuit of the historical development of the problem formulation is presented, aiming to uncover conceptual and methodological inconsistencies that may have been overlooked in the evolution of current standards.

A conceptual incongruence is highlighted regarding the dynamic overpressure in the case of circular silos: the current formula provided by EN1998-4 corresponds to an extremely high effective mass, up to twice the value of the full mass of the silo content (depending on the filling aspect ratio of the silo), which seems to be physically unjustified. This issue has also been encountered, but not explained, in several recent works by other researchers. In this short paper, a methodological explanation of this incongruence is provided, starting from the early dynamic pressure models and offering a comprehensive discussion of several related aspects. Moreover, a qualitative and quantitative evaluation of this incongruence is presented, and two possible scenarios explaining its origin are discussed. Finally, a proposal for a physically based correction is suggested, retaining the current assumptions made by EN1998-4.

Finally, the impact of the highlighted incongruence has been addressed through a list of the possible consequences in the design and manufacturing processes for both steel and reinforced concrete silos accounting for their technical differences.

Author Contributions: Both authors listed have made a substantial, direct, and intellectually equal contribution to the work and approved it for publication. All authors have read and agreed to the published version of the manuscript.

Funding: This research received no external funding.

Data Availability Statement: No new data were created or analyzed in this study. Data sharing is not applicable to this article.

Conflicts of Interest: The authors declare no conflicts of interest.

References

1. Khalil, M.; Ruggieri, S.; Uva, G. Assessment of Structural Behavior, Vulnerability, and Risk of Industrial Silos: State-of-the-Art and Recent Research Trends. *Appl. Sci.* **2022**, *12*, 3006. [CrossRef]
2. Khalil, M.; Ruggieri, S.; Tateo, V.; Nascimbene, R.; Uva, G. A numerical procedure to estimate seismic fragility of cylindrical ground-supported steel silos containing granular-like material. *Bull. Earthquake Eng.* **2023**, *21*, 5915–5947. [CrossRef]
3. Mansour, S.; Pieraccini, L.; Palermo, M.; Foti, D.; Gasparini, G.; Trombetti, T.; Silvestri, S. Comprehensive Review on the Dynamic and Seismic Behavior of Flat-Bottom Cylindrical Silos Filled with Granular Material. *Front. Built Environ.* **2022**, *7*, 805014. [CrossRef]
4. Janssen, H.A. Versuche uber getreidedruck in silozellen. *Z. Ver. Deut. Ing.* **1895**, *39*, 1045.
5. Sperl, M. Experiments on corn pressure in silo cells—Translation and comment of Janssen's paper from 1895. *Granular Matter.* **2006**, *8*, 59–65. [CrossRef]
6. Wood, J.H. Earthquake Induced Pressures on Rigid Wall Structure. Ph.D. Thesis, California Institute of Technology, Pasadena, CA, USA, 1973.
7. Wood, J.H. Earthquake induced pressures on rigid wall structure. *Bull. New Zealand Soc. Earthq. Eng.* **1975**, *8*, 175–186. [CrossRef]
8. Trahair, N.S.; Abel, A.; Ansourian, P.; Irvine, H.M.; Rotter, J.M. *Structural Design of Steel Bins for Bulk Solids*; Australian Institute of Steel Construction: Sydney, Australia, 1983; ISBN 0909945357.
9. Rotter, J.M.; Hull, T.S. Wall loads in squat steel silos during earthquakes. *Eng. Struct.* **1989**, *11*, 139–147. [CrossRef]
10. Federal Emergency Management Agency. *NEHRP Recommended Seismic Provisions for New Buildings and Other Structures*; FEMA P-2082-1 (Volume I: Parts 1 & 2, Provisions and Commentary); FEMA, Department of Homeland Security: Washington, DC, USA, 2020.
11. *EN 1998-4:2006*; Eurocode 8. Design of Structures for Earthquake Resistance, Part 4-Silos, Tanks and Pipelines. CEN: Brussels, Belgium, 2001.
12. Kurata, M.; Li, X.; Fujita, K.; Yamaguchi, M. Piezoelectric dynamic strain monitoring for detecting local seismic damage in steel buildings. *Smart Mater. Struct.* **2013**, *22*, 115002. [CrossRef]
13. Huang, H.; Chen, G.; Men, J.; Lai, E. Dynamic responses of steel cylindrical tanks subjected to sequential loadings of earthquake and explosion. *Eng. Fail. Anal.* **2023**, *143*, 106920. [CrossRef]
14. Hosseini, S.E.A.; Beskhyroun, S. Dynamic performance investigations of a full scale unanchored thin walled steel fluid storage tank via shake table tests. *Structures* **2024**, *70*, 107548. [CrossRef]
15. Zhang, H.; Wang, Y.; Liu, Q. Dynamic response analysis of vertical hybrid structure-equipment system under seismic excitation. *Case Stud. Constr. Mater.* **2023**, *19*, e02594. [CrossRef]
16. Jing, H.; Chen, H.; Yang, J.; Li, P. Shaking table tests on a small-scale steel cylindrical silo model in different filling conditions. *Structures* **2022**, *37*, 698–708. [CrossRef]
17. Yang, J.; Feng, G.; Jing, H.; Zhang, F. Seismic response and theoretical analysis of grain bulk material-steel silo structure under earthquake action. *J. Constr. Steel Res.* **2023**, *211*, 108207. [CrossRef]
18. Benkhellat, S.; Kadri, M. Seismic horizontal forces exerted by granular material on flat bottom silos: Experimental and analytical results. *Earthq. Eng. Struct. Dyn.* **2024**, *53*, 3983–4005. [CrossRef]
19. Benkhellat, S.; Kadri, M.; Seghir, A. Numerical investigation of the effects of soil-structure and granular material-structure interaction on the seismic response of a flat-bottom reinforced concrete silo. *Earthq. Eng. Eng. Vib.* **2024**, *23*, 609–623. [CrossRef]

20. Chen, J.; Ding, Y.; Xu, Q.; Liu, Q.; Zhou, Y. Evaluation method and experimental study on seismic performance of column-supported group silos. *Struct. Eng. Mech.* **2024**, *90*, 577–590. [CrossRef]
21. Wang, G.; Xu, Q.; Ding, Y.; Li, J.; Liu, Q. Vulnerability analysis of column supported reinforced concrete silo structures. *Appl. Sci.* **2025**, *15*, 2041. [CrossRef]
22. Chen, J.; Ding, Y.; Xu, Q.; Cheng, X. Research on seismic response of elevated silo by coupling discrete–finite element method. *Soil. Dyn. Earthq. Eng.* **2025**, *192*, 109287. [CrossRef]
23. Xu, Y.; Quan, C.; Li, R.; Wang, W.; Hua, Y.; Yang, H. Measurement of static and dynamic normal stress of granular materials in silos. *Measurement* **2025**, *252*, 117335. [CrossRef]
24. Qin, X.; Zeng, C.; Lu, Y. Seismic performance of wheat-silo-pile-soil interaction system implemented by a time staggered coupling of FEM and BEM. *Soil. Dyn. Earthq. Eng.* **2025**, *199*, 109660. [CrossRef]
25. Silvestri, S.; Gasparini, G.; Trombetti, T.; Foti, D. On the evaluation of the horizontal forces produced by grain-like material inside silos during earthquakes. *Bull. Earthq. Eng.* **2012**, *10*, 1535–1560. [CrossRef]
26. Mansour, S.; Foti, D.; Silvestri, S. Further refinements on Silvestri's theory for the dynamic overpressures prediction of filled silo systems under seismic excitation. In Proceedings of the World conference of Earthquake Engineering (WCEE2024), Milan, Italy, 30 June–5 July 2024.
27. Silvestri, S.; Ivorra, S.; Di Chiacchio, L.; Trombetti, T.; Foti, D.; Gasparini, G.; Pieraccini, L.; Dietz, M.; Taylor, C. Shaking table tests of flat-bottom circular silos containing grain-like material. *Earthq. Eng. Struct. Dyn.* **2016**, *45*, 69–89. [CrossRef]
28. Silvestri, S.; Mansour, S.; Marra, M.; Distl, J.; Furinghetti, M.; Lanese, I.; Hernández-Montes, E.; Neri, C.; Palermo, M.; Pavese, A.; et al. Shaking table tests of a full-scale flat-bottom manufactured steel silo filled with wheat: Main. results on the fixed-base configuration. *Earthq. Eng. Struct. Dyn.* **2022**, *51*, 169–190. [CrossRef]
29. Furinghetti, M.; Mansour, S.; Marra, M.; Silvestri, S.; Lanese, I.; Weber, F.; Pavese, A. Shaking table tests of a full-scale base-isolated flat-bottom steel silo equipped with curved surface slider bearings. *Soil. Dyn. Earthq. Eng.* **2024**, *176*, 108321. [CrossRef]
30. Mansour, S.; Silvestri, S.; Sadowski, A.J. The 'miniature silo' test: A simple experimental setup to estimate the effective friction coefficient between the granular solid and a horizontally-corrugated cylindrical metal silo wall. *Powder Technol.* **2022**, *399*, 117212. [CrossRef]
31. Nateghi, F.; Yakhchalian, M. Seismic behavior of silos with different height to diameter ratios considering granular material-structure interaction. *Int. J. Eng.* **2012**, *25*, 25–35. [CrossRef]
32. Djelloul, Z.; Mohammed, D. Contribution to the seismic behaviour of steel silos: Full finite-element analysis versus the Eurocode approach. *Asian J. Civ. Eng.* **2018**, *19*, 757–773. [CrossRef]
33. Mehretehran, A.M.; Maleki, S. Thin-Walled Structures 3D buckling assessment of cylindrical steel silos of uniform thickness under seismic action. *Thin-Walled Struct.* **2018**, *131*, 654–667. [CrossRef]
34. Jing, H.; Wang, X.; Yang, J.; Chen, H. Static and seismic pressure of cylindrical steel silo model with granular materials. *J. Constr. Steel Res. Nov.* **2022**, *198*, 107515. [CrossRef]
35. Khalil, M.; Ruggieri, S.; Tateo, V.; Butenweg, C.; Nascimbene, R.; Uva, G. Assessment of the seismic overpressure in flat bottom steel silos based on advanced FE modelling approach. *Soil. Dyn. Earthq. Eng.* **2025**, *190*, 109149. [CrossRef]
36. *EN 1991-4:2006*; Eurocode 1. Actions on structures, Part 4—Silos, tanks and pipelines. CEN: Brussels, Belgium, 2006.
37. Reimbert, M.L.; Reimbert, A.M. *Silos, Theory and Practice*; Trans Tech Publications: Baech, Switzerland, 1976.

designs

MDPI

Article

Influence of the Pattern of Coupling of Elements and Antifriction Interlayer Thickness of a Spherical Bearing on Structural Behavior

Anna A. Kamenskikh *, Anastasia P. Bogdanova, Yuriy O. Nosov and Yulia S. Kuznetsova

Department of Computational Mathematics, Mechanics and Biomechanics, Perm National Research Polytechnic University, 614990 Perm, Russia; anstasia_pankova@mail.ru (A.P.B.); ura.4132@yandex.ru (Y.O.N.); suhodolchik@mail.ru (Y.S.K.)

* Correspondence: anna_kamenskih@mail.ru; Tel.: +7-(342)239-15-64

Abstract: In this study, the behavior of the spherical bearing component of the L-100 bridge part (AlfaTech LLC, Perm, Russia) is considered within the framework of a finite element model. The influence of the pattern of the coupling of the antifriction interlayer with the lower steel plate on the operation of the part is examined in terms of ideal contact, full adhesion, and frictional contact. The thickness of the antifriction interlayer varied from 4 to 12 mm. The dependencies of the contact parameters and the stress–strain state on the thickness were determined. Structurally modified polytetrafluoroethylene (PTFE) without AR-200 fillers was considered the material of the antifriction interlayer. The gradual refinement of the behavioral model of the antifriction material to account for structural and relaxation transitions was carried based on a wide range of experimental studies. The elastic–plastic and primary viscoelastic models of material behavior were constructed based on a series of homogeneous deformed-state experiments. The viscoelastic model of material behavior was refined using data from dynamic mechanical analysis over a wide temperature range [−40; +80] °C. In the first approximation, a model of the deformation theory of plasticity with linear elastic volumetric compressibility was identified. As a second approximation, a viscoelasticity model for the Maxwell body was constructed using Prony series. It was established that the viscoelastic model of the material allows for obtaining data on the behavior of the part with an error of no more than 15%. The numerical analog of the construction in an axisymmetric formulation can be used for the predictive analysis of the behavior of the bearing, including when changing the geometric configuration. Recommendations for the numerical modeling of the behavior of antifriction layer materials and the coupling pattern of the bearing elements are given in this work. A spherical bearing with an antifriction interlayer made of Arflon series material is considered for the first time.

Keywords: bridge bearing; geometric configuration; polymer; antifriction interlayer; FEM; elastic–plastic; viscoelasticity; Prony; contact interaction

1. Introduction

1.1. Research Objectives

This study aimed to create a spherical bearing using relatively simple numerical analogs, taking into account the antifriction interlayer material's nonlinear behavior. This study includes an analysis of the structural behavior with different element matting patterns, as well as the thickness of the antifriction interlayer. Ensuring efficient operation of bridge bearings is an important task for the sustainable urbanization of territories.

Research tasks include the following:

1. The construction of an antifriction interlayer material model that takes into account the nonlinear effects of deformation behavior.
2. The influence analysis of the matting pattern of the lower steel plate and the antifriction interlayer in terms of structural behavior.
3. The influence analysis of the antifriction interlayer thickness in bearing operation.
4. The influence analysis of the material behavioral model for the contact stress–strain state of the bearing.
5. The analysis of the bearing behavior with structurally modified polytetrafluoroethylene (PTFE) without an AR-200 filler interlayer.

A digital analog of the bearing will allow us to predict the structural behavior under various conditions with high accuracy. Relatively simple structural models with a relatively small error from the real study object are of particular interest. They are necessary to minimizing the cost of computing resources and time for evaluating engineering solutions. This work aimed to create axisymmetric models of a spherical bearing to analyze the influence of the structural elements' geometric configuration on the contact stress–strain state.

1.2. Problem Context and Description

The design and construction of bridges is quite a complex process. Social, economic, resource, technical, and geographical factors determine the complexity of bridge structural design [1–3]. Building bridges in hard-to-reach areas, such as Arctic territories, mountain gorges, and coastal areas, requires the adaptation of important structural components to extreme conditions. The use of modern materials and the adoption of new design and technological solutions to ensure the long-term trouble-free operation of the structure are also important [1,4]. The high demand for such critical components requires more detailed comprehensive engineering and technical research. Preliminary analysis includes an assessment of materials and existing design solutions [5–7]. The final analysis includes the technical diagnostics and monitoring of the structure and its supporting elements [8,9]. Every detail of a bridge is an important component that affects the durability and safety of the structure as a whole.

The bearing is one of the key elements of the bridge [10]. Bearings redistribute the load from the span structures, and they bear dynamic impacts from vehicles and various natural and climatic phenomena [4,11,12]. Bridge bearings have many structural types. But spherical bearings have become widespread due to their high reliability and durability [13,14]. However, now, various factors that affect the stability, strength, and durability of the bearing must be taken into account when designing bridges [15–17]. This is because bridges are being built in difficult geographical and climatic conditions with extreme weather events and critical temperature fluctuations. For example, in [18], scientists assessed the impact of spherical steel bearings on train safety under seismic activity. Studies of the wind load on a long-span bridge during typhoons were conducted in order to further accurately assess fatigue damage to structural components [19]. The joint operation of the bearing and the span under complex loading, while also taking into account the means of transport, is of interest [17].

New, innovative designs of spherical bearings are also being developed [20–22], enabling increased load limits and fatigue strength. Changes in the geometry of the bearing elements significantly affect the deformation behavior of the structure [23,24]. The geometric parameters of the antifriction interlayer and its position relative to the steel plates have a significant impact on the contact characteristics [25]. Introducing an interlayer with convex and rounded edges allows for a slight increase in stresses and deformations compared to canonical geometry under increased vertical load [26]. Texturing the working surfaces also affects the effective coupling of elements [27–29]: it reduces friction, improves wear resistance, and increases the service life of the components. Studying bearing design is important for understanding the behavior of the elements, as well as for optimizing the structure to meet the required operating conditions.

The issues related to spherical bearing materials are also important. The primary influence on bearing performance is exerted by the antifriction interlayer material [25]. Scientists have made significant contributions by improving the physicomechanical and tribological characteristics of these materials [30–32]. A wide range of modern polymers and composites are currently available for use as bearing antifriction interlayers [33–35]. Polytetrafluoroethylene (PTFE), polyetheretherketone (PEEK), polymethylmethacrylate (PMMA), polyurethane (PU), polyamide (PA), ultra-high-molecular-weight polyethylene (UHMWPE), and composites based on these polymers are considered promising materials for antifriction interlayers [36]. The use of ceramics as sliding surfaces is also being explored [37,38]. Structurally modified PTFEs, both without additives and with nanofillers, manufactured by the Scientific and Production Enterprise Arflon LLC (Moscow, Russia) are also recommended for use as antifriction materials [39]. Arflon materials are characterized by high structural homogeneity, absence of porosity, and a wide range of operating temperatures. The material AR-200 is chemically analogous to PTFE but significantly surpasses it in physical, mechanical, and antifriction properties. In this study, AR-200 is considered as the antifriction interlayer. However, analyzing the behavior of the material within the bearing requires the development of numerical models. The digital model must accurately represent the physicomechanical and thermomechanical properties observed in experimental results. This necessitates a large number of empirical studies, as well as the application of mathematical tools for model development.

The mathematical description of the behavior of antifriction materials is of particular interest. The main research areas include rheology [40], viscoelasticity [41,42], viscoplasticity [43], viscoelastic-plasticity [44–46], phase transitions [47,48], etc. The complex nonlinear behavior of modern antifriction materials is important to note, as confirmed by numerous field experiments [49–51]. Experimental studies, combined with numerical modeling, make it possible to study the behavior of materials in greater detail. The processes occurring within the structural elements are analyzed, and the stress–strain state of parts and assemblies is evaluated based on key characteristics and parameters. Numerical methods help minimize the volume of experimental research. Numerical analogs of materials enable large-scale structural performance analysis within the framework of digital engineering. The choice of antifriction interlayer material and geometric design configuration can contribute to the optimization of spherical bearings for bridges. Numerical modeling allows for the prediction of structural deformation behavior under various loads, including the contact interaction between elements [52–54]. It is crucial to develop digital design models that take into account the nonlinear behavior of antifriction interlayer materials.

As part of the current study, a finite element model of the L-100 spherical bearing (AlfaTech LLC, Perm, Russia) with an antifriction interlayer made from Arflon line material was developed. A gradual refinement of the behavioral model for the antifriction interlayer material was carried out. Commonly used models, both elastic–plastic and viscoelastic, were considered. A high-quality digital analog of the material is essential for creating a spherical bearing model that closely reflects the actual structure. Using the bearing model with a nonlinear material representation of the antifriction interlayer, we investigated the influence of the element coupling pattern and the thickness of the antifriction interlayer on the bearing's contact interaction parameters. Multifactorial analysis of the structure's behavior makes it possible to obtain data relevant to designers when developing bearings for specific operating conditions.

2. Materials and Methods

2.1. The Design of the Spherical Bearing

The influence of the thickness of the polymer antifriction interlayer and its coupling pattern with the lower steel plate of the spherical bearing of the bridge span is examined in this study. The geometry of the contact unit of the spherical bearing includes an upper (1) and a lower (2) steel plate, which interact through a spherical antifriction interlayer made of a polymer or composite material (3) (Figure 1).

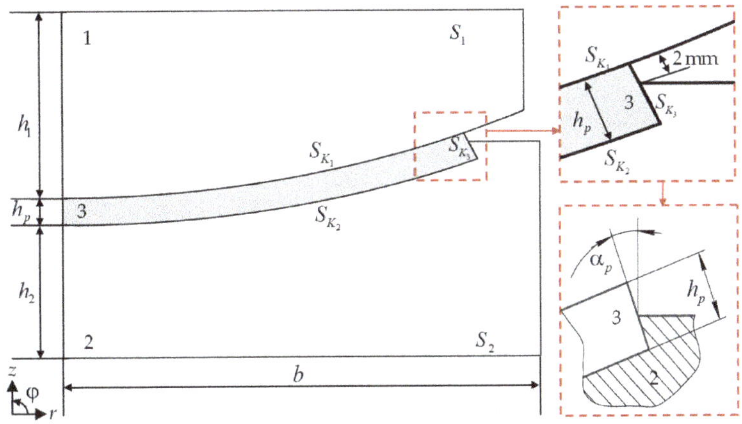

Figure 1. Calculation scheme of the spherical bearing.

The main geometric parameters of the L-100 bearing (AlfaTech LLC, Perm, Russia) are provided in Table 1. As part of this study, the effect of increasing the antifriction interlayer thickness to 6, 8, 10, and 12 mm on the performance of the structure will be examined. The end face of the antifriction interlayer S_{K_3} (3) is partially in contact with the steel structural elements. A 2 mm portion of the interlayer remains free from contact with the lower plate (a protrusion of the antifriction interlayer) to prevent metal-to-metal contact during construction work. The polymer interlayer is press-fitted into the bottom steel plate. The nominal vertical load is 1000 kN.

Table 1. Main geometric characteristics of the bearing.

Parameter	Designation on the Diagram	Meaning
The height of the structure	h	54 mm
The height of the upper steel plate with a polished spherical segment	h_1	20 mm
The height of the lower steel plate with a spherical cutout	h_2	22 ÷ 30 mm
The standard interlayer thickness	h_p	4 mm
The standard inclination angle of the antifriction interlayer end face	α_p	30°
The width	$B = 2b$	155 mm

A simplified geometry of the antifriction interlayer is considered, excluding the lubricant hole. The problem is analyzed in an axisymmetric formulation using the finite element method in ANSYS Mechanical APDL 2021R2 (Livermore, CA, USA).

A frictional contact S_{K_1} is implemented on the contact surface, with a previously unknown distribution pattern of contact status states.

- Sliding

$$u_n^1 \neq u_n^2,\ u_\tau^1 \neq u_\tau^2,\ \sigma_n^1 = \sigma_n^2,\ \sigma_{n\tau}^1 = \sigma_{n\tau}^2, \tag{1}$$

when $|\sigma_{n\tau}| = \mu(\sigma_n)|\sigma_n|$;

- No contact

$$\left|u_n^1 - u_n^2\right| \geq 0,\ \sigma_{n\tau} = \sigma_n = 0; \tag{2}$$

- Full attachment (adhesion)

$$\vec{u}^1 = \vec{u}^2,\ \sigma_n^1 = \sigma_n^2,\ \sigma_{n\tau}^1 = \sigma_{n\tau}^2, \tag{3}$$

where $\mu(\sigma_n)$ is the coefficient of friction, τ is the designation of the axis that lies in the plane tangent to the contact surface, u_n is displacement along the normal to the corresponding contact boundary, u_τ is displacement in the tangent plane, σ_n is stress along the normal to the contact boundary, $\sigma_{n\tau}$ are tangential stresses at the contact boundary, and 1–2 are conditional numbers of interface surfaces that belong to different bodies.

The friction contact corresponds to the polished surface of the steel in the interface area. On the contact surfaces S_{K_2}–S_{K_3}, three conjugation patterns are implemented:

1. An ideal contact, created by modeling a finite element node-to-node grid, which makes it possible to obtain uniform fields of displacements, stresses, and deformations at the interface:

$$u_n^1 = u_n^2,\ u_\tau^1 = u_\tau^2,\ \sigma_n^1 = \sigma_n^2,\ \sigma_{n\tau}^1 = \sigma_{n\tau}^2. \tag{4}$$

2. Full adhesion, which corresponds to the processing of the conjugation surface of the "torn thread" type (3).
3. Frictional contact (1)–(3), which includes different statuses with a previously unknown pattern of their distribution over the surface: full adhesion, sliding, and no contact.

Minimal roughness is required for the spherical sliding surface in the interface area between the upper balancer and the antifriction interlayer. This is achieved by using a polished surface on the spherical segment of the balancer. The bed of the balancer (the spherical cutout in the lower steel plate) can be treated in various ways: left unprocessed, given a rough "torn thread"-type surface to maximize adhesion, or polished. The surface treatment pattern can significantly influence the contact parameters and the stress–strain state of the structure. This paper considers the effect of the sliding antifriction interlayer's coupling pattern with the balancer bed on the performance of the bridge bearing.

2.2. *Antifriction Interlayer Material*

Structurally modified PTFE without fillers, AR-200 (Scientific and Production Enterprise Arflon LLC, Moscow, Russia), is considered as the material for the antifriction interlayer. The material is produced through compaction and high-temperature physico-chemical modification. It features a spherulitic macromolecular structure and is white in color. AR-200 is used in the manufacturing of sliding components.

To develop a material behavioral model, a series of experiments was initially conducted under homogeneous deformation conditions using cylindrical samples measuring 20 × 20 mm. Testing was performed with a Zwick Z100SN5A material testing machine (Zwick Roell AG, Ulm, Germany) and a Multisens extensometer (Zwick Roell AG, Ulm, Germany). The experimental studies were carried out at the Ural Branch of the Russian Academy of Sciences (ICMM UB RAS, Perm, Russia), under the supervision of Dr. A.A. Adamov. The maximum deformation level for the samples was set at 10%, based on the estimated maximum deformation of the bearing's antifriction interlayer. This limit represents the maximum allowable level for applying small deformation theory. A multi-stage testing program was implemented up to the 10% strain level, including relaxation, unloading, and recovery phases. The traverse movement speed was 3 mm/min. All experimental studies were conducted at room temperature.

The experiment and the processed results of empirical studies are shown in Figure 2.

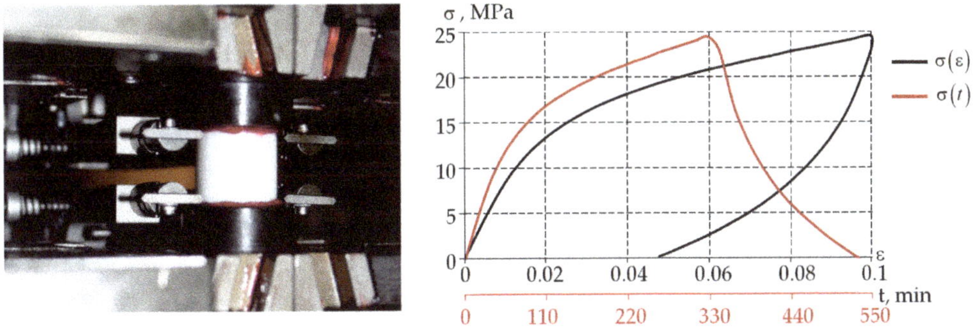

(a) (b)

Figure 2. Experiment on free compression of cylindrical samples: (**a**) the experiment; (**b**) the equilibrium experimental curves.

It was established that the material is a viscoplastic, compressible body with nonlinear viscoplastic volumetric compressibility and stress-induced phase transitions.

In the first approximation, a deformation theory of a plasticity model with linear elastic volumetric compressibility was identified (Figure 3a). This model was developed for the case of active loading and did not account for unloading.

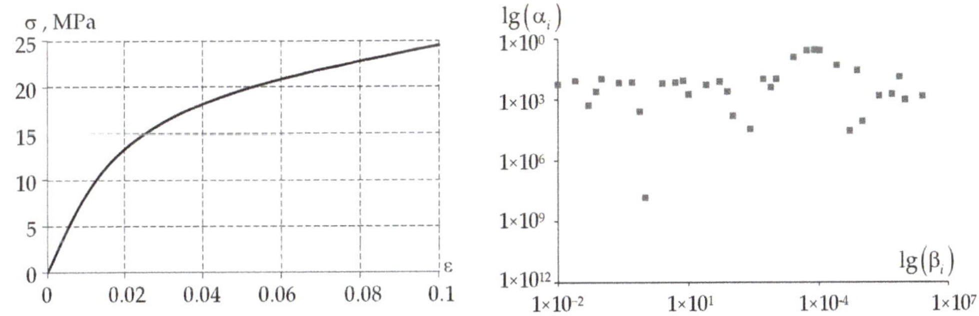

(a) (b)

Figure 3. Material models based on experimental data on the free compression of cylindrical samples: (**a**) elastic-plasticity; (**b**) viscoelasticity.

In the second approximation, a viscoelastic model for a Maxwell body was constructed using the Prony series (Figure 3b). The mathematical model was previously described in [55] and is based on the results of a series of experimental studies under uniaxial stress state (USS) conditions.

The dependence of the weighting coefficients α_i on the relaxation times β_i for 34 terms of the Prony series was obtained within the viscoelastic description of the material behavioral model. The empirical constants of the Williams–Landel–Ferry (WLF) temperature–time superposition $C_1 = 0.934$ and $C_2 = 1.231$, as well as the constant base temperature $T_r = 2.18\ ^\circ K$, were also determined. The model parameters correspond to room temperature. However, due to limited experimental data, additional testing is required to fully describe the behavior of AR-200 across a broader range of temperatures and deformation rates.

To further refine the viscoelastic behavioral model, a series of empirical studies was conducted using three-point bending of rectangular specimens. A dynamic mechanical analyzer DMA Q800 (TA Instruments, New Castle, DE, USA) was used (Figure 4). The average dimension of samples was 60 × 12 × 3 mm (Figure 4). The temperature range of the experiment was [−40; +80] °C. The research was conducted at the plastics laboratory of Perm National Research Polytechnic University (Perm, Russia).

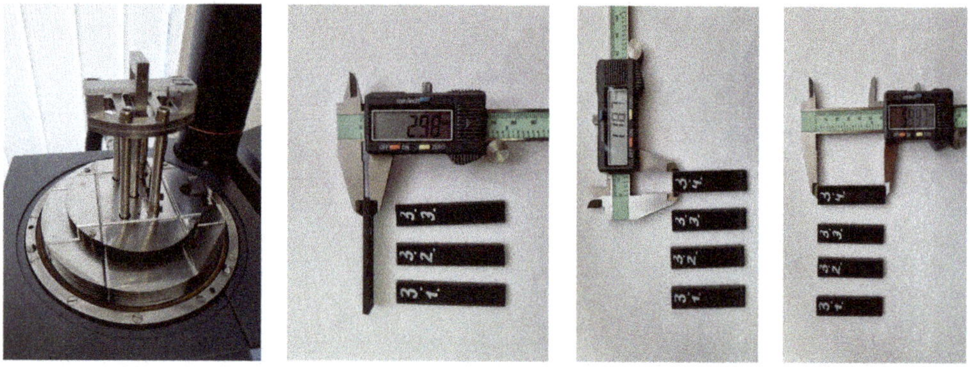

(**a**) (**b**) (**c**) (**d**)

Figure 4. Experiment on DMA: (**a**) three-point bend; (**b**–**d**) samples.

The dependencies of the storage modulus, loss modulus, and loss tangent (tan δ) on temperature were established (Figure 5). The glass transition temperature 8 °C of the AR-200 was determined. The material exhibited a distinct peak in the loss tangent, which corresponds to the transition region from a glassy to an elastic state near the glass transition temperature.

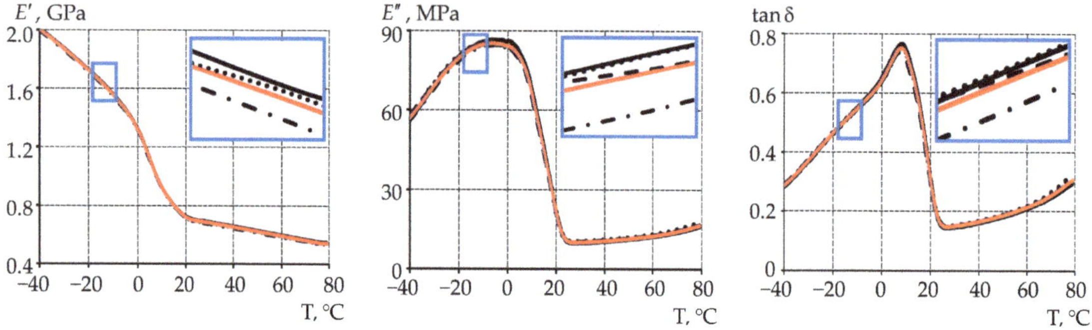

(**a**) (**b**) (**c**)

Figure 5. Temperature dependencies of the material parameters: (**a**) the storage modulus; (**b**) the loss modulus; (**c**) the loss angle tangent; black lines are the experimental data for a set of samples; red line is the averaged values of the experimental data.

The storage modulus reflects structural changes in the material. A decrease in the storage modulus is observed with increasing temperature, which is associated with material softening and a reduction in the elastic modulus. Changes in the loss modulus are related to relaxation transitions within the material. In the region where the loss modulus increases, the material is in a glassy state. As the temperature continues to rise, the loss modulus decreases, and the material transitions into a highly elastic state.

The results of the DMA experiments formed the basis for a refined model of AR-200 behavior within the framework of viscoelasticity theory, using the temperature–time superposition principle based on the WLF model (Figure 6).

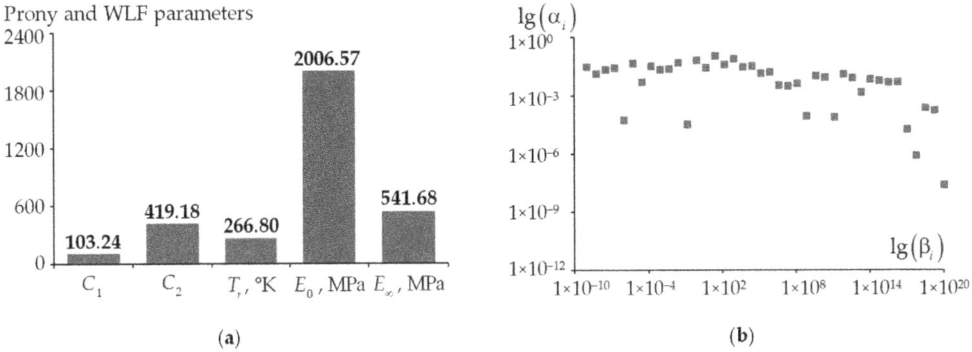

Figure 6. A viscoelastic Maxwell-type model based on Prony series: (**a**) Prony and WLF parameters; (**b**) the dependence of the weighting coefficients on the relaxation time.

The long-term modulus of elasticity at the initial and final moments in time E_0 and E_∞ was determined. The constant reference temperature T_r and the empirical constants of the temperature–time superposition (WLF) C_1 and C_2 were refined, taking into account a wide temperature range of [−40; +80] °C. For the description of the viscoelastic behavioral model of materials, 40 pairs of coefficients α_i, β_i were defined.

For all AR-200 models, Poisson's ratio was determined experimentally and was equal to 0.461.

Experimental studies included at least three repetitions to ensure the statistical significance of the results.

Viscoelastic models of the material behavior were constructed using the author's numerical identification procedure, with an error of less than 5% [55,56].

2.3. The Numerical Model

The numerical implementation of the problem was carried out in the ANSYS Mechanical APDL 2021R2 software package using PLANE182 axisymmetric quadrilateral elements with Lagrangian approximation. Surface-to-surface contact interaction was modeled using the contact element pair CONTA171 and TARGE169. The primary procedures for constructing finite element analogs were based on the Galerkin method, with basic functions selected to have compact support, as per the finite element method. A step-by-step loading algorithm was used, with iterative refinement of the distribution of contact zones, penetration values, and contact forces. The iterative procedure continued until the results of two consecutive iterations produced contact force and penetration values within the specified accuracy.

Frictional contact was modeled using the basic Coulomb friction model, incorporating the condition of critical tangential stress in the contact zone, i.e., exceeding the yield strength of the material beneath the surface (an expanded classical Coulomb friction model). The coefficient of friction was set as a constant value of 0.04, in accordance with manufacturer data for the steel–polymer contact pair used in bridge bearings.

To describe the elastic–plastic behavior of the antifriction interlayer material, the MISO (multilinear isotropic hardening) model was used. For the viscoelastic behavior, the Prony series was employed. The steel structural elements were modeled within the framework of linear elasticity theory, with an elastic modulus of 2×10^{11} and a Poisson's ratio of 0.3.

Previously, the convergence of the numerical solution with respect to the degree of system discretization was investigated [57]. It was determined that 16 finite elements through the thickness of the interlayer are required for a standard thickness of 4 mm. The finite element mesh in the contact area of the steel plates was aligned with the mesh division of the antifriction interlayer. Beyond the contact area, the mesh size increases gradually (in a gradient manner). Finite element models with antifriction interlayer thicknesses of 4 mm and 12 mm are characterized by approximately 50,000 and 63,000 nodal unknowns, respectively.

3. Results

3.1. Analysis of the Effect of the Interface Pattern of the Antifriction Interlayer with the Lower Steel Plate at an Interlayer Standard Thickness

In the first stage of this study, the effect of the interface pattern between the antifriction interlayer (of standard thickness) and the lower steel plate was assessed. The behavior of the antifriction interlayer material was modeled within an elastic–plastic formulation.

The levels and distribution patterns of the contact parameters on the sliding surface of the spherical balancer S_{K_1}, as well as the deformation of the antifriction interlayer near the protrusion, are of particular interest.

The pattern of the contact pressure and contact tangential stress distribution on S_{K_1}, obtained by modeling different types of conjugation on S_{K_2}–S_{K_3}, is shown in Figure 7.

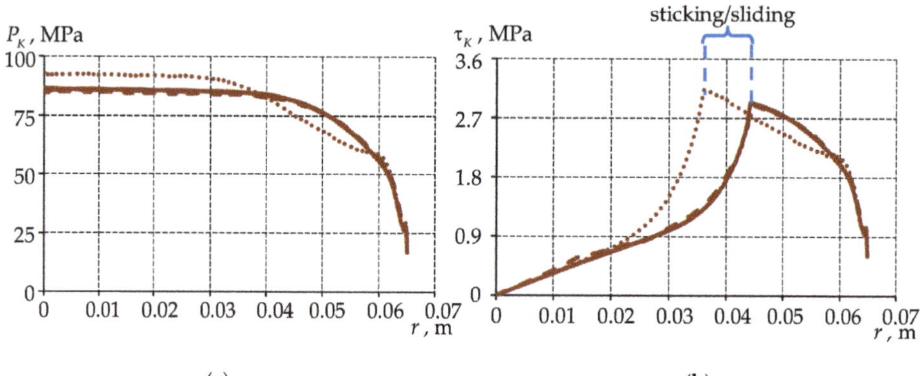

Figure 7. Contact parameters S_{K_1} with different conjugation S_{K_2}–S_{K_3}: (**a**) contact pressure; (**b**) contact tangential stress; the solid line is ideal contact; the dashed line is full adhesion; dots are frictional contact.

The contact surfaces did not separate under any coupling pattern S_{K_2}–S_{K_3}, i.e., the "no contact" status did not occur during structural deformation. The interface pattern between the lower steel plate and the antifriction interlayer has the greatest influence on the distribution of zones with different contact statuses S_{K_1}. With ideal contact and full adhesion, the transition zone between full adhesion and sliding shifts closer to the edge of the antifriction interlayer. The full adhesion area S_{K_1} accounts for 30.97% of the initial contact area with ideal contact and full adhesion S_{K_2}–S_{K_3}, and 16.48% of the initial contact area with frictional contact S_{K_2}–S_{K_3}. The total adhesion area S_{K_2} comprises 24.09% of the initial contact area for the frictional contact S_{K_2}–S_{K_3}. The maximum level of contact pressure on the surfaces S_{K_1} and S_{K_2} is the same for all types of coupling of the antifriction interlayer with the lower steel plate. The maximum level of P_K and τ_K at frictional contact S_{K_2}–S_{K_3} is 7–9% higher than at ideal contact and full adhesion. This effect is associated with a decrease in the area of full adhesion at the interface with the spherical balancer (the area of adhesion is almost two times smaller). The maximum levels of P_K and τ_K at full adhesion S_{K_2}–S_{K_3} are lower than at ideal contact by 1.41 and 0.84%, respectively.

Plastic flow of the material is observed near the protrusion of the antifriction interlayer. The deformation of the interlayer is overestimated compared to that of the actual structure, due to the fact that the elastic–plastic model does not account for material relaxation. The influence of the element connection pattern S_{K_2}–S_{K_3} on the normal displacements of the antifriction interlayer's end face is shown in Figure 8.

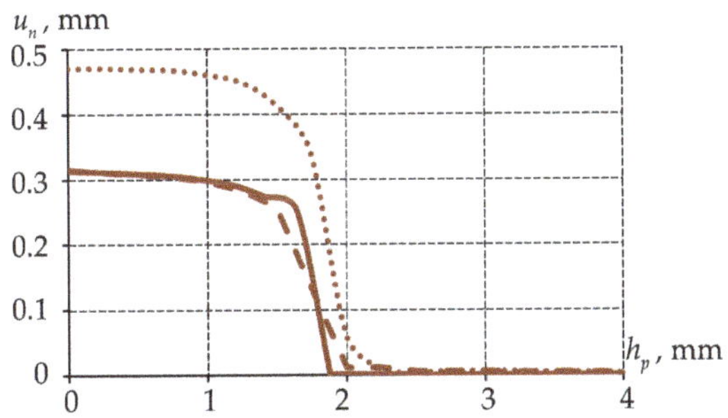

Figure 8. Distribution of normal displacements along the end of the interlayer S_{K_3} at different interfaces S_{K_2}–S_{K_3}: the solid line is ideal contact; the dashed line is full adhesion; the dots are friction contact.

The maximum displacement of the antifriction interlayer is observed near the sliding surface S_{K_1}. Under the standard load during indentation by the spherical balancer, the interlayer undergoes significant deformation. In the protrusion zone of the antifriction interlayer, the material enters a state of plastic flow. This deformation can be more accurately modeled by incorporating viscosity into the material behavioral model. Under frictional contact, the antifriction interlayer material can slip relative to the lower steel plate, resulting in a 49.8% increase in the maximum normal displacement. In contrast, the difference in maximum displacement between ideal contact and full adhesion S_{K_2}–S_{K_3} does not exceed 0.1%. The distribution pattern of displacements along the normal direction at the interlayer's end face shows only minor differences.

The differences in the average values ΔP_K, $\Delta\tau_K$, and Δu_n are considered for better clarity (Figure 9). ΔP_K, $\Delta\tau_K$, and Δu_n are the mean arithmetic errors of the parameters across the surface. The parameters obtained during ideal contact S_{K_2}–S_{K_3} were selected as the reference values: the model has only one contact surface and requires less computational power to assess changes in the geometry of the structure during engineering surveys.

The greatest differences between the solutions are observed in the normal displacement parameter. The difference between frictional contact and ideal contact is approximately 50%, while the difference with full adhesion does not exceed 7.5%. As expected, the smallest differences occur in contact pressure, which is the least sensitive to variations in the numerical implementation parameters. Nevertheless, the difference between frictional contact and ideal contact still exceeds 6.5%. Contact shear stress is more sensitive to contact settings: the difference between full adhesion and ideal contact is 5.64%, whereas for frictional contact, it reaches 23.47%.

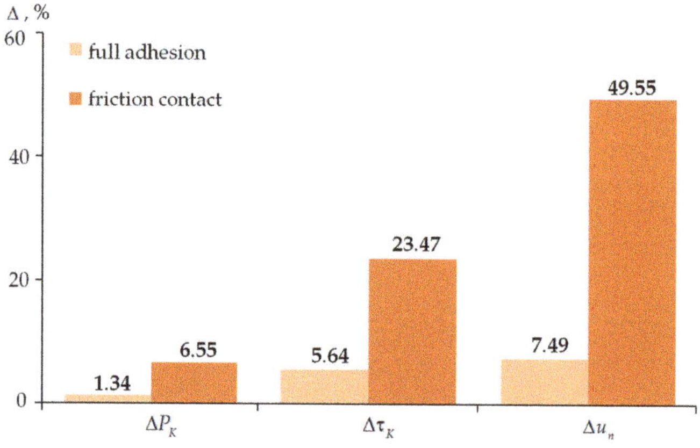

Figure 9. The arithmetic mean differences in the contact parameters S_{K_1} and normal displacement on S_{K_3}.

Figure 10 shows the isofields of the stress–strain state parameters (intensity of stresses and deformations) of the bearing antifriction interlayer with different contact patterns S_{K_2}–S_{K_3}.

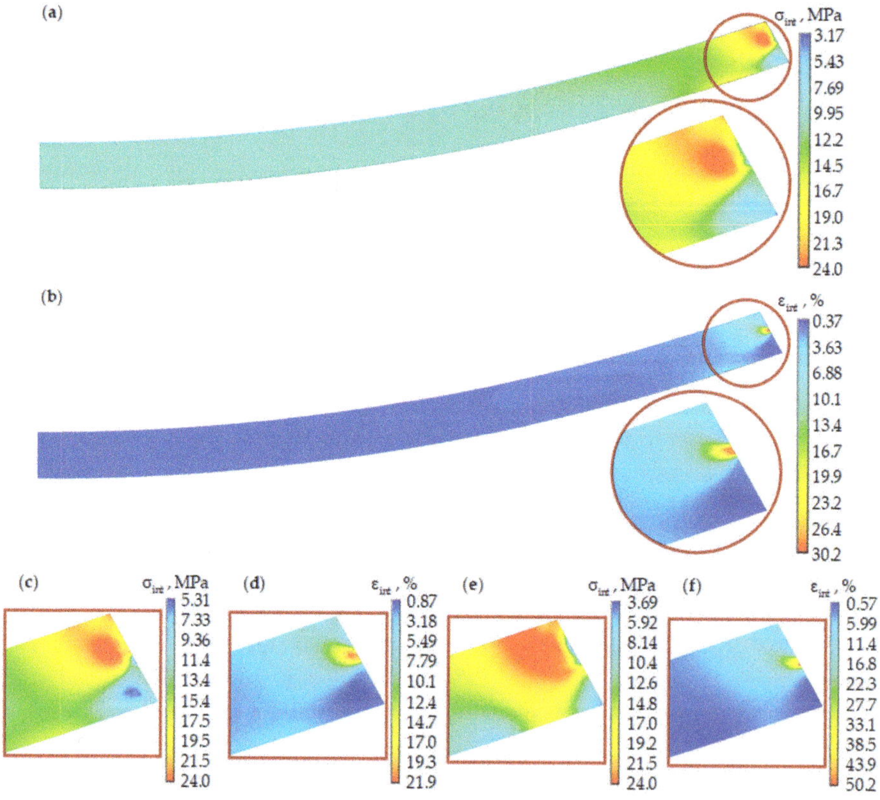

Figure 10. The parameters of the stress–strain state of the antifriction interlayer: (**a**,**c**,**e**) σ_{int}; (**b**,**d**,**f**) ε_{int}; (**a**,**b**) ideal contact S_{K_2}–S_{K_3}; (**c**,**d**) full adhesion S_{K_2}–S_{K_3}; (**e**,**f**) frictional contact S_{K_2}–S_{K_3}; the red circle is the area of the antifriction layer near the edge.

The maximum levels of stress and strain are observed near the edge of the antifriction interlayer, in the protrusion area. The maximum strain level exceeds 10%, and the stress level reaches the peak value from the material's deformation curve σ–ε. This is attributed to limitations of the current material model. The maximum area σ_{int} is observed during frictional contact S_{K_2}–S_{K_3}. This occurs because, in this formulation, the antifriction interlayer material undergoes greater deformation due to the possibility of sliding along the contact surfaces. The maximum intensity of deformations is observed near the collar of the spherical cutout of the lower steel plate, where a concentrator forms. To improve accuracy, the material behavioral model should be refined to account for relaxation effects. Implementing full adhesion S_{K_2}–S_{K_3} allows for a more accurate representation of the stress–strain distribution, as it avoids inconsistencies in stress and strain fields between the mating surfaces of the antifriction layer and the bottom plate. The minimum levels of stress and deformation parameters are observed with full adhesion S_{K_2}–S_{K_3}, which demonstrates the effectiveness of processing the contact surface of the lower steel plate using the "torn thread" method.

The "torn thread" surface interaction can be modeled as ideal contact when it is necessary to reduce computational cost. In this case, the differences in the numerical solution for the standard thickness of the antifriction interlayer do not exceed 10%.

This subsection presents the results of analyzing the influence of the conjugation pattern on the stress–strain state of the antifriction interlayer under constant temperature and pressure. However, factors such as temperature, load magnitude, and load periodicity also have a significant impact on the structure's behavior. These effects are currently being investigated. An increase in temperature leads to a decrease in contact pressure and an increase in both the sliding contact area and contact penetration. This behavior is attributed to the material's softening under elevated temperatures, caused by structural and phase transitions. Conversely, when the temperature decreases, the opposite effects are observed—associated with the glass transition of the antifriction interlayer material. Similar trends were confirmed by the DMA experimental results.

3.2. *Analysis of the Effect of the Thickness of the Bearing Antifriction Interlayer on the Structure's Behavior*

In the second stage of this study, the effect of the antifriction interlayer thickness was assessed under constant protrusion and various coupling patterns between elements.

The distribution patterns of contact parameters and the stress–strain state of the structure do not significantly differ across the different coupling patterns between the antifriction interlayer and the lower steel plate. The influence of interlayer thickness on structural behavior is illustrated using a model with ideal steel–polymer contact S_{K_2}–S_{K_3}.

Subsequently, the effect of interlayer thickness on the maximum parameter levels is analyzed for the different coupling patterns.

3.2.1. Ideal Contact Along the Interface Surfaces of the Antifriction Interlayer with the Lower Steel Plate

The distribution of contact parameters on S_{K_1} at different h_p is shown in Figure 11.

The distribution pattern of contact pressure and tangential contact stress shows only minor differences with changes in the thickness of the antifriction interlayer. As the interlayer thickness increases, there is a tendency for the contact parameters to approach zero near the edge of the interlayer. This effect is associated with increased deformation of the antifriction interlayer material and a larger volume of material exhibiting plastic behavior. When the interlayer thickness > 12 mm or under altered deformation conditions, regions of complete detachment (i.e., loss of contact) may appear. The maximum level of contact parameters S_{K_1} changes slightly at ideal contact S_{K_2}–S_{K_3}: an increase in the antifriction interlayer thickness up to 8 mm increases the parameter level by a maximum of 1.2%, followed by a decrease. The maximum tangential contact stress occurs in the transition zone between sticking and sliding contact statuses. A general trend is observed—a decrease in the percentage of the contact area that remains in full adhesion. At the same time, a larger area in the state of adhesion corresponds to a greater volume of material operating under hydrostatic compression, exhibiting primarily elastic properties.

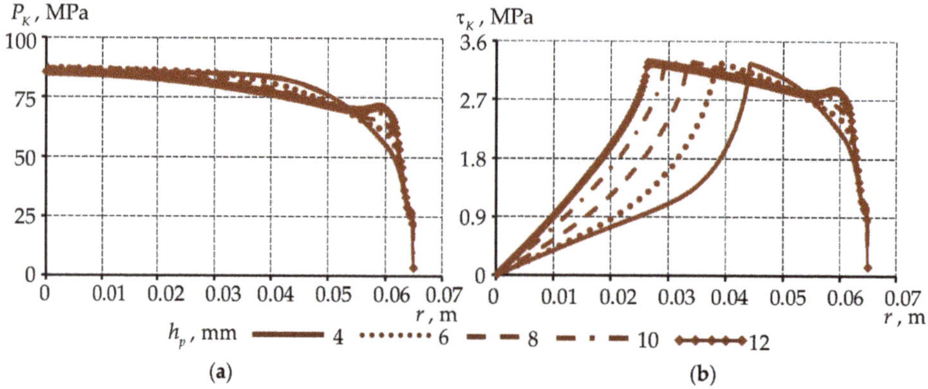

Figure 11. Distribution of contact parameters on S_{K_1} at different h_p: (**a**) contact pressure; (**b**) tangential contact stress.

The distributions of the normal end face displacement are shown in Figure 12.

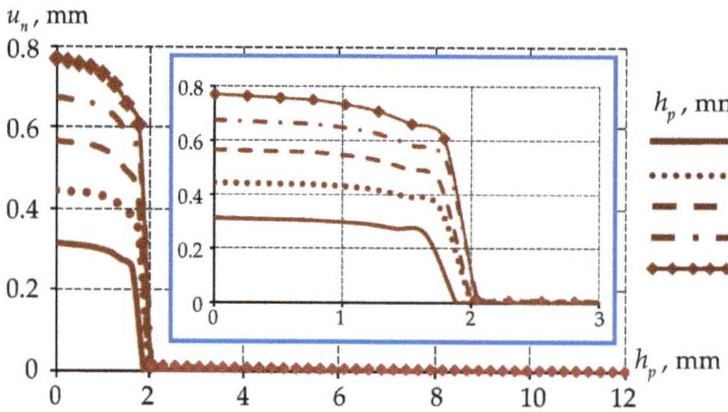

Figure 12. Distributions of normal displacement along the end of the interlayer S_{K_3} at different thicknesses of the antifriction interlayer.

An increase in the thickness of the antifriction interlayer leads to a significant rise in the maximum normal displacement at the surface of the protrusion. Specifically, increasing the thickness from 4 mm to 12 mm results in a more than 2.5-fold increase in the maximum normal displacement. The interlayer thickness is regulated by the Federal Road Agency "Rosavtodor" (Moscow, Russia). Therefore, it is necessary to consider the effect of interlayer thickness across all sizes and configurations of spherical bearings. These structures vary in the radius of curvature of the spherical segment, as previously discussed in [56]. Additionally, different approaches are used to determine the size of the antifriction material protrusion. One potential area for further research is a comprehensive assessment of how the projection geometry influences the overall behavior of the structure. To better capture the deformation behavior of the antifriction interlayer's end face, it is essential to account for the relaxation properties of the polymer material. This was addressed in subsequent research through the implementation of a viscoelastic material behavioral model.

3.2.2. Comparative Analysis of the Structure's Behavior at Different Thicknesses and Antifriction Interlayer Interfaces

The next stage of this study focused on analyzing the influence of the coupling pattern between the lower steel plate and the antifriction interlayer across all variants of the bearing's geometric configuration.

Figure 13 shows the dependencies of the maximum level of contact pressure S_{K_1}, tangential contact stress S_{K_1}, and normal displacements S_{K_3}, as well as the percentage of the surface area S_{K_1}, in the state of full adhesion on the antifriction interlayer thickness with different interface patterns S_{K_2}–S_{K_3}.

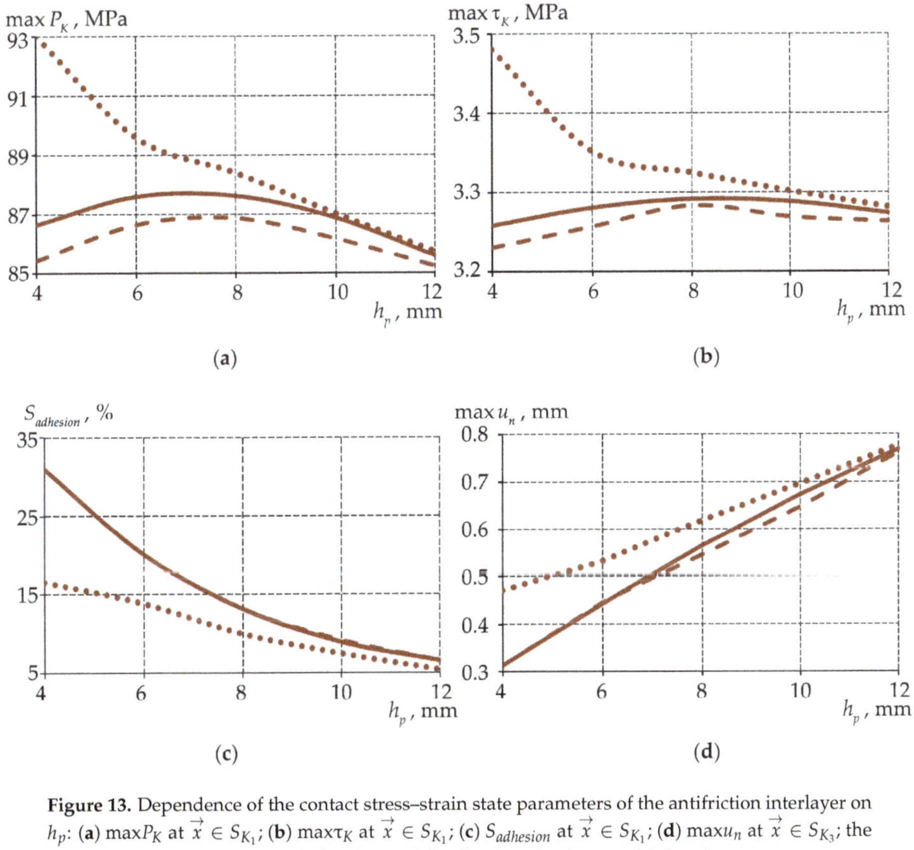

Figure 13. Dependence of the contact stress–strain state parameters of the antifriction interlayer on h_p: (**a**) maxP_K at $\vec{x} \in S_{K_1}$; (**b**) maxτ_K at $\vec{x} \in S_{K_1}$; (**c**) $S_{adhesion}$ at $\vec{x} \in S_{K_1}$; (**d**) maxu_n at $\vec{x} \in S_{K_3}$; the solid line is ideal contact; the dashed line is full adhesion; the dots are frictional contact.

It can be noted that the spherical bearing deformation at ideal contact and full adhesion S_{K_2}–S_{K_3} has small differences: the field of stress and strain distribution is specified, especially near the antifriction interlayer, as well as contact parameters. The model with a single contact surface (ideal contact on S_{K_2}–S_{K_3}) can be used to increase the speed of evaluating the deformation behavior of the structure during engineering surveys.

The level of contact parameters S_{K_1} at ideal contact and full adhesion S_{K_2}–S_{K_3} is lower than that of a frictional contact with an antifriction interlayer thickness of up to 10 mm. As the antifriction interlayer thickness increases, the differences in the parameters of the contact stress–strain state of the spherical bearing with a different type of interface (surface treatment) with the lower steel plate decrease. With thicknesses of more than 10 mm, the differences are less than 2%.

The percentage of the surface area S_{K_1} in the state of full adhesion decreases with increasing thickness of the interlayer. As the sliding area increases, there is a significant rise in the normal displacements at the end face of the interlayer, particularly in the protrusion zone. The "no contact" status between the mating surfaces is not observed on S_{K_1} for any of the contact options S_{K_2}–S_{K_3}.

An increase in interlayer thickness leads to a greater volume of material exhibiting nonlinear behavior, which affects the overall performance of the bearing. A larger portion of the interlayer experiences plastic deformation; however, plastic flow does not increase substantially due to the state of all-around compression in the fully adhered contact zones. Nonetheless, the normal displacements at the end face of the interlayer increase by several times—this is considered a negative effect. These results correspond to static loading conditions at constant temperature. Under dynamic, cyclic, or temperature-variable loading, it becomes necessary to use a viscoelastic or viscoelastoplastic material behavioral model for accurate simulation of the bearing's performance.

3.3. Influence Analysis of the Behavioral Model of the Antifriction Interlayer Material

The influence of the antifriction interlayer material behavioral model on the results of structural deformation needs to be evaluated. This is necessary to minimize the error of the bearing model from the real object. The evaluation of the model was performed with an antifriction interlayer thickness of 4 mm.

A gradual refinement of the material behavioral model was performed to account for structural and relaxation transitions as part of the work. This is necessary to move on to dynamic tasks, including at a wide range of temperatures, to assess the life of the structure and its operation under cyclic and seismic influences.

Three models of behavior are considered:

1. The elastic–plastic model of the material behavior is based on experimental data on the free compression of cylindrical samples. The model is implemented only for the case of active loading.
2. The viscoelasticity (USS) model is based on data from multi-stage tests for free compression of cylindrical samples to a maximum strain level of 10% with relaxation, unloading, and recovery areas. The model is limited within the range of temperatures close to room temperatures of 22–23 °C. The plasticity of the material and a number of phenomena and effects of the viscoelastic pattern are not taken into account.
3. The viscoelasticity (DMA) model is based on the entire set of experimental data, taking into account the DMA study of material behavior over a wide temperature range. The model does not take into account the plasticity of the material.

Figure 14 shows a comparative analysis of the three models of antifriction material behavior within the parameters of the stress–strain state of the elements of the spherical bearing with a standard antifriction interlayer thickness of 4 mm.

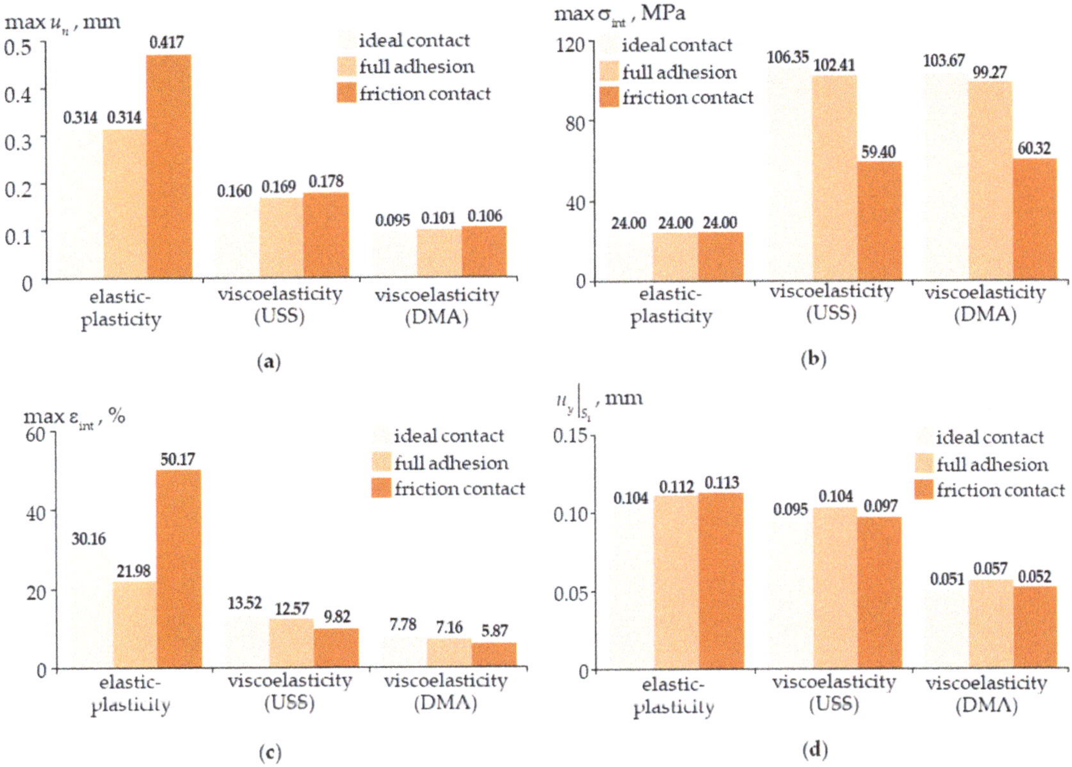

Figure 14. The parameters of the stress–strain state of the bearing: (**a**) maxu_n at $\vec{x} \in S_{K_3}$; (**b**) maxσ_{int} at $\vec{x} \in V_3$; (**c**) maxε_{int} at $\vec{x} \in V_3$; (**d**) the settlement of the bearing.

The elastic–plastic behavioral model of the material in the antifriction interlayer has a significant limitation associated with an asymptotic stress level when the deformation level exceeds 10%. Additionally, the model does not account for material relaxation, which can be observed even with iterative loading application. This results in overestimated deformation characteristics at significantly underestimated stress levels. The modeling error of the deformation of the spherical bearing compared to the data from field tests is from 57.7 to 71.9%. The maximum error is observed during frictional contact S_{K_2}–S_{K_3}.

The viscoelasticity (USS) model makes it possible to refine data on the behavior of the antifriction interlayer material. This leads to a significant reduction in the deformation of the antifriction interlayer, while clarifying the level of stresses and deformations of the material. The model makes it possible to effectively describe the multi-stage cyclic deformation of samples [56]. The error in modeling the deformation of the spherical bearing is from 44.1 to 57.4% of the experimental test data. The results of numerical modeling are overestimated. The maximum error is observed at full adhesion S_{K_2}–S_{K_3}.

The viscoelasticity (DMA) model makes it possible to obtain a numerical analog of the spherical bearing, as close as possible to the real design. The error of modeling the deformation of the spherical bearing is from 13.8 to 21.1% of the experimental test data. The minimum error is observed with full adhesion S_{K_2}–S_{K_3}.

The results of experimental tests of the bearing draft contained multi-stage loading with load holding. These effects are not taken into account in the current model of deformation of the bearing. The pattern of the processing of the bottom plate balancer is "torn thread", which is confirmed by the minimum error of the model with full adhesion to the interface surfaces S_{K_2}–S_{K_3}. The level of settlement of the spherical bearing model with a refined viscoelastic model of polymer material behavior is as close as possible to the actual design. It should be noted that the model is implemented in an axisymmetric formulation and allows one to quickly evaluate engineering solutions for changing the geometric configuration of elements.

The behavioral model of the antifriction interlayer material can be refined after conducting a series of numerical experiments on the dependence of the thermal expansion coefficients on temperature, as well as a series of experiments on uniaxial stress over a wide temperature range. It also requires consideration of the plasticity of the material, which can be carried out when switching to the Anand model.

Comparative Analysis of Viscoelastic and Elastic–Plastic Models of Material Behavior

We will perform a comparative analysis of the bearing behavior in the case of elastic–plastic and viscoelasticity (DMA) models of the behavior of the antifriction interlayer material.

The pattern of the distribution of contact parameters when modeling the antifriction interlayer material behavior within the framework of elastic–plastic and viscoelastic settings is shown in Figure 15.

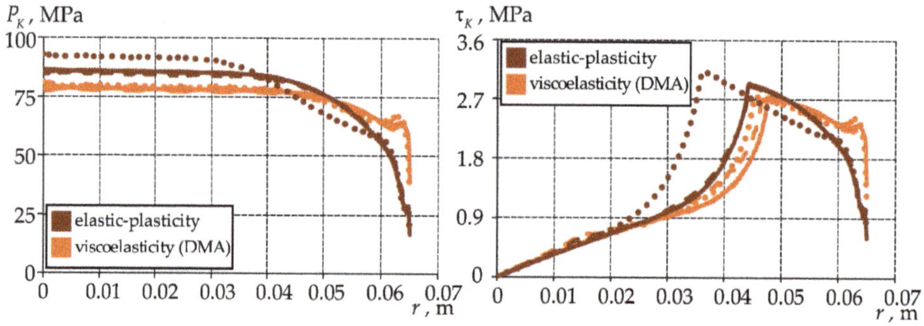

(**a**) (**b**)

Figure 15. Comparative analysis of the contact parameters S_{K_1} at different conjugation S_{K_2}–S_{K_3} and different models of the antifriction interlayer material behavior: (**a**) contact pressure; (**b**) tangential contact stress; the solid line is ideal contact; the dashed line is full adhesion; the dots are frictional contact.

The behavioral model of the antifriction interlayer material does not affect the pattern of the distribution of contact parameters. The viscoelasticity (DMA) model of the antifriction interlayer material shows a decrease in the level of contact parameters compared to elastic-plasticity by 6–8% with ideal contact and full adhesion S_{K_2}–S_{K_3}, and by 13–15% with frictional contact. The percentage area of the contact surface S_{K_1} in the state of full adhesion is greater for the viscoelastic material behavior.

Figure 16 shows the normal displacements of the end of the antifriction interlayer when modeling the material behavior in the framework of elastic–plastic and viscoelastic settings.

The maximum level of normal displacement of the end of the antifriction interlayer with viscoelastic behavior of the material is 3–5 times lower than with elastic-plasticity.

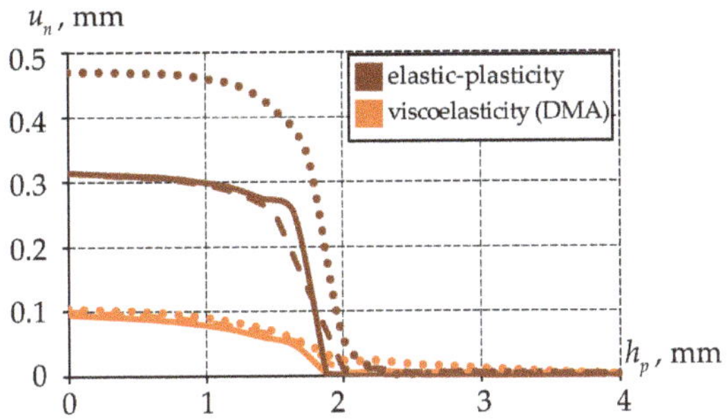

Figure 16. Comparative analysis of the distribution of displacements along the normal of the end face of the interlayer S_{K_3} at different conjugation S_{K_2}–S_{K_3} and different models of behavior of the antifriction interlayer material: the solid line is ideal contact; the dashed line is full adhesion; the dots are frictional contact.

Figure 17 shows the dependencies of the maximum level of contact pressure S_{K_1}, tangential contact stress S_{K_1}, and normal displacements S_{K_3}, as well as of the bearing draft on the antifriction interlayer thickness in the case of an elastic–plastic and viscoelastic behavioral material model.

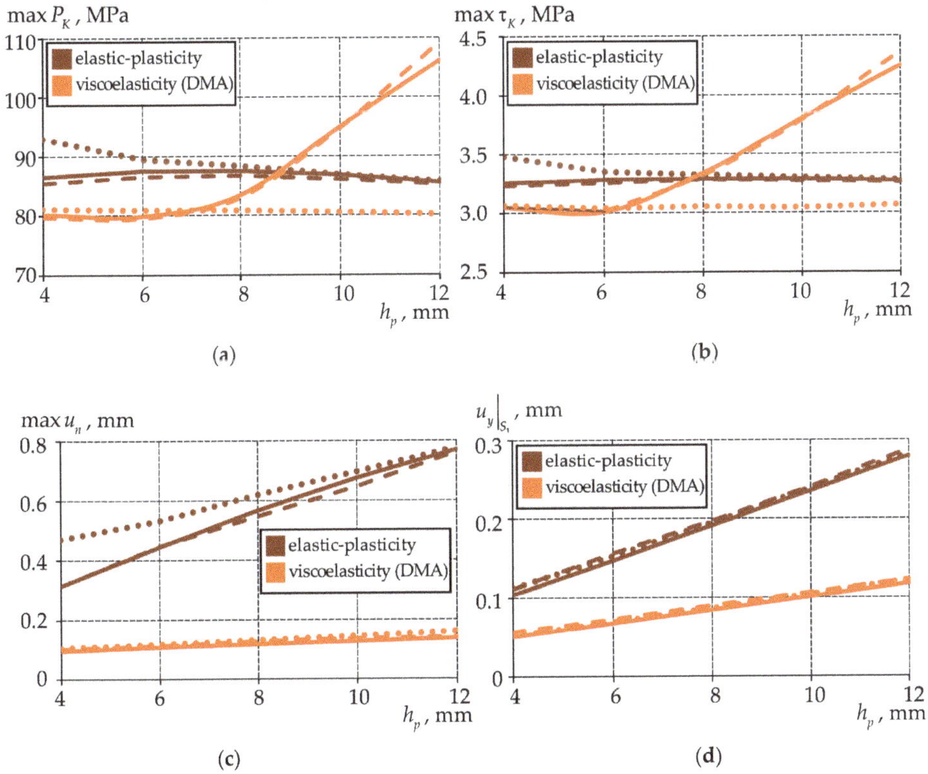

Figure 17. The dependence of the parameters of the contact stress–strain state of the antifriction interlayer on h_p for elastic–plastic and viscoelastic models of the antifriction interlayer material behavior: (**a**) $\max P_K$ at $\vec{x} \in S_{K_1}$; (**b**) $\max\tau_K$ at $\vec{x} \in S_{K_1}$; (**c**) $\max u_n$ at $\vec{x} \in S_{K_3}$; (**d**) settlement of the bearing; the solid line is ideal contact; the dashed line is full adhesion; the dots are frictional contact.

A significant increase in the level of contact pressure and tangential contact stress is observed with ideal contact and full adhesion S_{K_2}–S_{K_3} with an increase in the thickness of the antifriction interlayer in the viscoelastic behavioral material model. These effects are associated with rather harsh deformation conditions of the structure with an increase in the volume of a more malleable material with the possibility of structural and relaxation transitions. The decrease in contact pressure and a small change in the level of tangential contact stress are observed with frictional contact. In general, the contact parameters of the spherical bearing with viscoelastic behavior of the polymer material are lower, with an antifriction interlayer thickness of 10 mm less than that with elastic–plastic behavior.

The deformation of the antifriction material in particular and the spherical bearing as a whole is less with the viscoelastic behavior of the antifriction interlayer than with the elastic–plastic material behavior.

The parameters of the contact stress–strain state of the bearing with a standard thickness of the antifriction interlayer depend little on the coupling pattern.

Of interest is the change in the percentage of the contact surface area in the state of full adhesion, since in this case, hydrostatic compression of the antifriction interlayer is observed, and the material operates within the framework of the theory of elasticity.

Figure 18 shows the dependencies of the percentage of surface area S_{K_1} and S_{K_2} (for the case of frictional contact along the interface surfaces of the antifriction interlayer and the lower steel plate) in the state of full adhesion on the antifriction interlayer thickness with elastic–plastic and viscoelastic models.

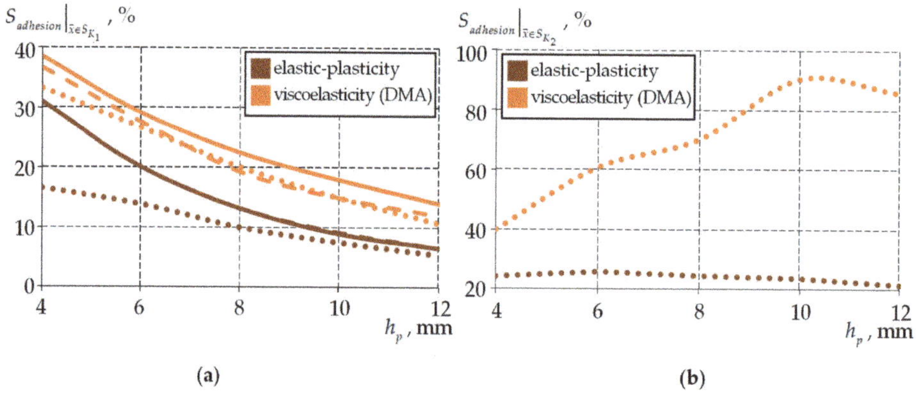

Figure 18. The percentage of the contact surface area in the state of full adhesion depends on the thickness of the antifriction interlayer: (**a**) S_{K_1}; (**b**) S_{K_2}; the solid line is ideal contact; the dashed line is full adhesion; the dots are frictional contact.

The viscoelastic material behavior and its deformation affect the distribution of contact statuses on the sliding surface of the spherical balancer. The percentage of the surface area S_{K_1} in the state of full adhesion is greater with the viscoelastic behavior of the antifriction interlayer material. As the thickness increases, the percentage of the surface area S_{K_1} in the state of full adhesion decreases with all the considered models of the material behavior.

It can be noted that with an increase in the antifriction interlayer thickness with viscoelastic material behavior on the interface surface S_{K_2} in the case of frictional contact, an increase in the percentage of the surface area in the state of full adhesion is observed. This means that with sufficiently large antifriction interlayers, additional surface treatment of the interface of the lower steel plate with a "torn thread" interlayer is not required, thus reducing the time and economic costs of manufacturing the structure. The damping properties will also be improved. This study will be useful in analyzing the contact interaction of the bridge bearing elements under the influence of more complex loads, taking into account the time factor, as well as temperature.

4. Discussion

Spherical bearings are a type of high-load multirotational bridge bearings that compensate for various types of loads from the bridge span and the environment. The effective operation of bearings is related to the gaps between the steel structural elements and the antifriction interlayer thickness [58]. Increasing the antifriction interlayer thickness can reduce the resistance to rotation, internal wear, and damage, and it can also increase the bearing capacity of the structure. However, the restrictions on the antifriction interlayer thickness have not been determined so far. In the current study, the dependencies of the contact parameters and the stress–strain state of the spherical bearings on the antifriction interlayer thickness were determined. These studies were transferred to the manufacturers of bridge bearings and taken into account in engineering methods for calculating the geometric configuration of the structure.

The results of the bearing deformation correlate with data from other studies and the results of field experiments [20,59]. The displacement level of the bearing at 1000 kN is in good agreement with the data of field experiments presented in [20]. The maximum stress level of a standard spherical bearing is also commensurate with the data obtained in our study. The pattern of the stress distribution in the antifriction interlayer corresponds to the results obtained by Deng et al. [59]. The maximum stresses of the antifriction interlayer are also observed near the protrusion. This confirms the possibility of using models in the framework of simulation modeling for a multifactorial analysis of the behavior of structures in statics and dynamics.

Mechanical treatment of the sliding surface can ensure efficient and long-term operation of the bearing [58]. In this case, the influence of the pattern of the surface treatment of the lower steel plate in the area of contact with the antifriction interlayer can also affect the operation of the spherical bearing. As part of this study, it was found that by ensuring full adhesion at the interface, the interface of the antifriction interlayer with the lower steel plate reduced the deformation of the protrusion of the polymer interlayer and the settlement of the structure. At the same time, studies of the behavior of bearings most often consider only one type of coupling of elements [22,59]. Studies often do not take into account the friction on the interface surfaces [59]. Modeling the behavior of the antifriction interlayer material does not take into account the viscoelastic, creep, and plasticity of materials. However, in the framework of field tests, viscous deformation of the antifriction interlayer and plastic flow of materials were observed [58,60]. Nonlinear effects in spherical bearings related to the behavior of materials and the contact state of the elements are complex phenomena and an interesting area of research for understanding the work of the structure [61]. The influence of the coupling pattern of antifriction material and steel structural elements, including taking into account frictional properties, is necessary to evaluate the operation and rationalize the bearings [26]. The current work analyzed, as a first approximation, the influence of the behavioral model of the antifriction interlayer material and the pattern of its interface with steel structural elements. A model of a spherical bearing with viscoelastic behavior of an antifriction material with full adhesion on the interface surfaces of the antifriction interlayer with the lower steel plate made it possible to obtain a discrepancy with experimental data of less than 15%. A comparison with data from experimental studies by other authors shows an error of less than 10% in the model [20]. A large percentage of the difference between the settlement of the model and the experiment is due to the multi-stage experimental program, as well as the limitations of the geometric model of the bearing in the numerical implementation (grooves with lubricant and a flat antifriction interlayer are not taken into account). The model error can be justified by the accepted assumptions on the set of physical, mechanical, and thermal properties of the polymer material: the thermal expansion coefficient and the friction coefficient are constants and do not depend on other characteristics (time, temperature, load value, etc.). This introduces limitations in the application of the constructed models, from the point of view of the quantitative assessment of the parameters of the stress–strain state of the structure.

Let us consider the multi-stage loading of a model with exposure sections, according to a full-scale experiment (Figure 19).

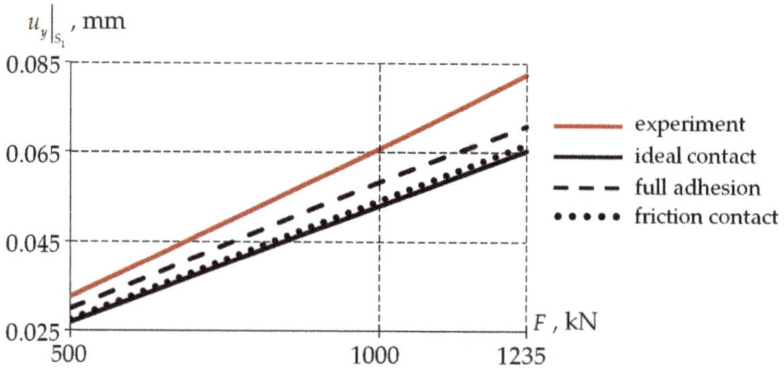

Figure 19. Settlement of the supporting part during multi-stage loading.

The discrepancy between the results of the numerical simulation and the experiment does not exceed 20% for all variants of the interface of the antifriction interlayer with the lower steel plate of the bearing. The maximum error is observed when modeling an ideal contact S_{K_2}–S_{K_3}. The minimum error of numerical simulation at a nominal load of 1000 kN is observed at full adhesion S_{K_2}–S_{K_3} (~11.37%). This is due to the implementation of the specified type of coupling, due to the equality of the fields of the displacement vector, and due to the components of the stress and strain tensor on the elements' coupling surfaces. This is not quite "physical" for the considered "steel–polymer" contact. The experiment involved bearings with the character of processing the metal surfaces of the interface of the lower steel plate according to the "torn thread" type, which allows for maximum adhesion to the antifriction interlayer of the material.

5. Conclusions

As part of the first stage of this study, the influence of the coupling pattern on the deformation behavior of spherical bearings of a bridge structure at a standard thickness of the antifriction interlayer was analyzed. Three coupling options were considered: ideal contact, full adhesion, and frictional contact. The behavior of the antifriction interlayer material was modeled within the framework of an elastic–plastic formulation. A number of patterns of the deformation behavior of bridge bearings were established:

(1) The change in the coupling conditions S_{K_2}–S_{K_3} has the greatest effect on the area of full adhesion S_{K_1}. In the case of ideal contact and full adhesion S_{K_2}–S_{K_3}, the total adhesion area is 30.97% of the initial contact area, and in the case of frictional contact, it is 16.48% of the initial contact area.

(2) The maximum differences between solutions with the modeling of a contact pair of elements for different types of contact are observed in the normal movement of the end face. The difference between frictional contact and ideal contact is about 50%; full adhesion is no more than 7.5%. Contact pressure: the difference between full adhesion and ideal contact is 5.64%, and for frictional contact is 23.47%.

(3) The effects of plastic flow of the material in the vicinity of the protrusion zone of the antifriction interlayer were observed, which needs to be clarified within the framework of applying the viscoelastic behavioral model in numerical implementation.

A quick qualitative assessment of the characteristics of the stress–strain state of the structure and the parameters of contact interaction in a static formulation of the problem is the main important point of using the elastic–plastic model of material behavior. The "torn thread" conjugation can be modeled as an ideal contact to save computing resources according to the results obtained.

In the second stage of this study, an assessment was made of the effect of the antifriction interlayer thickness at a constant protrusion and the different patterns of the coupling of the elements. The pattern of the distribution of contact pressure and tangential contact stress has small differences when the antifriction interlayer thickness changes. However, an increase in the antifriction interlayer thickness leads to an increase in the maximum level of normal displacement on the surface of the protrusion, which is also confirmed by the results obtained when changing the pattern of the interface. This effect is associated with a decrease in the surface area S_{K_1} in the state of full adhesion.

An assessment of the influence of the behavioral patterns of the antifriction interlayer material on the results of structural deformation was performed to form a model of the spherical bearing that is closest to reality. The MISO model was used to describe the elastic–plastic behavior of the antifriction interlayer material, and the Prony series was used for viscoelastic behavior. The viscoelastic material behavior and its deformation affect the distribution of contact statuses on the sliding surface of the spherical balancer. The percentage of the surface area S_{K_1} in the state of full adhesion is greater with the viscoelastic behavior of the antifriction interlayer material. The percentage of the surface area S_{K_1} in the state of full adhesion decreases with all the considered models of the behavior of the antifriction interlayer material as the thickness increases. The viscoelastic behavioral model of the antifriction material makes it possible to minimize the deviation of the numerical model of the bearing from the actual structure. This proves the necessity of using modern numerical models to describe the behavior of materials in engineering practice.

Several directions are being considered for further development of the work:

- The consideration of the experimentally obtained friction properties of the steel–polymer pair in the analysis of the deformation behavior of the structure;
- The analysis of the effects of cyclic and seismic loads, including temperature and time considerations;
- The transition to a viscoelastic–plastic model of antifriction material behavior.

Author Contributions: Conceptualization, A.A.K., Y.O.N., and A.P.B.; methodology, A.A.K.; software, A.A.K., Y.O.N., Y.S.K., and A.P.B.; validation, A.A.K.; writing—original draft preparation, A.A.K., Y.S.K., and Y.O.N.; writing—review and editing, Y.S.K. and A.P.B.; visualization, A.A.K., Y.S.K., Y.O.N., and A.P.B.; funding acquisition, Y.O.N. All authors have read and agreed to the published version of the manuscript.

Funding: This study was supported by the Russian Science Foundation grant no. 25-29-00470, https://rscf.ru/project/25-29-00470/ (accessed on 30 August 2025).

Institutional Review Board Statement: Not applicable.

Informed Consent Statement: Not applicable.

Data Availability Statement: The original contributions presented in this study are included in the article. Further inquiries can be directed to the corresponding author.

Conflicts of Interest: The authors declare no conflicts of interest.

References

1. Gonzalez, A.; Wiener, M.; Valdez salas, B.; Mungaray, A. Bridges: Structures and Materials, Ancient and Modern. In *Infrastructure Management and Construction*; IntechOpen: London, UK, 2020. [CrossRef]
2. Presta, F.; Gibbens, B.; Turner, J. Sydney Gateway's Twin Network Arches: A Case Study in Complex Bridge Design and Construction. In Proceedings of the Environmentally Friendly Technologies and Structures: Focusing on Sustainable Approaches, Tokyo, Japan, 18–21 May 2025; pp. 1630–1638. [CrossRef]
3. Zhao, W.; Yang, B.; Nian, Y. Study on economic design and construction program of steel-hybrid combined girder bridge. *Adv. Comput. Eng. Technol. Res.* **2024**, *1*, 52. [CrossRef]
4. Jalaei, F.; Zhang, J.; Mcneil-Ayuk, N.; McLeod, C. Environmental life cycle assessment (LCA) for design of climate-resilient bridges—A comprehensive review and a case study. *Int. J. Constr. Manag.* **2024**, *25*, 99–114. [CrossRef]
5. D'Amato, M.; Ranaldo, A.; Rosciano, M.; Zona, A.; Morici, M.; Gioiella, L.; Micozzi, F.; Poeta, A.; Quaglini, V.; Cattaneo, S.; et al. The Development and Statistical Analysis of a Material Strength Database of Existing Italian Prestressed Concrete Bridges. *Infrastructures* **2025**, *10*, 203. [CrossRef]
6. Ranaldo, A.; Lo Monaco, A.; Palmiotta, A.; D'Amato, M.; Lippolis, A.; Vacca, V.; Sarno, R. A preliminary investigation on material properties of existing prestressed concrete beams. *Procedia Struct. Integr.* **2024**, *62*, 145–152. [CrossRef]
7. Immanuel, Y.; Rifai, A.I.; Saputra, A.J. Bridge Structural Design Simulation: Case Study of Nongsa Pura Bridge. *OPSearch Am. J. Open Res.* **2024**, *3*, 268–276. [CrossRef]
8. Xiong, C.; Shang, Z.; Wang, M.; Lian, S. Dynamic Monitoring of a Bridge from GNSS-RTK Sensor Using an Improved Hybrid Denoising Method. *Sensors* **2025**, *25*, 3723. [CrossRef] [PubMed]
9. Xi, R.J.; He, Q.Y.; Meng, X.L. Bridge monitoring using multi-GNSS observations with high cutoff elevations: A case study. *Measurement* **2021**, *168*, 108303. [CrossRef]
10. Eggert, H.; Kauschke, W. *Structural Bearings*; Ernst & Sohn: Berlin, Germany, 2002.
11. Taghipour, A.; Zakeri, J.; Ghozat, A.; Mosayebi, S. Dynamic Behavior Assessment of a Railway Bridge in Isfahan under Over-Height Vehicle Collision Loads and Proposing Maintenance Strategies to Enhance Its Performance. *Int. J. Railw. Res.* **2025**, *12*, 34–39. [CrossRef]
12. Lin, Z.; Xia, D.; Jiang, Y.; Yuan, Z.; Wang, H.; Lin, L. Experimental and Numerical Investigation of Localized Wind Effects from Terrain Variations at a Coastal Bridge Site. *J. Mar. Sci. Eng.* **2025**, *13*, 1223. [CrossRef]
13. Kang, D.H.; Hyun, J.H. Evaluation of the durability of spherical bridge bearing using polyamide engineering plastic middle plate. *J. Korean Soc. Urban Railw.* **2021**, *9*, 1021–1031. [CrossRef]
14. Yang, Y.; Zhang, Y.; Ju, J. Study on the mechanical properties of a type of spherical bearing. *J. Theor. Appl. Mech.* **2021**, *59*, 539–550. [CrossRef]
15. Borisov, A.I.; Gnatyuk, G.A. Assessment of transport accessibility of the Arctic regions of the Republic of Sakha (Yakutia). *Transp. Res. Procedia* **2022**, *61*, 289–293. [CrossRef]
16. Abarca, A.; Monteiro, R.; O'Reilly, G.J. Seismic risk prioritisation schemes for reinforced concrete bridge portfolios. *Struct. Infrastruct. Eng.* **2023**, *21*, 49–69. [CrossRef]
17. Zhang, P.; Zhao, H.; Shao, Z.; Jiang, L.; Hu, H.; Zeng, Y.; Xiang, P. A rapid analysis framework for seismic response prediction and running safety assessment of train-bridge coupled systems. *Soil Dyn. Earthq. Eng.* **2024**, *177*, 108386. [CrossRef]
18. Wei, B.; Chen, M.; Jiang, L.; Li, C.; Chen, Y.; Lu, A.; Yang, Z.; Li, S. The influence of spherical bridge bearings on train running safety during earthquakes considering train-track–bridge interaction and soil specification. *Arch. Civ. Mech. Eng.* **2025**, *25*, 143. [CrossRef]
19. Ye, X.W.; Xia, P.S.; Su, Y.H.; Chen, B. Analysis and probabilistic modeling of wind characteristics of an arch bridge using structural health monitoring data during typhoons. *Struct. Eng. Mech. Int. J.* **2017**, *63*, 809–824. [CrossRef]
20. Yan, L.; Gou, X.Y.; Zhang, X.; Jiang, Y.; Ran, X.W.; Zhang, P. Experimental and Numerical Investigations on the Spherical Steel Bearing Capacity of New Anti-Separation Design. *KSCE J. Civ. Eng.* **2024**, *28*, 889–903. [CrossRef]
21. Ding, M.; Li, C.X.; Liang, Z.M.; Ju, J.S. Research on the Failure Characteristics of a Spherical Bearing. *Strength Mater.* **2022**, *54*, 536–544. [CrossRef]
22. Shi, X.; Liu, Z.; Guo, T.; Cai, C.S.; Jiang, C. Investigation on contact stress calculation method of spherical hinge structures for swivel construction. *Structures* **2024**, *69*, 107290. [CrossRef]
23. Lin, S.-C. Friction and Lubrication of Sliding Bearings. *Lubricants* **2023**, *11*, 226. [CrossRef]
24. Gajewski, M.D.; Miecznikowski, M. Assessment of the Suitability of Elastomeric Bearings Modeling Using the Hyperelasticity and the Finite Element Method. *Materials* **2021**, *14*, 7665. [CrossRef] [PubMed]
25. Adamov, A.A.; Kamenskikh, A.A.; Pankova, A.P.; Strukova, V.I. Comparative Analysis of the Work of Bridge Spherical Bearing at Different Antifriction Layer Locations. *Lubricants* **2022**, *10*, 207. [CrossRef]
26. Han, O.; Kwark, J.-W.; Lee, J.-W.; Han, W.-J. Analytical Study on the Frictional Behavior of Sliding Surfaces Depending on Ceramic Friction Materials. *Appl. Sci.* **2023**, *13*, 234. [CrossRef]

27. Hu, Q.; Pei, Q.; Li, P. Reducing the Friction Coefficient of Heavy-Load Spherical Bearings in Bridges Using Surface Texturing—A Numerical Study. *Lubricants* **2025**, *13*, 180. [CrossRef]
28. Zhang, H.; Liu, Y.; Wang, W.; Qin, L.; Dong, G. Surface texture design and its tribological application. *J. Mech. Eng.* **2019**, *55*, 85–93. [CrossRef]
29. Lu, P.; Wood, R.J. Tribological performance of surface texturing in mechanical applications—A review. *Surf. Topogr. Metrol. Prop.* **2020**, *8*, 043001. [CrossRef]
30. Wang, H.; Sun, A.; Qi, X.; Dong, Y.; Fan, B. Experimental and analytical investigations on tribological properties of PTFE/AP composites. *Polymers* **2021**, *13*, 4295. [CrossRef]
31. Deshwal, D.; Belgamwar, S.U.; Bekinal, S.I.; Doddamani, M. Role of reinforcement on the tribological properties of polytetrafluoroethylene composites: A comprehensive review. *Polym. Compos.* **2024**, *45*, 14475–14497. [CrossRef]
32. Berladir, K.; Antosz, K.; Ivanov, V.; Mitaľová, Z. Machine Learning-Driven Prediction of Composite Materials Properties Based on Experimental Testing Data. *Polymers* **2025**, *17*, 694. [CrossRef]
33. Yi, X.; Du, S.; Zhang, L. *Composite Materials Engineering, Volume 1: Fundamentals of Composite Materials*; Springer: Singapore, 2018. [CrossRef]
34. Wang, Q.J.; Chung, Y.W. *Antifriction Materials and Composites*; Springer: Boston, MA, USA, 2013. [CrossRef]
35. Lin, Z.; Zhang, K.; Ye, J.; Li, X.; Zhao, X.; Qu, T.; Liu, Q.; Gao, B. The effects of filler type on the friction and wear performance of PEEK and PTFE composites under hybrid wear conditions. *Wear* **2022**, *490–491*, 204178. [CrossRef]
36. Abdul, S.M. Recent Advances in UHMWPE/UHMWPE Nanocomposite/UHMWPE Hybrid Nanocomposite Polymer Coatings for Tribological Applications: A Comprehensive Review. *Polymers* **2021**, *13*, 608. [CrossRef]
37. Park, J.-H.; Lee, J.-W. Friction Behavior of Ceramic Materials for the Development of Bridge-Bearing Friction Materials. *Appl. Sci.* **2025**, *15*, 152. [CrossRef]
38. Park, J.H.; Lee, J.W.; Kwark, J.W.; Han, W.J.; Han, O. Characteristics of Friction Behavior of Ceramic Friction Materials according to Surface Materials. *J. Korean Recycl. Constr. Resour. Inst.* **2023**, *11*, 535–541.
39. Bokhoeva, L.A.; Rogov, V.E.; Chermoshentseva, A.S. Antifriction fluoroplastic materials for sliding layers in bridge supports. *Mag. Civ. Eng.* **2024**, *17*, 13104. [CrossRef]
40. Lenk, R.S. *Polymer Rheology*; Springer Science & Business Media: Dordrecht, The Netherlands, 2012. [CrossRef]
41. Kraus, M.; Niederwald, M.; Siebert, G.; Keuser, M. Rheological modelling of linear viscoelastic materials for strengthening in bridge engineering. In Proceedings of the 11th German Japanese Bridge Symposium, Osaka, Japan, 30–31 August 2016; pp. 30–31.
42. Nguyen, T.Q.; Nguyen, T.T.; Nguyen, P.T. Nguyen Analysis of vibration characteristics of bridge spans based on the viscoelastic material model: Investigating the relationship between material properties and dynamic parameters. *Structures* **2025**, *7*, 108788. [CrossRef]
43. Wang, D. Modelling the Contact Behaviour in the Presence of Viscoelasticity. Ph.D. Thesis, University of Leeds, Leeds, UK, 2025.
44. Doh, J.; Hur, S.H.; Lee, J. Viscoplastic parameter identification of temperature-dependent mechanical behavior of modified polyphenylene oxide polymers. *Polym. Eng. Sci.* **2019**, *59*, E200–E211. [CrossRef]
45. Liu, E.; Wu, J.; Li, H.; Liu, H.; Xiao, C.; Shen, Q.; Kong, L.; Lin, J. Research on viscoelastic behavior of semi-crystalline polymers using instrumented indentation. *Polym. Sci.* **2021**, *59*, 1795. [CrossRef]
46. Alam, M.I.; Khan, D.; Mittal, Y.; Kumar, S. Effect of crack tip shape on near-tip deformation and fields in plastically compressible solids. *J. Braz. Soc. Mech. Sci. Eng.* **2019**, *41*, 441. [CrossRef]
47. Persson, B.N.J. *Sliding Friction: Physical Principles and Applications*; Springer Science & Business Media: Berlin/Heidelberg, Germany, 2013. [CrossRef]
48. Li, Y.; Guo, W.; Huang, X.; Chen, Z.; Gao, Y. Friction Characteristics and Lubrication Properties of Spherical Hinge Structure of Swivel Bridge. *Lubricants* **2024**, *12*, 130. [CrossRef]
49. Adamov, A.A.; Keller, I.E.; Ivanov, Y.N.; Utev, N.V. Basic Tests and Identification of a Model of Viscoelastic Behavior of Elastomers under Finite Deformations. *Mech. Solids* **2024**, *59*, 3831–3843. [CrossRef]
50. Ilyin, S.O. Structural Rheology in the Development and Study of Complex Polymer Materials. *Polymers* **2024**, *16*, 2458. [CrossRef]
51. Adamov, A.A.; Kamenskikh, A.A. The Deformation Behavior of Modern Antifriction Polymer Materials in the Elements of Transport and Logistics Systems with Frictional Contact. In *Digital Science 2019. DSIC 2019. Advances in Intelligent Systems and Computing*; Antipova, T., Rocha, Á., Eds.; Springer: Cham, Switzerland, 2020; Volume 1114. [CrossRef]
52. Chen, W.W.; Wang, Q.J. Thermomechanical analysis of elastoplastic bodies in a sliding spherical contact and the effects of sliding speed, heat partition, and thermal softening. *J. Tribol.* **2008**, *130*, 041402. [CrossRef]
53. Fang, X.; Zhang, C.; Chen, X.; Wang, Y.; Tan, Y. A new universal approximate model for conformal contact and non-conformal contact of spherical surfaces. *Acta Mech.* **2015**, *226*, 1657–1672. [CrossRef]
54. Tao, L.; Wang, Q.; Qi, Z.; Wu, H.; Zhu, H.; Huang, J. Analysis of the Rolling Interface Contact Characteristics in Mixed Lubrication Based on Gaussian Distribution Theory. *Materials* **2023**, *16*, 5220. [CrossRef]

55. Nosov, Y.O.; Kamenskikh, A.A. Experimental Study of the Rheology of Grease by the Example of CIATIM-221 and Identification of Its Behavior Model. *Lubricants* **2023**, *11*, 295. [CrossRef]
56. Kamenskikh, A.A.; Nosov, Y.O.; Bogdanova, A.P. The Study Influence Analysis of the Mathematical Model Choice for Describing Polymer Behavior. *Polymers* **2023**, *15*, 3630. [CrossRef]
57. Kamenskih, A.A.; Trufanov, N.A. Regularities interaction of elements contact spherical unit with the antifrictional polymeric interlayer. *J. Frict. Wear* **2015**, *36*, 170–176. [CrossRef]
58. Roeder, C.W.; Stanton, J.F.; Campbell, T.I. Rotation of High Load Multirotational Bridge Bearings. *J. Struct. Eng.* **1995**, *121*, 747–756. [CrossRef]
59. Deng, N.; He, M.; Gu, N.; Liang, H. Design and Performance Research of a New Type of Spherical Force-Measuring Bearing of Bridges Based on Button Type Microsensor. *KSCE J. Civ. Eng.* **2024**, *28*, 5066–5076. [CrossRef]
60. Adamov, A.A.; Keller, I.E.; Petukhov, D.S.; Kuzminykh, V.S.; Patrakov, I.M.; Grakovich, P.N.; Shilko, I.S. Evaluation of the Performance of PTFE-Composites as Antifriction Layers in Supporting Parts with a Spherical Segment. *J. Frict. Wear* **2023**, *44*, 127–134. [CrossRef]
61. Zhao, L.; Sun, X.; Wu, Z.; Chen, Y.; Liu, J.; Wang, Y. Nonlinear Static Analysis of Spherical Hinges in Horizontal Construction of Bridges. *Buildings* **2024**, *14*, 3726. [CrossRef]

designs

MDPI

Article

An Innovative Design of Railroad Crossties

Moses Karakouzian *, Maple Crow, William Van Vlerin, Patrick Whitton and Mehrdad Karami *

Department of Civil and Environmental Engineering, University of Nevada, Las Vegas, NV 89154-4015, USA; crowm1@unlv.nevada.edu (M.C.); willvanvlerin@gmail.com (W.V.V.); whittp1@unlv.nevada.edu (P.W.)
* Correspondence: mkar@unlv.nevada.edu (M.K.); mkarami@unlv.nevada.edu (M.K.)

Abstract: This study presents an initial feasibility concept paper for a proposed crosstie system, an innovative railroad crosstie reinforcement system designed to reduce the stresses transmitted to the underlying ballast. While not developed for a specific industry client, the proposed crosstie system lays the groundwork for patent application and potential commercialization, offering a novel alternative to conventional railroad construction. Finite Element Analysis demonstrated that this system can reduce effective stress on the ballast by up to 24%, effectively making train loads appear lighter to the substructure. The design of the proposed system focuses on mitigating the excessive stresses transmitted from crossties to the ballast layer in heavy axle load (HAL) freight rail operations. The goal was to create a reinforcement mechanism that is modular, compatible with existing track infrastructure, and capable of reducing maintenance costs by distributing loads more effectively across the ballast and subgrade. The findings indicate that this system is not only the most cost-effective and sustainable solution but also holds promise for reducing fixed stock investment, minimizing downtime for track maintenance, and enabling expanded rail network connectivity. These results support continued research and investment in the system's development and deployment.

Keywords: railroad crosstie reinforcement; crosstie design; Finite Element Analysis (FEA); ballast stress; sustainability

1. Introduction

Railroads efficiently transport goods and passengers along established routes but have significant initial and maintenance costs associated with track and railbed infrastructure (also called fixed stock). Given the importance of environmental and economic efficiency, engineers are obligated to reduce costs in any way they can to reduce the impact of heavy industry on the natural world while also saving their clients' money. A significant contributor to the costs of railroads is the enormous dynamic and cyclic load that must be diffused from the wheel, through the track, across the subgrade, and finally into the surrounding native soil [1–3].

The proposed crosstie system (Figure 1) reduces the cost and environmental impact by making better use of existing track structure and subgrade. This product does so by spreading the applied load of a typical track-wheel loading scenario across many ties, spreading the peak wheel loads over a longer length of track, and reducing the magnitude of stress conveyed to the ballast under the crossties.

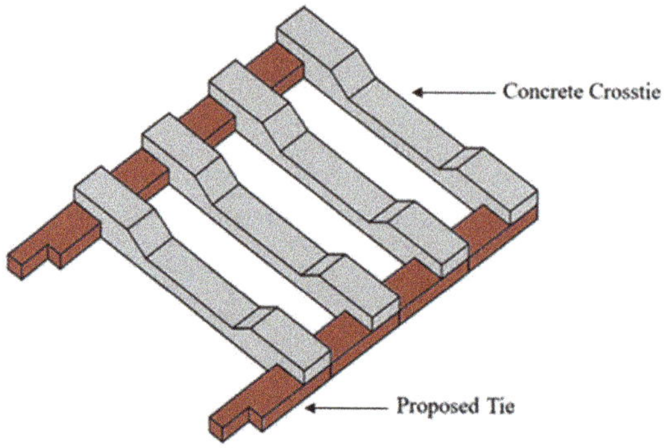

Figure 1. Proposed Tie system configuration. The tessellating tiles distribute loads across larger ballast areas.

This technique allows railroads to save money in reduced excavation for track structure and subgrade improvement, easier remediation of heaving, and more route options by enabling freight routes to be constructed in what are considered "less ideal" soils. Under high-stress cyclic loading, cohesive soils will tend to flow away from the centerline, as shown in Figure 2. At the onset of instability (Figure 2a), a localized zone of remolded clay forms within the clay subgrade, indicating the initiation of shear failure. With continued deformation, this instability propagates upward, resulting in surface manifestation in the form of heave (Figure 2b). At this stage, the displaced remolded clay pushes the track substructure upward, causing significant distortion and potential serviceability issues [4]. This sequence highlights how subgrade failure in cohesive soils transitions from a hidden internal mechanism to a visible surface deformation that compromises track stability. The void left behind by the heave weakens the track, requiring increased maintenance costs or a "go-slow" order on the track, reducing commercial throughput. The objective of this research is to provide an initial feasibility assessment for the New-Tie system that will reduce the maximum stress transmitted from crosstie to ballast, saving railroad companies money and reducing the impact on the surrounding environment.

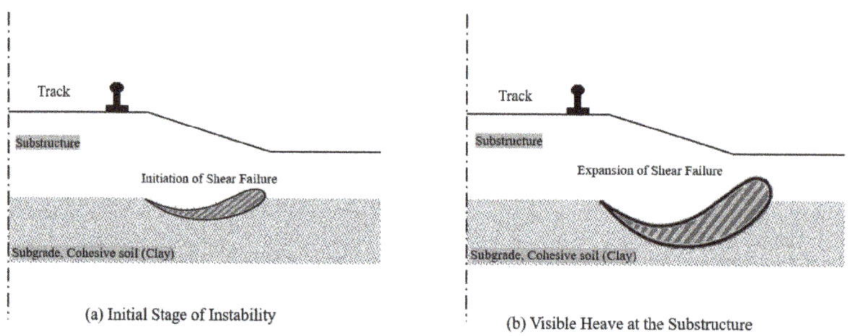

Figure 2. Heaving, visualized.

New-Tie can be used in place of traditional crossties to reduce the stress applied to the existing ballast, thereby bringing a high-risk system into acceptable tolerances. By reducing the maintenance impact of heaving remediation from total replacement of subgrade to a simple crosstie replacement, New-Tie will save railroad operators money by eliminating the need for costly changes in operations, allowing necessary maintenance to be performed faster and with less heavy equipment than currently required. Regular, non-heaving maintenance cycles are driven by the rate of ballast fouling. No matter how tough ballast rock is, eventually enough will be down to fill up the spaces between rocks (fouling). Losing the air voids between rocks reduces track drainage and decreases the layer's elasticity, causing more force to be transmitted into the subgrade beneath. By reducing the peak loads transmitted into and through the ballast layer, New-Tie installation will reduce the need for the frequency of maintenance from regular rail operations.

2. Methodology

The present research presents an initial feasibility concept analysis for a novel railroad crosstie system (New-Tie) that reduces the environmental and economic costs of railroads. A numerical model simulation using Finite Element Analysis (FEA) is conducted to compare the difference in the stress and displacement in both the ballast layer and subgrade. The results of this analysis determine the magnitude of benefit for construction with New-Tie. The remainder of the research is a case study of track remediation in the Black Rock Desert of Northern Nevada. This case study indicates the relative economic viability of the New-Tie system compared to other typical alternatives. All design parameters were cross-referenced with the AREMA Manual for Railway Engineering to ensure code compliance [5]. Subgrade strength parameters and train-track load assumptions were validated using published soil data.

2.1. Materials and Geometries

A comprehensive list of terms to use as a glossary for the analysis is presented in Table 1. Typical rail, tie, and fastener, granular and subgrade layer, and wheel load parameters were utilized to determine a baseline for deviator stress analyses used to compare a system with and without New-Tie. The input parameters for the track design include the properties of the rails, ties, fasteners, ballast, and subgrade, along with the applied wheel load. The rail is defined by its elastic modulus (E = 207 GPa), moment of inertia (I = 3.95 × 10^{-6} m^4), cross-sectional area (8.61 × 10^{-3} m^2), gauge (1.5 m, distance between two rails), and mass per unit length (68 kg/m). The tie and fastener system is characterized by a base width of 0.273 m, base length of 2.6 m, cross-sectional area of 5.6 × 10^{-2} m^2, elastic modulus of 31 GPa, moment of inertia of 2.42 × 10^{-4} m^4, tie mass of 363 kg, and spacing of 0.61 m, with the fastening system providing a stiffness of 175 kN/mm. The granular ballast layer is represented by a density of 1.76 Mg/m^3, Poisson's ratio of 0.3, and modulus of 280 MPa, while the subgrade layer has a density of 1.92 Mg/m^3, Poisson's ratio of 0.35, modulus of 55 MPa, and an at-rest earth pressure coefficient (K_0) of 1.0. Finally, the wheel load is applied vertically with a magnitude of 230 kN, representing the axle load of a heavy freight train.

Table 1. Parameters used for Comparative Roadbed Analysis [1,5,6].

Track Variable	Value
Rail Properties	
E (GPa)	207
I (m^4)	3.95 × 10^{-6}
Cross area (m^2)	8.61 × 10^{-3}
Gauge (m)	1.5
Mass (kg/m)	68
Tie and Fastener	
Base width (m)	0.273
Base length (m)	2.6
Cross area (m^2)	5.6 × 10^{-2}
E (GPa)	31

Table 1. *Cont.*

Track Variable	Value
I (m^4)	2.42×10^{-4}
Mass (kg)	363
Spacing (m)	0.61
Fastener stiffness (kN/mm)	175
Granular Layer	
Density (Mg/m^3)	1.76
Poisson's ratio	0.3
Modulus (MPa)	280
Subgrade Layer	
Density (Mg/m^3)	1.92
Poisson's ratio	0.35
Modulus (MPa)	55
K_0	1.0
Wheel load	
Magnitude vertically (kN)	230

Figure 3, "Railroad Tie", defines the railroad crosstie to be used in the presented research. The material properties for the crosstie are listed in Table 1.

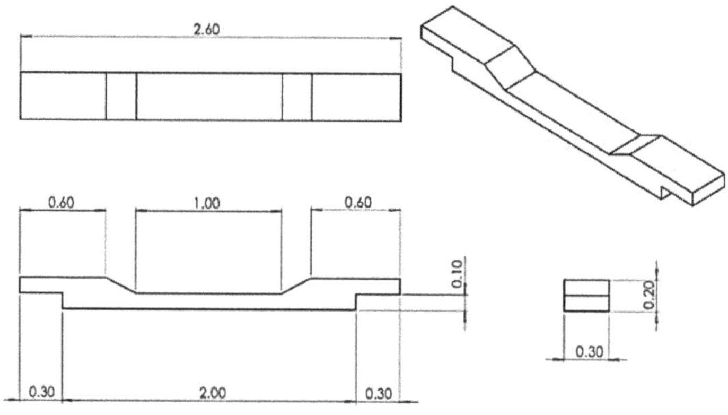

Figure 3. Original concrete crosstie (units are in meters).

Figure 4, "New-Tie," defines the proposed cross-tie modification. The dimensions of the New-Tie were selected to have the same width and material properties as the original crosstie. Consistent width of New-Tie and crossties improves constructability by enabling installation and maintenance with existing equipment. The elongated 'Z' shape of the New-Tie allows for even overlap with the crosstie that rests on top, as shown in the "Assembly" drawing.

Figure 5, "Assembly," shows the arrangement of New-Tie and crosstie. This drawing shows four New-Ties in contact with each crosstie, allowing for maximum stress dissipation from each crosstie through the New-Tie into the ballast. This configuration was intended not only to provide additional rigidity between adjacent crossties but also to allow for greater surface area for the axle load to be transferred into the ballast.

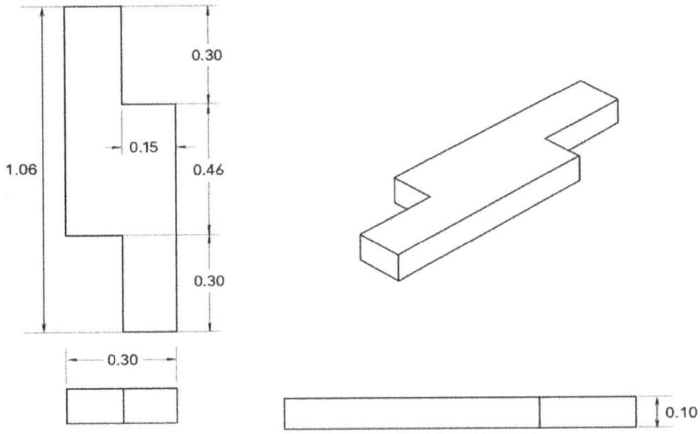

Figure 4. Proposed "New-Tie" geometry (units are in meters).

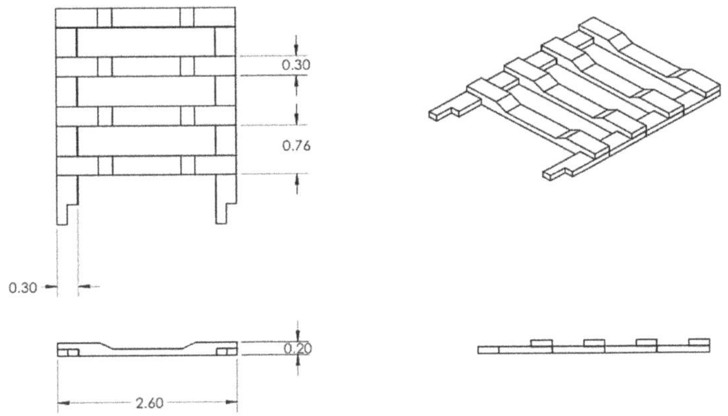

Figure 5. Proposed "New-Tie" system assembly (units are in meters).

2.2. Numerical Modeling

Comparative analysis of stress and displacement in ballast and subgrade soil lays the groundwork for this research. The FEA will use a selected set of soil, ballast, rail, and crosstie parameters to determine the general difference between standard crossties and the New-Tie system when loaded with a single axle load. It is noted that the assumptions and model setup are not a perfect representation of reality; however, this analysis is meant to indicate whether further analysis and investment are warranted for this technology, not to give functional design parameters. The general design of the New-Tie system was developed through several iterations based on preliminary model results. The location and geometry of the new crosstie system are adjusted to improve results, but material properties are kept constant, and the fine design of the crosstie-New-Tie interface is considered beyond gravity. This analysis will also only be valid for vertical loading on a straight, level track to simplify the inputs to the model and the generated results. The FEA model did not include dynamic loading, lateral forces, or cyclic fatigue effects and assumed idealized material behavior.

Figure 6 demonstrates the existing condition of the conventional ballast and crosstie system used in freight railroad operations and the proposed New-Tie system, showing the configuration of rails, crossties, ballast, and subgrade layer used in FEA modeling. The model consists of rails, ties, ballast, and subgrade, discretized using three-dimensional, eight-node reduced integration solid elements (C3D8R). The boundary conditions were applied such that the bottom of the subgrade was fully constrained in all three directions (X, Y, and Z), representing a fixed support condition. The vertical sides of the model were restrained in the X and Y to prevent lateral movement, while remaining free in the vertical direction (Z) to allow settlement. The ties were connected to the underlying soil through a hard contact interaction to ensure realistic load transfer, while the rail was rigidly fixed to the ties to simulate the fastening system. This formulation prevents any penetration between the contacting surfaces and allows the transfer of compressive stresses only, ensuring a realistic representation of the rail–soil interface. The penalty enforcement method was adopted as the default approach, allowing minimal numerical penetration while maintaining solution stability. The rail was assigned as the major surface due to its higher stiffness, and the soil as the minor surface. This contact definition effectively simulates the support and load transfer mechanism between the rail and the underlying soil.

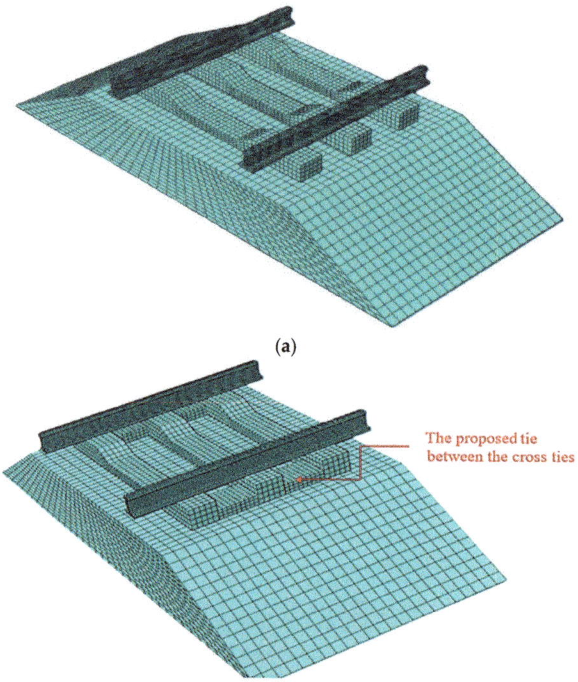

Figure 6. FEA model of (**a**) standard freight railbed and (**b**) proposed New-Tie system.

3. Numerical Modeling Results

The two models, including the original crosstie and the proposed New-Tie, were run in a stress analysis finite element software. For the original crosstie, Figure 7a shows the stress at the crosstie–ballast interface with a maximum compression value in dark blue of 1.094 MPa and a general ballast stress in orange of 45.1 KPa. Figure 8a shows the displacement at the crosstie–ballast interface with a maximum compression of 1.233 mm, represented by the dark blue color. Figure 7b shows the stress at the crosstie–ballast interface with a maximum compression value in dark blue of 1.018 MPa. Figure 8b shows the displacement at the crosstie–ballast interface with a maximum compression of 1.090 mm, represented by the dark blue color.

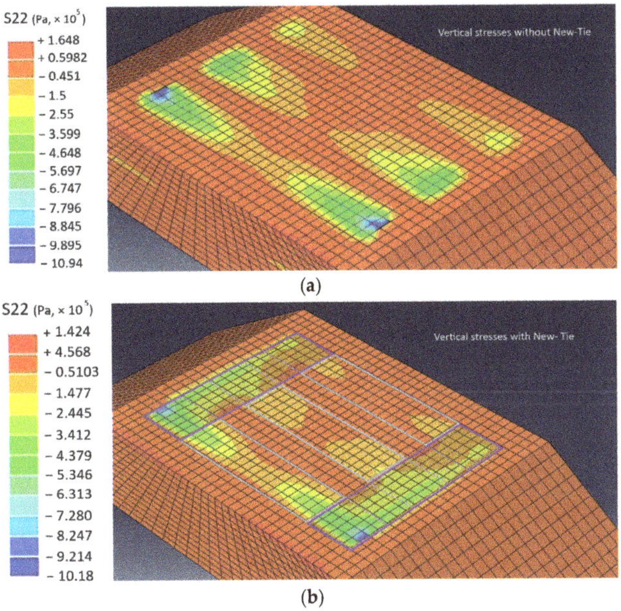

Figure 7. Model of vertical stress at ballast–crosstie interface (**a**) without New-Tie and (**b**) with New-Tie.

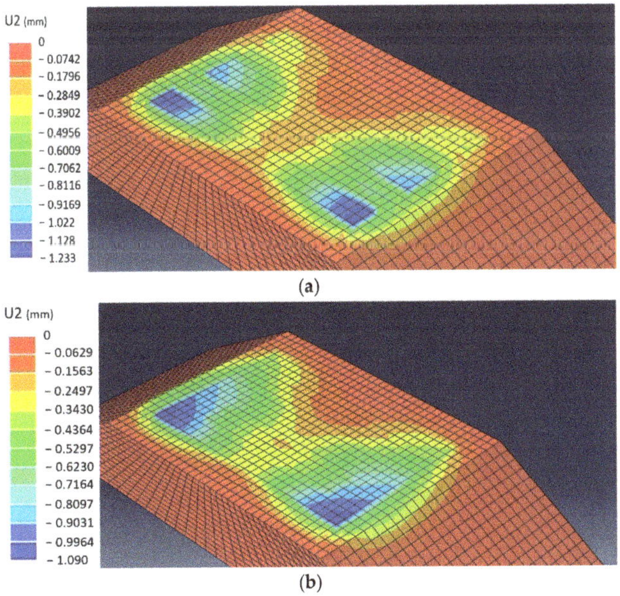

Figure 8. Model of vertical displacement at ballast–crosstie Interface (**a**) without New-Tie and (**b**) with New-Tie.

The crosstie and New-Tie were modeled using the same very stiff material, which resulted in significant load transmitted directly through the crosstie into the ballast, with relatively less spread through the New-Tie. Additional research and modeling are warranted to determine whether a reduction in the stiffness (and therefore cost) of the crosstie relative to the New-Tie would provide acceptable performance while also further improving the performance of the stress-dissipation effect of the New-Tie. This could enable wood or composite crossties to be used with concrete New-Tie or other material optimizations to reduce cost. The nearly 24% reduction in stress in the ballast shown in Table 2 was significant, indicating improved load distribution and long-term ballast and subgrade durability. This is due to the fact that stress, defined as force over area, is spread over a substantial area in the ballast. Table 2 summarizes the change in ballast values between the two models.

Table 2. Comparative model heaving values.

Stress and Deflection Reduction Calculations		
(In the Ballast)	**Stress (Pa)**	**Deflection (m)**
Without New-Tie	59,820	0.00003109
With New-Tie	45,680	0.00003037
% Difference	23.64	2.32

When this stress in the ballast was reduced, this reduced the chance of deformation and fouling, increasing the lifespan of the track. The reduction in stress in terms of the maximum values was calculated as roughly 15%. This was substantial when considering that a reduction in stress overall leads to longer service life and reduced maintenance cycles. Table 3 shows the change in maximum magnitude values.

Table 3. Comparative model maximum magnitude values.

Stress and Deflection Reduction Calculations		
(In the Ballast)	**Stress (Pa)**	**Deflection (m)**
Without New-Tie	59,820	0.001233
With New-Tie	51,030	0.00109
% Difference	14.69	11.60

Although the change in maximum deflection was small (around 0.2 mm), any reduction in the magnitude of cyclic loading of the ballast granite could have a significant impact on the fatigue and overall life cycle of the ballast. For the sake of brevity, this report focuses on the change in maximum stress and leaves the analysis of cyclic loading to further research.

Change in ballast stress is summarized by Figure 9, which shows the stress profile across the midline of the crosstie with and without New-Tie. This plot shows the greatest reduction is at the ends of the crosstie, where the snowshoe effect of the New-Tie is the greatest. Figure 9 illustrates the distribution of vertical stress beneath the rail tie with and without the proposed New-Tie system. The results show that the inclusion of the proposed Tie leads to a more uniform stress distribution across the tie width. In the case without the proposed Tie, the stress is concentrated near the tie edges, indicating higher local deformation in the subgrade. When the New-Tie system is used, these edge concentrations are reduced, and the peak stress shifts slightly toward the center, demonstrating improved load dispersion and a more stable contact condition between the tie and the supporting soil. This indicates that the proposed system enhances the stress transfer mechanism and can potentially reduce localized settlement under repeated train loads.

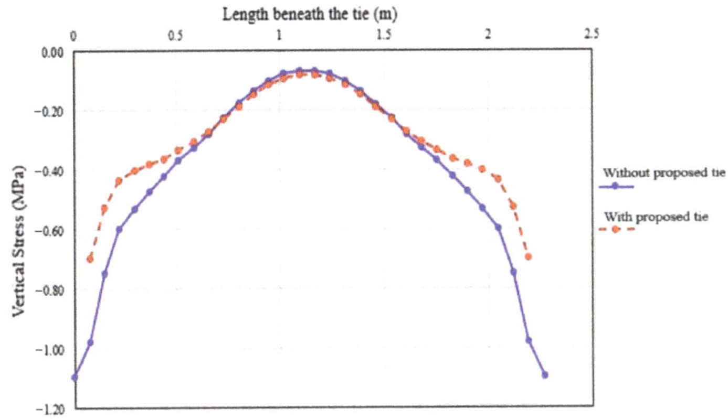

Figure 9. Comparison of vertical stress beneath the crosstie centerline.

Since ballast remediation may still be performed easily with the New-Tie system, as they were designed to be modular and easily installable and removable. With New-Tie, the decreases in the magnitude of stress and deflection would lead to a reduction in maintenance cycles and increase serviceability. The decreases in stresses and deflections in the ballast layer would increase the structural integrity of the system overall.

4. Case Study Design Criteria

The present study considers the viability of the New-Tie system by comparing the most promising alternatives (slab track and geotechnical improvement) with the New-Tie system. The chosen case study area was the Black Rock Desert, just east of Gerlach, Nevada (Figure 10). This site was chosen due to its silty sand, which is typically considered to be unfit for railroad construction. Despite this soft soil, a Union Pacific freight line runs just south of Burning Man, within the endorheic basin that encompasses it. This 3 km stretch of rail line was selected based on three key factors. First, the site is underlain by soft soil with characteristics that typically demand significant subgrade improvements and pose the highest risk of heaving among soils in Nevada. Second, the location already contains an existing rail line, making it suitable for evaluation and comparison. Finally, the rail alignment is flat and straight, providing consistent conditions for analysis.

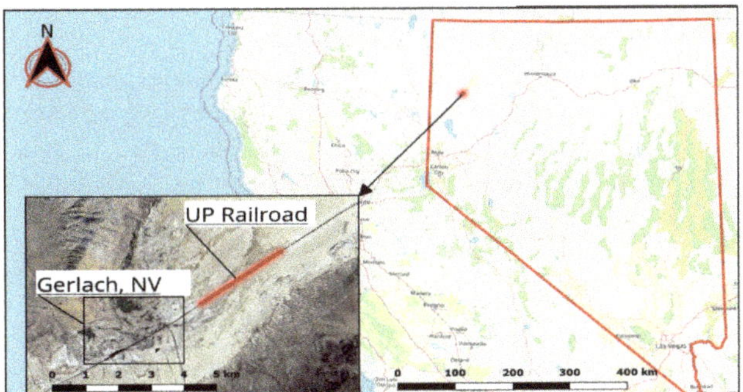

Figure 10. An outline of the selected site in the Black Rock Desert, NV.

Three kilometers was chosen as the length of analysis because of the relative "flatness" of this section of the rail line within the Black Rock Desert. This "flatness" was crucial to compare alternatives based on the differences between New-Tie and the two chosen alternatives of subgrade improvement and slab track construction, without accounting for slope and superelevation improvements.

The classification of the soil of the Black Rock Desert site was determined from UC Davis' California Soil Resource Soil Web application [7]. This interactive map provided the Black Rock Desert's United States Classification System soil classification of SM—Typic Udorthents, which is a fine-grained, silty sand-type soil. The bearing capacity of typic udorthents decreases by 25% when moist [8]; however, due to the comparative nature of this case study, no accommodations were made for the moisture content of the soil. With this classification, typical soil characteristic values were determined (as necessary for the analysis of "typical" soil for railroad construction). Table 4 presents the mechanical properties of the existing subgrade in the area of the case study.

Table 4. Subgrade layer in the Black Rock Desert (for case study) [9].

Soil Type	SM
Density (g/m^3)	2.09
Poisson's ratio	0.25
Modulus (MPa)	16
Earth Pressure Coefficient K_0	1
Allowable Pressure (bearing capacity) (MPa)	13.8

The three design alternatives considered in this case study are geotechnical improvement, replacement of the existing track with slab track, and installation of New-Tie.

With these characteristics of the soil determined and the volumes of necessary materials required, a comparison between New-Tai and the alternative improvements of subgrade improvements and slab track construction was made. All models were performed considering the same load of 333 kN. The major differences between the analysis of New-Tie, subgrade improvements, and slab track inclusion were the geometric and material properties of the slab track and the geometry, or increased depth, of the subgrade improvements.

4.1. Geotechnical Improvement of Existing Soil

This design alternative consists of removing the existing track and improving the ballast and soil layers so that the stress transmitted to the subgrade is within an acceptable 3-factor of safety of the bearing capacity of 25 psi [10]. The existing track must be removed, and the subgrade excavated and compacted to a depth of 1.2 m. Crossties were replaced with prestressed concrete crossties for consistency with the New-Tie alternative. All installation was performed according to standard rail construction procedures. Table 5 shows the designed subgrade improvement for the case study.

Table 5. Excavation volumes for the case study.

Subgrade Improvement Volumes—Subgrade Excavation over 3 km:	
Length total (m)	3000
Total width of subballast layer (m)	5.18
Minimum depth (m)	1.2
Cross-sectional area (m^2)	6.22
Volume—Subgrade excavation (m^3)	18,645
Total vol. of ballast + subballast (m^3)	4068

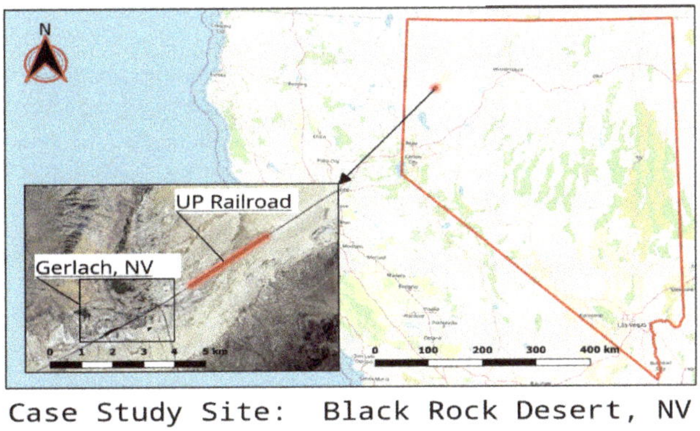

4.2. Replacement with Slab Track

This design alternative consists of removing the existing track and excavating the subgrade to install the underlayment and formwork required for slab track. The subgrade was excavated to a depth of 1.2 m with a section width of 5.18 m. The underlayment for the slab track was 0.15 m of granite [11–13]. Table 6 shows the computed dimensions and volume for the slab track for the case study.

Table 6. Slab track volume computations.

Slab Track Volume over 3 km of Case Study Area:	
Length total (m)	3000
Width (m)	2.4
Thickness (m)	0.19
Volume (m^3)	1368
Slab Track Supporting Layer Volume over 3 km of Case Study Area:	
Length total (m)	3000
Width (m)	2.95
Thickness (m)	0.3
Volume (m^3)	2656

This gives a total concrete volume for slab-track construction of 4024 m^3.

4.3. Installation of New-Tie

This design alternative preserved the rail and replaced the crossties and 0.19 m of ballast depth so that the top of the rail elevation was preserved. The New-Tie system was installed on top of the ballast layer according to standard rail construction procedures.

The volumes of materials required for the construction of suitable alternatives were determined from the geometries of each improvement. The slab track volumes were determined from the standard width, depth, and length of the case study, 3 km. It is important to note, however, that the case study length of 3 km was not divisible evenly by the standard slab section length of 4.96 m, so a total length of 3000 m was required for slab-track construction. For subgrade improvements, the material properties of the Black Rock Desert site were used to determine the subgrade excavation depth, as the purpose of subgrade improvements is to "fix" the existing soil by an increased depth of compaction or outright soil replacement.

Figures 11 and 12 show the dimension reduction cross-section of the railroad by using the New-Tie system versus subgrade improvement.

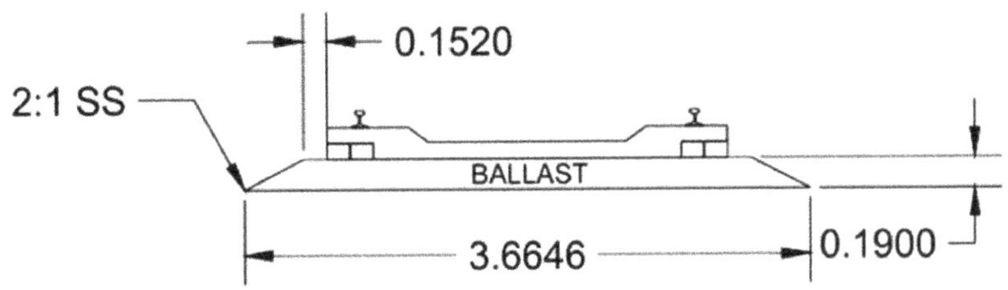

Figure 11. Cross-section of the railroad with the New-Tie system.

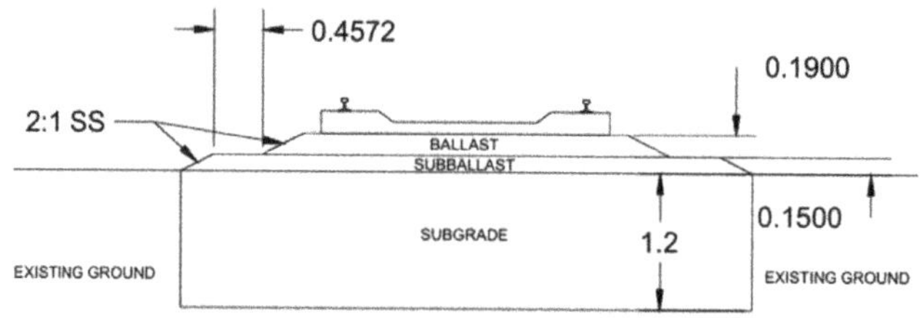

Figure 12. Cross-section of the railroad with subgrade improvement.

4.4. Performance and Economic Evaluation

The New-Tie System not only reduces costs but also enhances performance. It decreases the maximum displacement of concrete crossties by 12% and reduces ballast displacement by 2%. In terms of stresses, the system lowers the maximum vertical stress by 7% and ballast stress by 24%.

Environmentally, the system reduces the overall carbon footprint of granular layer reinforcement in railroads. While the exact reductions in deflection, stress, maintenance cycles, construction cost, and environmental impact may vary depending on site-specific material properties and layouts, the case study conducted in the Black Rock Desert, Nevada, demonstrates consistent benefits. The results confirm that neither construction nor maintenance parameters are negatively impacted, even in soft soils with low bearing capacity.

For these reasons, further research is recommended to test different crosstie and New-Tie materials, as well as to optimize designs for curved track sections and turnouts, where stress and deflection behavior differ from straight alignments. Overall, the New-Tie System provides a practical and innovative middle ground between subgrade replacement and slab track construction and presents a strong candidate for broader industry implementation.

5. Conclusions

This study presented an initial feasibility concept assessment of the New-Tie system as an innovative reinforcement method for railroad crossties. Through finite element modeling and a comparative case study, the system was assessed against conventional alternatives, including subgrade improvement and slab track construction. The findings demonstrate that New-Tie not only reduces stresses and displacements in the ballast layer but also offers potential economic and environmental benefits. The conclusions of this research are summarized as follows:

- The New-Tie System effectively reduces stresses transmitted to the ballast, achieving up to a 24% reduction in ballast stress and lowering crosstie displacement by 12%.
- Finite Element Analysis confirmed that stress distribution in the ballast is improved, leading to greater track durability and extended maintenance cycles.
- The case study in the Black Rock Desert, Nevada, demonstrated that New-Tie is suitable for both "standard" soils and softer soils with low bearing capacity, without negatively impacting performance.
- Environmentally, the system contributes to a reduced carbon footprint by minimizing the need for extensive subgrade replacement or large-scale concrete use.
- The New-Tie system represents a practical middle ground between subgrade replacement and slab track construction, offering an efficient solution that enhances load distribution while maintaining ease of installation and adaptability for existing railway infrastructures.

- Further research is recommended to test different materials and geometries for New-Ties and crossties, especially in curved track sections and turnouts, to optimize performance under diverse loading conditions.

Author Contributions: Conceptualization, M.K. (Moses Karakouzian); methodology, M.K. (Moses Karakouzian), M.C., W.V.V., P.W. and M.K. (Mehrdad Karami); software, M.K. (Mehrdad Karami); writing—original draft preparation, M.C., W.V.V. and P.W.; writing—review and editing, M.K. (Moses Karakouzian) and M.K. (Mehrdad Karami); supervision, M.K. (Moses Karakouzian). All authors have read and agreed to the published version of the manuscript.

Funding: This research received no external funding.

Data Availability Statement: The raw data supporting the conclusions of this article will be made available by the authors on request.

Conflicts of Interest: The authors declare no conflict of interest.

References

1. Li, D.; Selig, E.T. Method for railroad track foundation design. I: Development. *J. Geotech. Geoenviron. Eng.* **1998**, *124*, 316–322. [CrossRef]
2. Wang, Y.; Dimitrovová, Z.; Yau, J.D. Dynamic responses of vehicle ballasted track interaction system for heavy haul trains. In *MATEC Web of Conferences*; EDP Sciences: London, UK, 2018; Volume 148, p. 05004.
3. Selig, E.; Waters, J. *Track Geotechnology and Substructure Management*; Thomas Telford Publications: London, UK, 1994.
4. Sussmann, T.R.; Hyslip, J.P. Track Substructure Design Methodology and Data. In Proceedings of the Joint Rail Conference, Urbana, IL, USA, 27–29 April 2010; Volume 49064, pp. 369–376.
5. *Manual of Railroad Engineering Ch 1 (Version 2014)*; American Railway Engineering and Maintenance-of-Way Association: Lanham, MD, USA, 2014
6. Tutumluer, E.; Dombrow, W.; Huang, H. Laboratory characterization of coal dust fouled ballast behavior. In Proceedings of the AREMA 2008 Annual Conference & Exposition, Salt Lake City, UT, USA, 21–24 September 2008; pp. 21–24.
7. SoilWeb: An Online Soil Survey Browser | California Soil Resource Lab. (n.d.). Available online: https://casoilresource.lawr.ucdavis.edu/gmap/(accessed on 4 April 2025).
8. Kererat, C. Bearing capacity determination of sandy soil layer using light weight penetration test. *J. Appl. Eng. Sci.* **2016**, *14*, 290–295. [CrossRef]
9. Cheng, W.C.; Li, G.; Zhou, A.; Xu, J. Rethinking the water leak incident of tunnel Luo09 to prepare for a challenging future. *Adv. Civ. Eng.* **2019**, *2019*, 4695987. [CrossRef]
10. Li, D.; Selig, E.T. Evaluation of railway subgrade problems. *Transp. Res. Rec.* **1995**, *1489*, 17.
11. Luo, J.; Zhu, S.; Zhai, W. An advanced train-slab track spatially coupled dynamics model: Theoretical methodologies and numerical applications. *J. Sound Vib.* **2021**, *501*, 116059. [CrossRef]
12. Ye, Q.; Luo, Q.; Feng, G.; Wang, T.; Xie, H. Stress distribution in roadbeds of slab tracks with longitudinal discontinuities. *Railw. Eng. Sci.* **2023**, *31*, 61–74. [CrossRef]
13. Gao, Z.; Dersch, M.S.; Qian, Y.; Csenge, M.V.; Edwards, J.R. Use of ballast support condition back-calculator for quantification of ballast pressure distribution under concrete sleepers. In *Bearing Capacity of Roads, Railways and Airfields*; CRC Press: Boca Raton, FL, USA, 2017; pp. 1955–1960.

MDPI AG
Grosspeteranlage 5
4052 Basel
Switzerland
Tel.: +41 61 683 77 34

Designs Editorial Office
E-mail: designs@mdpi.com
www.mdpi.com/journal/designs

www.ingramcontent.com/pod-product-compliance
Lightning Source LLC
Chambersburg PA
CBHW061834210926
51850CB00006B/891
* 9 7 8 3 7 2 5 8 7 9 9 5 3 *